PROTEIN METABOLISM DURING INFANCY

The 33rd Nestlé Nutrition Workshop, Protein Metabolism During Infancy, was held in Magaliesburg, South Africa, May 10–12th, 1993.

Workshop participants: *(left to right, from front to back)*: I. E. M. Axelsson, E. Fern, A. Latronico, A. Fazzolari, K. Pengsaa, Lay Sian Soh, E. Eggermont, J. K. Mbuthia, P. Tienboon, A. Pandit, S. Mantagos, G. Schöch, D. Beatty, A. Dhansay, A. Marini, G. Biscatti, S. Zaverio, A. diComite, K. J. Tracey, P. R. Guesry, C. Schübl, S. Kashyap, A. Quinto, G. Lan Kuan, P. Pengsaa, A. Wasunna, P. J. Garlick, A. Priolisi, D. Labadarios, V. Tripodi, I. Corthouts, L. Vandenbossche, C. Househam, N. C. R. Räihä, M. Bowie, D. F. Wittenberg, R. Uauy, O. J. Ransome, B. A. Wharton, H. J. Bremer, Y. Yamashiro, Per H. Finne, N. G. Partington, P. Hesseling, H. Hoffman, G. Kirsten, B. Lönnerdal, D. K. Rassin, W. Heine, R. Cronjé, J. Pettifor, P. A. Cooper, F. Macagno.

Nestlé Nutrition Workshop Series
Volume 33

PROTEIN METABOLISM DURING INFANCY

Editor

Niels C. R. Räihä, M.D., Ph.D.
University of Lund
Department of Pediatrics
Malmö General Hospital
Malmö, Sweden

NESTLÉ NUTRITION SERVICES

RAVEN PRESS ■ NEW YORK

Nestec Ltd., 55 Avenue Nestlé, CH-1800 Vevey, Switzerland
Raven Press, Ltd., 1185 Avenue of the Americas, New York,
New York 10036

© 1994 by Nestec Ltd. and Raven Press, Ltd. All rights reserved. This book is protected by copyright. No part of it may be reproduced, stored in a retrieval system, or transmitted, in any form or by any means, electronical, mechanical, photocopying, or recording, or otherwise, without the prior written permission of Nestec and Raven Press.

Made in the United States of America

Library of Congress Cataloging-in-Publication Data

Protein metabolism during infancy / editor Niels Räihä.
    p. cm.—(Nestlé nutrition workshop series ; v. 33)
  Based on the 33rd Nestlé Nutrition Workshop held in Magaliesburg, South Africa, 10–12th May, 1993.
    "Nestlé nutrition services."
  Includes bibliographical references and index.
  ISBN 0-7817-0215-1
  1. Proteins in human nutrition—Congresses. 2. Infants—Nutrition—Requirements—Congresses. 3. Proteins—Metabolism—Congresses. I. Räihä, Niels. II. Nestlé Nutrition Workshop (33rd : 1993 : Magaliesburg, South Africa) III. Series.
  RJ216.P758 1994
  612.3'98'0832—dc20
                                                                                             94-1490

The material contained in this volume was submitted as previously unpublished material, except in the instances in which credit has been given to the source from which some of the illustrative material was derived.

Great care has been taken to maintain the accuracy of the information contained in the volume. However, neither Nestec nor Raven Press can be held responsible for errors or for any consequences arising from the use of the information contained herein.

9 8 7 6 5 4 3 2 1

Preface

Optimal growth, development, and functional maturation of the infant and child clearly depend on the provision of an adequate intake and balance of more than 50 essential macro- and micro-nutrients. Of all these, *protein* is by far the most important determinant. Protein in the diet provides the essential amino acids necessary for protein synthesis and thus not only protein quantity but also quality is important. However, protein cannot be considered alone since there is a close interrelation between protein and energy metabolism with reference to growth.

In many areas of the world protein-energy malnutrition of children is still an acute problem and understanding the complicated relationships of early malnutrition, brain development and mental functions later in life has become an urgent problem not only in unpriviledged areas but also in the industrial countries where extremely small and sick infants are surviving in increasing numbers.

On the other hand, excessive protein intakes during critical periods of development have also been associated with an increased incidence of reduced intellectual outcome later in life, increased frequency of childhood diabetes, and possible long-term effects on renal function.

Thus, it is evident that optimal protein-energy nutrition is desirable during critical periods of development, which in the human may last for as long as the first three years.

The investigators gathered at this meeting have made excellent scientific contributions to the understanding of protein metabolism and to its correlation to protein requirement during infancy. Readers should find this book useful in their attempts to feed infants optimally or in their efforts to further study protein metabolism during infancy.

NIELS C. R. RÄIHÄ, M.D., Ph.D.
University of Lund
Department of Pediatrics
Malmö General Hospital
Malmö, Sweden

Foreword

It was important for many reasons to hold the Nestlé Nutrition Workshop on Protein Metabolism: the first, and probably least important, was to complete the cycle of Workshops in which we dealt with carbohydrates in 1989 (1) and lipids in 1990 (2), also trace elements in 1989 (3) and vitamins in 1986 (4) and 1988 (5). A much stronger reason to favor this topic was the hot debate that has been ongoing for many years about the real protein needs for both premature infants and term babies during their first year of life. During the 1960s premature infants were given very large amounts of protein; later on the trend was to decrease the quantity of protein for both low birthweight and term infants. However, healthy term newborns do not have the same requirements as premature or small for date infants who need to grow at a faster pace, so excessively decreasing protein intake must be avoided if a comparable speed of postnatal and intrauterine growth is to be observed.

It is now generally agreed that for the term baby, especially after eight weeks of life, the protein intake should be decreased, but it is clear that needs can vary from day to day, from one baby to another, and depending on pathological conditions such as fever, vomiting, and diarrhea. Since the danger of giving 10% more protein than needed is smaller than giving 10% less than needed, the value considered as adequate in a protected environment cannot be recommended for all babies. In particular, a safety margin has to be added for less favored infants. It was our concern for less developed countries, where the "protein gap" has not been closed but is quite often on the increase, which prompted us to organize the Workshop on protein needs in South Africa. We should not forget that Professor Hansen was a pioneer in the field of kwashiorkor management and prevention in this country.

Protein scarcity cannot be compensated simply by acquiring more precise information about true protein needs per se, but wider knowledge should help us to make better usage of what is available. If this can be achieved we shall consider the 33rd Nestlé Nutrition Workshop to have been a success.

PIERRE R. GUESRY, M.D.
Vice President, Nestec Ltd.
Vevey, Switzerland

REFERENCES

1. Gracey M, Kretchmer N, Rossi E, eds. *Sugars in nutrition.* Nestlé Nutrition Workshop Series, vol 25. New York: Raven Press, 1991.
2. Bracco U, Deckelbaum RJ, eds. *Polyunsaturated fatty acids in human nutrition.* Nestlé Nutrition Workshop Series, vol 28. New York: Raven Press, 1992.
3. Chandra RK, ed. *Trace Elements in Nutrition of Children—II.* Nestlé Nutrition Workshop Series, vol 23. New York: Raven Press, 1991.
4. Glorieux FH, ed. *Rickets.* Nestlé Nutrition Workshop Series, vol 21. New York: Raven Press, 1991.
5. Berger H, ed. *Vitamins and minerals in pregnancy and lactation.* Nestlé Nutrition Workshop Series, vol 16. New York: Raven Press, 1988.

Contents

The Biochemistry and Physiology of Protein and Amino Acid
    Metabolism, with Reference to Protein Nutrition 1
*Vernon R. Young, Antoine E. El-Khoury, Melchor Sánchez
and Leticia Castillo*

Isotopic Methods for Studying Protein Turnover 29
Peter J. Garlick and Margaret A. McNurlan

Interrelations between the Degradation Rates of RNA and Protein
    and the Energy Turnover Rates 49
Gerhard Schöch and Heinrich Topp

Digestibility and Absorption of Protein in Infants 53
Bo Lönnerdal

International Recommendations on Protein Intakes in Infancy:
    Some Points for Discussion 67
Brian A. Wharton

Protein Content of Human Milk, from Colostrum to Mature Milk .. 87
Niels C. R. Räihä

Nutritional Importance of Non-protein Nitrogen 105
Bo Lönnerdal

Qualitative Aspects of Protein in Human Milk and Formula: Amino
    Acid Pattern 121
Willi E. Heine

Protein Requirements of Low Birthweight, Very Low Birthweight,
    and Small for Gestational Age Infants 133
Sudha Kashyap and William C. Heird

Protein Requirement of Healthy Term Infants during the First Four
    Months of Life 153
Niels C. R. Räihä

CONTENTS

Protein Needs during Weaning 165
Irene E. M. Axelsson

Essential and Non-essential Amino Acids in Neonatal Nutrition ... 183
David Keith Rassin

Significance of Nucleic Acids, Nucleotides, and Related
   Compounds in Infant Nutrition 197
Ricardo Uauy-Dagach and Richard Quan

Inborn Errors of Metabolism: A Model for the Evaluation of
   Essential Amino Acid Requirements 211
*Jean-Louis Bresson, Francoise Rey, Florence Poggi, Eliane Depondt,
   Véronique Abadie, Jean-Marie Saudubray, Jean Rey*

Role of Tumor Necrosis Factor in Protein Metabolism 229
Kevin J. Tracey

Subject Index 243

Contributors

Irene E. M. Axelsson
Department of Pediatrics
University of Lund
Malmö General Hospital
21401 Malmö, Sweden

Peter J. Garlick
State University of New York
Department of Surgery
Health Sciences Center
Stony Brook, New York 11794-8191,
USA and The Rowett Research
Institute Aberdeen AB2 9SB,
Scotland UK

Willi E. Heine
Kinderklinik
Medizinische Fakültät
Universität Rostock
Rembrandt Strasse 16/17
1855 Rostock, Germany

Sudha Kashyap
Department of Pediatrics
Columbia University College
 of Physicians and Surgeons
630 West 168th Street
New York, New York 10032, USA and
Babies Hospital (Presbyterian Hospital)

Bo Lönnerdal
Department of Nutrition
University of California
Davis, California 95616-8669, USA

Niels C. R. Räihä
Department of Pediatrics
University of Lund
Malmö General Hospital
21401 Malmö, Sweden

David K. Rassin
Department of Pediatrics
Children's Hospital C3T16
301 University Boulevard
University of Texas Medical Branch
Galveston, Texas 77555-0344, USA

Jean Rey
Département de Pédiatrie
Hôpital des Enfants Malades
149 rue de Sèvres
75743 Paris Cédex 15, France

Gerhard Schöch
Forschungsinstitut für Kinderernährung
Dortmund
Heinstück 11
44225 Dortmund, Germany

Kevin J. Tracey
Department of Surgery
Division of Neurosurgery
North Shore University Hospital
Cornell University Medical College
300 Community Drive
Manhasset, New York 11030, USA

Ricardo Uauy-Dagach
Instituto de Nutrición y Tecnología de
 los Alimentos
Universidad de Chile
Casilla 138-11 Santiago, Chile
and Hospital Sótero
 de Río

Brian A. Wharton
Old Rectory
Belbroughton
Worcestershire DY9 9TF, United
 Kingdom

Vernon R. Young
Massachusetts Institute of Technology
Laboratory of Human Nutrition
School of Science and Clinical Research
 Center
Cambridge, Massachusetts 02139, USA

Invited Attendees and Nestlé Participants

David Beatty / *Rondebosch, South Africa*
Giuliano Biscatti / *Cantu, Italy*
Malcolm Bowie / *Rondebosch, South Africa*
Hans J. Bremer / *Rondebosch, South Africa*
Antonio di Comite / *Taranto, Italy*
Peter A. Cooper / *Bertsham, South Africa*
Ivo Corthouts / *Dendermonde, Belgium*
Reinhold Cronjè / *Pretoria, South Africa*
Muhammad Ali Dhansay / *Tygerberg, South Africa*
Ephrem Eggermont / *Leuven, Belgium*
Angela Fazzolari Nesci / *Palermo, Italy*
Edward Fern / *Vevey, Switzerland*
Per H. Finne / *Oslo, Norway*
Pierre R. Guesry / *Vevey, Switzerland*
Peter Hesseling / *Tygerberg, South Africa*
Hercules Hoffman / *Panorama, South Africa*
Craig Househam / *Bloemfontein, South Africa*
Una Mac Intyre / *Medunsa, South Africa*
Geret Kirsten / *Parow, South Africa*
Geo k Lan Kuan / *Malacca, West Malaysia*
Demetre Labadarios / *Tygerberg, South Africa*
Alberto Latronico / *Milano, Italy*
Franco Macagno / *Udine, Italy*
Stephanos Mantagos / *Patras, Greece*
Antonio Marini / *Milano, Italy*
Joseph K. Mbuthia / *Nairobi, Kenya*
Etienne Nel / *Parow, South Africa*
Anand Pandit / *Pune, India*
Dave Parker / *Randburg, South Africa*
Nick G. Partington / *Randburg, South Africa*
Krisana Pengsaa / *Khon Kaen, Thailand*
John Pettifor / *Bertsham, South Africa*
Antonio Priolisi / *Palermo, Italy*
Anna Quinto / *Napoli, Italy*
Olliver J. Ransome / *Lonehill, South Africa*
Jean Rey / *Paris, France*
Alan Rothberg / *Johannesburg, South Africa*
Claudia Schübl / *Tygerberg, South Africa*
Martin Seip / *Oslo, Norway*
Lay Sian Soh / *Kuala Lumpur, Malaysia*
Prasong Tienboon / *Chiang Mai, Thailand*
Vittorio Tripodi / *Napoli, Italy*
Luc Vandenbossche / *Lommel, Belgium*
Aggrey Wasunna / *Nairobi, Kenya*
Dankwart F. Wittenberg / *Durban, South Africa*
Yuichiro Yamashiro / *Bunkyo-ky Tokyo, Japan*
Silvia Zaverio / *Voghera, Italy*

Nestlé Nutrition Workshop Series

Volume 33: Protein Metabolism During Infancy
Niels C. R. Räihä, Editor; 264 pp., 1994.
Volume 32: Nutrition of the Low Birthweight Infant
Bernard L. Salle and Paul R. Swyer; Editors, 240 pp., 1993.
Volume 31: Birth Risks
J. David Baum, Editor; 256 pp., 1993.
Volume 30: Nutritional Anemias
Samuel J. Fomon and Stanley Zlotkin, Editors; 232 pp., 1992.
Volume 29: Nutrition of the Elderly
Hamish N. Munro and Günter Schlierf, Editors; 248 pp., 1992.
Volume 28: Polyunsaturated Fatty Acids in Human Nutrition
Umberto Bracco and Richard J. Deckelbaum, Editor; 256 pp., 1992.
Volume 27: For a Better Nutrition in the 21st Century
Peter Leathwood, Marc Horisberger, and W. Philip T. James, Editors, 272 pp., 1993.
Volume 26: Perinatology
Erich Saling, Editor; 208 pp., 1992.
Volume 25: Sugars in Nutrition
Michael Gracey, Norman Kretchmer, and Ettore Rossi, Editors; 304 pp., 1991.
Volume 24: Inborn Errors of Metabolism
Jürgen Schaub, Françis Van Hoof, and Henri L. Vis, Editors; 320 pp., 1991.
Volume 23: Trace Elements in Nutrition of Children—II
Ranjit Kumar Chandra, Editor; 248 pp., 1991
Volume 22: History of Pediatrics 1850–1950
Buford L. Nichols, Jr., Angel Ballabriga, and Norman Kretchmer, Editors; 320 pp., 1991.
Volume 21: Rickets
Francis H. Glorieux, Editor; 304 pp., 1991.
Volume 20: Changing Needs in Pediatric Education
Cipriano A. Canosa, Victor C. Vaughan III, and Hung-Chi Lue, Editors; 336 pp., 1990.
Volume 19: The Malnourished Child
Robert M. Suskind and Leslie Lewinter-Suskind, Editors; 432 pp., 1990.
Volume 18: Intrauterine Growth Retardation
Jacques Senterre, Editor; 336 pp., 1989.

The Biochemistry and Physiology of Protein and Amino Acid Metabolism, with Reference to Protein Nutrition

Vernon R. Young, Antoine E. El-Khoury, Melchor Sánchez, and Leticia Castillo

Laboratory of Human Nutrition, School of Science and Clinical Research Center, Massachusetts Institute of Technology, Cambridge, Massachusetts 02139, USA

All of the relevant topics that could be included under the title of this chapter would cover an extraordinary range of biological knowledge; more than is possible to include in an article of restricted length. Because of this and because of our own limited area of understanding, we have adopted the strategy of covering a few selected areas without attempting to be comprehensive; further elaboration of some of these areas will be made in later chapters. The emphasis to be given here is toward physiology rather than biochemistry, since our own interests relate to the flows of metabolites through pathways and the impact of nutritional and other factors that affect these, as well as the *in vivo* mechanisms involved. A particular biochemical focus would involve giving more attention to the detailed cellular and subcellular pathways responsible for the formation of proteins and amino acids, their movement to particular sites, and their interconversion and degradation, together with details of the molecular mechanisms by which these processes take place. Our objective is to attempt to provide a better understanding of the metabolic basis of the protein and amino acid requirements under various pathophysiological conditions.

Since amino acids are the currency of the nitrogen and protein economy of the host, this introductory chapter begins with a statement about them, particularly those of quantitative dietary and nutritional significance. We shall then discuss the major metabolic systems responsible for the utilization of dietary amino acids and the maintenance of body protein and amino acid homeostasis.

THE NUTRITIONALLY RELEVANT AMINO ACIDS

There are hundreds of amino acids in nature (1) but only approximately 20 of these appear in proteins via charging by cognate tRNA with subsequent recognition of a codon on the mRNA, as for example a UGA codon which was identified recently for the uncommon amino acid, selenocysteine, which is made from serine on a unique tRNA (2). Other amino acids, such as hydroxyproline, hydroxylysine, N^τ-methylhistidine, and the rare amino acids 3-hydroxyaspartic acid and 3-hydroxyasparagine, to mention a few examples, are also present in proteins due to post-translational modification of specific amino acid residues. It is, however, these 20 amino acids, together with a few others not in peptide-bound form, such as ornithine, citrulline, and taurine, that are of physiologic and quantitative importance in the nitrogen economy and nutrition of the mammalian organism.

Three amino acids, tryptophan, lysine, and histidine, had been shown to be indispensable components of the diet in the growing rat by 1932; once an abundant supply of threonine became possible all of the dietary amino acids that were necessary for rat growth could be identified, and in 1948 Rose and his co-workers (e.g., see 3) had determined that these were valine, leucine, isoleucine, methionine, threonine, lysine, phenylalanine, tryptophan, histidine, and arginine. Further, by the middle of the twentieth century evidence had been obtained showing that except for histidine and arginine, these same amino acids were essential for establishment and preservation of body nitrogen equilibrium in human adults. A "final" classification of the dietary amino acids, according to their role in the maintenance of nitrogen equilibrium in healthy young men, was made by Rose et al. (3) and this is reproduced in Table 1 under the columns headed "1954." Research subsequent to the impressive series of metabolic balance studies by Rose and his co-workers has indicated that this earlier nutritional classification of amino acids is no longer satisfactory. Thus histidine is now considered to be an essential (indispensable) amino acid (4), and the amino acids glycine, cystine, tyrosine, proline, and arginine, together with glutamine and taurine, are more appropriately defined as "conditionally indispensable." In particular, active

TABLE 1. Changing view of the role of amino acids in human nutrition

Indispensable (essential)		Conditionally indispensable	Dispensable (non-essential)	
1954[a]	Present[b]	Present[b]	1954[a]	Present[b]
Valine	Valine	Glycine	Glycine	
Isoleucine	Isoleucine	Cystine	Cystine	
Leucine	Leucine	Glutamine	Glutamic acid	Glutamic acid
Lysine	Lysine	Tyrosine	Tyrosine	
Methionine	Methionine	Proline	Proline	
Phenylalanine	Phenylalanine	Arginine	Arginine	
Threonine	Threonine	Taurine	Alanine	Alanine
Tryptophan	Tryptophan		Serine	Serine
	Histidine		Aspartic acid	Aspartic acid
			Histidine	Asparagine
			Hydroxyproline	
			Citrulline	

[a] Based on "Final classification" by Rose et al. (3).
[b] A current interpretation.

areas of current research in amino acid metabolism and nutrition focus heavily on the role played by these various amino acids in the maintenance of organ and body protein homeostasis and function under various pathophysiological conditions. For example, an exciting and relatively recent development in this context is the role of arginine as a precursor of nitric oxide, which functions in cellular communication, signal transduction, and in defense against invading microorganisms (5). The nonessential (dispensable) and conditionally indispensable amino acids probably will continue to attract increasing research attention, particularly in terms of the potential therapeutic benefits to be achieved by manipulating their intakes and/or tissue availability in different disease states.

Amino Acid Functions

The nutritionally relevant amino acids (Table 1) serve multiple, diverse functions. Some of these are listed in Table 2, together with those functions met by the proteins

TABLE 2. *Some functions of amino acids and their products*[a]

Function	Example
Amino acids	
Substrates for protein synthesis	Those for which there is a codon
Regulators of protein turnover	Leucine, arginine
Regulators of enzyme activity (allosteric)	Arginine and NAG synthetase; Phe and PAH activation
Precursor of signal transducer	Arginine and nitric oxide
Methylation reactions	Methionine
Neurotransmitter	Tryptophan (serotonin); glutamate
Ion fluxes	Taurine; glutamate
Precursor of "physiologic" molecules	Arg (creatinine); glut(NH_2)-purines
Transport of nitrogen	Alanine; glut(NH_2)
Oxidation-reduction properties	Cystine; glutathione
Precursor of conditionally indispensable amino acids	Methionine (cys); phe (tyr)
Gluconeogenic substrate and fuel	Alanine; serine; glut(NH_2)
Proteins	
Enzymatic catalysis	BCKADH
Transport	B-12 binding proteins; ceruloplasmin; apolipoproteins
Messenger/signals	Insulin; growth hormone
Movement	Kinesin; actin
Structure	Collagens; elastin
Storage/sequestration	Ferritin; metallothionein
Immunity	Antibodies; TNF; interleukins
Growth; differentiation; gene expression	EGF; IGFs; transcription factors

[a] PAH, phenylalanine hydroxylase; BCKADH, branched-chain ketoacid dehydrogenase; TNF, tumor necrosis factor; NAG, *N*-acetylglutamate.

which are elaborated from the amino acid pools. The point we should make is that in the course of carrying out their physiologic and functional roles, the proteins and amino acids turn over and part of their nitrogen and carbon is lost via the excretory pathways, including CO_2 in expired air and urea and ammonia in urine. To maintain an adequate body protein and amino acid status, these losses must be balanced by an appropriate dietary supply of (a) a utilizable source of nitrogen to support cell and organ function and (b) the indispensable amino acids, and under specific states the "conditionally indispensable" amino acids, to replace those that are lost during their daily metabolic transactions and/or deposited in new tissues during growth, repletion, and/or secretion via milk during lactation. It is relevant, therefore, to consider the nature of the systems that underlie these dietary needs and the changes in these systems when physiological and/or pathological factors intervene.

THE MAJOR SYSTEMS

The principal metabolic systems responsible for the maintenance of body protein and amino acid homeostasis are shown in Fig. 1. They are: protein synthesis, protein degradation, amino acid oxidation and urea production, and amino acid synthesis, in the case of the nutritionally dispensable or conditionally indispensable amino acids. Changes in the rates of these systems lead to an adjustment in nitrogen balance and retention, the direction and extent of which depend on the factor(s) responsible for initiating an effect. For purposes of this brief introductory review, the biochemical and physiological processes that might be highlighted are schematically outlined in Fig. 2; other than gene transcription, which will not be discussed in detail here, these events include the translational phase of protein synthesis, post-translational events, particularly protein degradation, and subsequent intra- and intercellular transport of amino acids, amino acid oxidation, the production of urea, and the formation of conditionally indispensable and dispensable amino acids from carbon and nitrogen precursors. Each of these processes will be considered in the following sections, with the hope that this will provide a sufficient introduction to some aspects of the physiology and biochemistry of protein and amino acid metabolism to make subsequent discussions on various nutritional topics a bit more valuable.

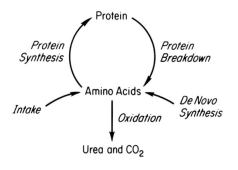

FIG. 1. Schematic representation of the major biochemical systems responsible for the homeostasis of protein and amino acid metabolism *in vivo*.

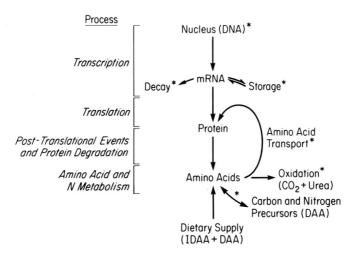

FIG. 2. Major cellular processes that account for the "biochemistry and physiology of protein and amino acid metabolism."

Translational Aspects of Protein Synthesis

Although the primary control of gene expression lies at the level of gene *transcription*, and very rapid changes in inducible gene expression can occur in response to changes in the cellular environment (6), a workshop that is devoted to the protein nutrition of infants and children suggests to us that it would be more useful to begin by giving an initial focus to the translational stage involved in formation of the gene product; this involves a readout of the codons in the mRNA and comprises three distinct phases: initiation, elongation, and termination, with each using different factors to catalyze specific interactions among the components of the protein-synthesizing machinery. The initiation step is the major phase of regulation following acute changes in the availability of amino acids (7). To facilitate a brief consideration of some recent findings, the scheme shown in Fig. 3 (8) depicts the steps and the factors involved in formation of the initiation complex (7). In summary, the initiation phase consists of (a) formation of the 43S initiation complex, (b) formation of the 48S initiation complex, consisting of mRNA binding to the 43S complex, followed by scanning for the initiation codon (AUG), and (c) formation of the 80S initiation complex. Following these major events the elongation phase proceeds via participation of the aminoacyl-tRNAs and elongation factors (9).

Protein synthesis is regulated in most cases by changes in either the cellular concentration or phosphorylation state of only a few of these initiation factors; modifications of eIF-2α and eIF-4E appear to be the most important sites of regulation (7), and global alterations in translation rates can be achieved by modifications in the phosphorylation state of either one of these general translation initiation factors (10).

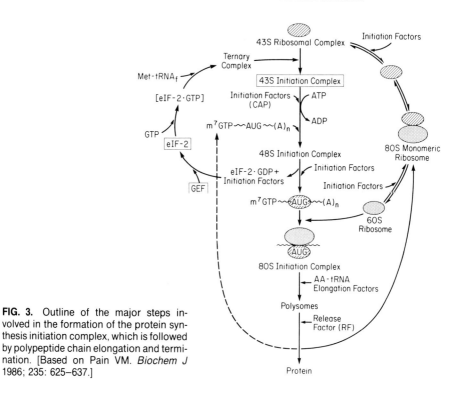

FIG. 3. Outline of the major steps involved in the formation of the protein synthesis initiation complex, which is followed by polypeptide chain elongation and termination. [Based on Pain VM. *Biochem J* 1986; 235: 625–637.]

Additionally, translational regulation of specific mRNAs can be achieved by reversible RNA/protein interactions in the 5-untranslated region of mRNA. Regulation of the iron storage protein ferritin in response to iron levels is a good example of this mechanism (10,11).

In the context of the nutritional supply of amino acids, a continued deficient intake level will later be accompanied by alterations at the transcriptional phase of protein synthesis. For example, the levels of mRNA for albumin, transferrin, transthyretin, the β-chain of fibrinogen, and apolipoprotein E are reduced in livers obtained from rats following 3 days of protein depletion (12); refeeding experiments revealed that changes in mRNA levels occur more slowly than changes in the actual rates of protein synthesis (12). Clearly, alterations in mRNA levels may be due to changes in transcription rates, mRNA splicing and its delivery to the cytoplasmic compartment, and/or mRNA stability, but much remains to be learned about the quantitative impact of different nutritional factors on these various sites and processes of potential control. A final point worth making here is that it is unlikely that the concentration of amino acids functions as the sole nutritional regulator of translation. Indeed, there is now considerable evidence that hormonal factors are intimately involved in the signaling of dietary change (see 13, 14 for reviews), via effects on transcription as well as on translation. However, changes in tissue amino acid levels are apparently of importance in bringing about their regulatory influence and, in consequence, achieving a so-called anabolic drive (15). The latter process is relevant to the discussion of oxidative losses of amino acids and their varying nutritional significance, to follow later in this chapter.

Protein Degradation

Tissue and organ protein composition and content are also determined by the relative rates of breakdown of proteins. Briefly, the regulation of protein breakdown is determined by incompletely defined innate properties of protein substrates (e.g., 16); by various factors, such as ubiquitination (17) or oxidation of amino acid residues (18), that interact with, or "prepare," proteins for degradation; and by the degradative machinery in the cell (19,20), which includes both lysosomal and non-lysosomal systems (Table 3).

With reference to nutritional aspects, two points are of particular interest here. First, proteins in eukaryotic cells are continually and extensively degraded and replaced, and this turnover serves several important functions: cells selectively degrade proteins with abnormal conformations, which could be harmful if these accumulated; the rapid degradation of regulatory peptides is essential for the control of metabolic pathways and the cell cycle; the breakdown of proteins, such as those in skin and muscle, during starvation or in response to the stress of infection or physical trauma provides amino acids for the synthesis of immunologically competent proteins, for glucose production, and for energy metabolism; and the digestion of protein antigens to small peptides is of importance in antigen presentation to T lymphocytes (21). This turnover of proteins has important implications for protein and amino acid requirements, since high turnover rates are probably responsible for higher dietary nitrogen and amino acid requirements. We shall refer again to this issue below.

Second, the degradation of most cell proteins requires metabolic energy (20). Further, it is now known that in the eukaryotic cytosol and nucleus, there are large multimeric proteolytic complexes (20–22) that require ATP hydrolysis in the course of their functions, which include selective intracellular proteolysis and cell cycle traverse (22). Although much remains to be learned about the enzymatic roles of ATP and precise functions of the large proteolytic complexes, it is important for us

TABLE 3. *Summary of the major systems involved in the degradation of proteins within cells*

Enzyme system	Function
Lysosomal proteinases	Nutrient deprivation
Ubiquitin/ATP-dependent proteases	Short-lived proteins; abnormal proteins
Ca-dependent calpains	Myofibrillar proteins
Multicatalytic/ATP-dependent protease	
Metalloproteinases	Peptides
Chymotrypsin- and trypsin-like proteinases	Pathological conditions

From Reeds PJ, & Davis TA. *Control of fat and lean disposition.* Oxford: Butterworth-Heinemann, 1992; 1–26.

to appreciate that the process of protein degradation is energy requiring, and that the multiple and mechanistically distinct pathways of proteolysis are subject to considerable *in vivo* regulation, especially via nutritional and hormonal factors (13,23,24). These matters have significance for considerations of protein and amino acid nutrition and requirements and for dietary protein–energy interactions, as we shall mention in a later section. A recent, nutritionally relevant example of this point is that while oxygen radical-mediated oxidation of enzymes is a marking step in protein turnover (18), there is an age-related accumulation of oxidized protein which may be involved in age-related loss of some physiological functions (18). However, this accumulation can be reduced, in rodents, by protein or energy restriction (25). It seems, therefore, that better knowledge of the molecular and cellular mechanisms responsible for protein degradation and of the nutritional factors that intervene will be likely to lead to effective, nutritionally based strategies for promoting nitrogen retention and net deposition under various pathophysiological conditions.

Amino Acid Oxidation

As indicated in Fig. 1, the process of amino acid oxidation plays an important role in determining body protein homeostasis. This topic has been reviewed (24,26), so we shall just highlight the importance of oxidative metabolism in relation to the maintenance of body protein and amino acid metabolism under conditions of altered protein and amino acid intakes.

An important biochemical feature to emphasize is the profound reduction in the rate of oxidation of specific indispensable amino acids when rats are given diets that are low in these amino acids (see 24,26). Furthermore, oxidation remains at these reduced levels until the approximate intake level of the amino acid that is necessary to meet the needs for maximum growth is exceeded. Similarly, studies in human subjects (e.g., 24) reveal that low intakes of protein or specific indispensable amino acids bring about an efficient conservation of amino acids, due in large part to a reduced rate of amino acid catabolism.

This conservation occurs with a decline in blood and tissue amino acid concentrations (27) and this directly affects the rate of amino acid oxidation. The reason for this is that the K_m values for the various amino acid degrading enzymes are relatively high (26,28,29), so they are rarely saturated within the physiological range of plasma and tissue amino acid concentrations (Table 4). In contrast, the K_m values of aminoacyl-tRNA synthetases are low (26). Hence, the latter enzymes would be expected to be functioning at high rates and their tRNAs essentially fully charged, even when amino acid intakes are low. The net effect of these differences in the kinetic properties of the various enzymes would be that amino acids are more efficiently channeled into the pathway of protein synthesis and the efficiency of dietary and endogenous amino acid retention would increase as the intake level is reduced. Concomitantly, the oxidation of amino acids would decline.

It follows, then, that the reduction in amino acid oxidation at low amino acid

TABLE 4. *Summary of Michaelis constants (K_m) for some enzymes of amino acid metabolism in relation to liver and plasma amino acid levels in the rat*

	K_m
Aminoacyl-tRNA synthetases	1–50 μM
Amino acid-degrading enzymes	
Lysine-1-oxoglutarate reductase	18 mM
Phenylalanine hydroxylase	660 μM
BCAA amino acid concentrations	4 mM
Liver amino acid concentrations (low protein diet)	30–50 μM
Plasma amino acid concentrations:	
Lysine	~0.5 mM
Phe	~0.06 mM
Other	0.03–1.1 mM

From Young VR, Marchini JS. Am J Clin Nutr 1990; 51: 570–589 and based on refs. 26, 28, and 29.

intakes would parallel changes in plasma amino acid concentrations. We have shown that this occurs in adult humans given diets that are low in leucine or lysine, for example (see 30 for a review). Because plasma and tissue amino acid concentrations are highly labile and promptly affected by acute changes in protein and amino acid intake, rates of amino acid oxidation would be expected to change rapidly in response to a dietary change. For example, as depicted in Fig. 4, we have been carrying out a series of 24 hour leucine kinetic studies in young adults who received different intakes of leucine from a generous intake (89 mg/kg/d) to a lower (38 mg/kg/d) and/or submaintenance (14 mg/kg/d) level (El-Khoury A, *et al.*, 1993, unpublished results). With feeding (which began at 6 A.M. as small, equal meals, at hourly intervals for 10 h) there was a prompt rise in leucine oxidation which was sustained when leucine intake was generous. However, oxidation declined after a few hours when intakes were low or even sooner when submaintenance levels were consumed (Fig. 4a). These patterns of prandial, oxidative loss of leucine were paralleled closely by the changes in plasma leucine (and presumably tissue) concentrations (Fig. 4b). Thus we conclude that substrate availability is an important factor in determining the rate of amino acid oxidation, and this is to be predicted from the foregoing discussion of K_m values. Furthermore, for generous protein intakes it now appears that there is a significant prandial oxidation of amino acids, which actually accounts for about one half of the daily oxygen consumption by the liver; on this basis it has been suggested that gluconeogenesis from amino acids is to be regarded as a normal prandial process (31).

The status of amino acid oxidation is also determined by the level and properties of tissue enzymes, and they are affected by alterations in protein and amino acid intakes. Involved are allosteric mechanisms, alterations in the amount of enzyme protein, and changes in the proportion of the enzyme in its active and inactive forms. An important example of the latter mechanism is the regulation of activity of the

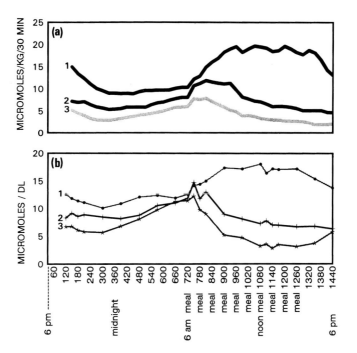

FIG. 4. Rate of leucine oxidation throughout the 24 h day (a) in healthy young men, in relation to the concentration of leucine in plasma (b). Lines 1, 2, and 3 refer to subjects receiving a generous (89 mg/kg/d, $n = 7$) maintenance (38 mg/kg/d, $n = 5$), or submaintenance (14 mg/kg/d, $n = 5$) intake of leucine with an otherwise fully adequate diet. [Unpublished data from the authors' laboratories (El-Khoury et al., 1993).]

second enzyme in the pathway of branched-chain amino acid catabolism, namely branched-chain keto acid dehydrogenase (BCKAD) (see 32 for a review). Here the activity of BCKAD is regulated by phosphorylation (inactive) and dephosphorylation (active) of the α-subunit of the E1 component of the complex, via the BCKAD kinase. The latter enzyme is inhibited by ketoleucine and, to a lesser extent, by the ketoanalogs of isoleucine and valine. Hence at low leucine intakes the kinase will be active, giving rise to a reduced activation of BCKAD and, in consequence, to a low rate of leucine oxidation.

Thus, in summary, amino acid oxidation changes rapidly in response to an altered amino acid intake, and changes in substrate availability appear to be a primary regulatory factor. The amino acid-degrading enzymes are also affected, with alterations in their specific activity, as well as in the amount of enzyme, when protein and amino acid intakes are changed. The net effect of these factors is a conservation of amino acids, and of nitrogen, when intake is low and, conversely, higher rates of loss of amino acid nitrogen and carbon when their intakes are abundant. Finally, this conservation of amino acids at low intakes may complicate interpretation of the nutritional significance of many of the earlier studies designed to establish the minimum physiological requirements for specific amino acids, as we have discussed elsewhere (22).

Urea Cycle Enzymes and Urea Production

Also with reference to the systems depicted in Fig. 1, the urea cycle enzymes and production of urea might be considered an important, final common pathway in the removal of amino nitrogen and possibly for adjusting nitrogen loss to nitrogen intake. As discussed by Cohen (33), the enzymes of urea biosynthesis provide a metabolic pathway for conversion of potentially toxic ammonia (34), as well as removal of excess amino acids via their oxidation and transfer of the nitrogen to arginine and ultimately, urea, when intakes of protein are high. Alternatively, the pathway might facilitate conservation of nitrogen when intakes are low.

Briefly, the enzymes of the ornithine-urea cycle are compartmentalized in the liver (see 33–35 for reviews) (Fig. 5), and detailed studies of their physical and chemical properties have been described. Additionally, some of these enzymes are abundant in the mitochondria of the intestinal mucosa and in the extramitochondrial fraction in the kidney; this topic will be considered later with reference to our discussion of the metabolism of arginine.

The classical studies by Schimke (e.g., 36) showed that metabolic adaptation to differing consumption levels of protein is associated with changes in the activities of urea cycle enzymes in the liver; these changes were found to be due to differences in the content of specific protein rather than to changes in specific activity. The mechanism underlying these alterations in the urea cycle enzyme levels appears to reside at a pretranslational level (37) for all of the enzymes and is largely coordinate. Finally, Das & Waterlow (38) showed that the activity in liver argininosuccinate synthetase was reduced in parallel with the decline in urinary nitrogen output when rats received a low protein diet. This parallelism is not necessarily reflective of a simple cause-and-effect relationship since, for example, the *in vitro* capacity of the rate-limiting enzyme, argininosuccinate synthetase, in liver of rats given a low protein

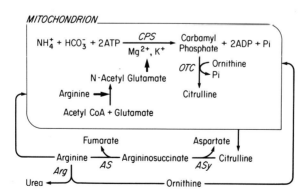

FIG. 5. Urea cycle enzymes and their intracellular distribution within the liver. CPS, carbamyl phosphate synthetase; OTC, ornithine transcarbamylase; ASy, argininosuccinic synthetase; AS, arginosuccinase; Arg, arginase. [From Cohen PP. *Nitrogen metabolism in man.* London: Applied Science Publishers, 1981; 215–228.]

diet is still in considerable excess of the activity that is theoretically needed to handle the flow of nitrogen through the urea cycle.

In any event, we still do not appreciate the extent to which the process of urea production "drives" the elimination of amino acids when they are consumed in excess of needs or "decelerates" their rates of loss when intakes are inadequate. This, then, raises the issue of urea production rates and how they change with protein (nitrogen) intake.

As we have discussed above, lowered intakes of indispensable amino acids bring about lowered rates of amino acid oxidation. There is also a reduced output of nitrogen when the intake of protein is restricted. Thus it would be reasonable to predict a proportionate change in urea production throughout a wide range with changes in the level of dietary nitrogen intake. On the other hand, Jackson (39) claims that in normal adults there is no demonstrable association between urea production and dietary intake over an adequate range of intakes in excess of about 35 g of protein per day. Further, he (39) points out that from his own studies, the relationship between urea excretion and intake does not necessarily arise through urea production and that there is a protein-intake-level-dependent salvaging of urea nitrogen, via intestinal hydrolysis of urea, which is again made available to the host as *indispensable* and *dispensable* amino acids in functionally significant amounts. His view is highly speculative and it appears to us that the available evidence does not support this hypothesis (40,41). Thus the physiological role and importance of urea nitrogen salvaging is still a matter of great uncertainty, but perhaps this process might serve as a mechanism for conserving nitrogen when intakes are very much below physiological needs. Assuming this to be the case, a biochemical consideration of the urea cycle enzymes and of the mechanism of urea nitrogen formation has to be made finally in relation to the *in vivo* context if the role of this system in the protein and amino acid economy of the host is to be fully appreciated.

Amino Acid Transport

The cellular sites of protein synthesis, amino acid oxidation, and urea production require that the nutritional substrates be transported across cell membranes. Further, the role of membrane transport in the regulation of substrate utilization and interorgan metabolic flows has received increased attention in recent years (e.g., 42), and *in vivo* models are now being developed to quantify amino acid transport rates (e.g., 43). In summary, it can be stated that the amino acid transport systems are affected by, and respond to, changes in the supply of amino acids as well as other factors, including hormones (44,45). The multiplicity of specific transport systems and the changes in these due to diet are complex (see 46 for a review). Hence, it is difficult for us to grasp how they operate and are integrated in the intact organism to facilitate the conservation and effective distribution among organs of amino acids when their supply falls below requirements for meeting cellular and metabolic functions. Nevertheless, the regulation of alanine and aromatic amino acid metabolism in the rat liver and of arginine metabolism in hepatocytes appears to be determined by the membrane transport phase of utilization of these amino acids (see 24). The relationships between changes in transport, amino acid uptake into proteins, and amino acid catabolism will have to be examined more extensively in the intact organism under various conditions before the quantitative importance and role played by membrane transport in the maintenance of amino acid homeostasis under various nutritional, physiological, and pathological states is established. Finally, and in a similar context to the earlier discussion on protein degradation, the transport systems are energy-dependent, and this point will be considered again when the basic aspects of protein–energy interrelationships are discussed briefly in a later section.

AN INTEGRATION OF THESE SYSTEMS

The biological level of primary focus at this workshop is the intact organism. Hence, given the type of biochemical and physiological detail discussed above, in combination with *in vivo* studies of various kinds, a general picture of protein and amino acid metabolism at the whole body level has now emerged. This picture reveals the cooperative participation of different organs, such as the intestines, liver, kidney, and muscle, in determining the fate of free amino acids entering the body pools from exogenous and endogenous sources. Therefore, the major pathways of amino acid flow at the whole body level can be presented, in schematic form, as shown in Fig. 6; the key involvement of glutamine and alanine and their interrelationships with the nutritionally indispensable, branched-chain amino acids in the economy of whole body nitrogen metabolism are also depicted here.

The fluxes or rates of transport of specific amino acids within the circulation have been measured in human subjects, as summarized by Bier (48), and these will be reviewed in greater detail elsewhere in this volume with reference to protein turnover at the whole body and regional level (see the chapter by Garlick). However, as Christensen (49) has stated succinctly, the fluxes of amino acids are not simply a consequence of an indiscriminate and intermittent inflow of amino acids arising from protein-containing meals but are regulated at specific organ sites and by various mechanisms. The quantitative significance of these fluxes, together with an understanding of the *in vivo* mechanisms and points of control of specific amino acid metabolism in subjects under different nutritional conditions, is necessary if we are to define fully the metabolic basis of the requirements for protein and amino acids. Hence, we turn now from a somewhat reductionist level to a number of examples of the integration, coordination, and quantitative nature of protein and amino acid metabolism in the *in vivo* setting.

Arginine Kinetics as an Example of Metabolic Cooperativity

Because this chapter must, inevitably, be strictly limited in length, we shall use the results obtained from a recent series of studies, conducted in our laboratories in young adults, on plasma arginine kinetics, to illustrate both the complexity and coordination of amino acid metabolism in the whole organism. This example will

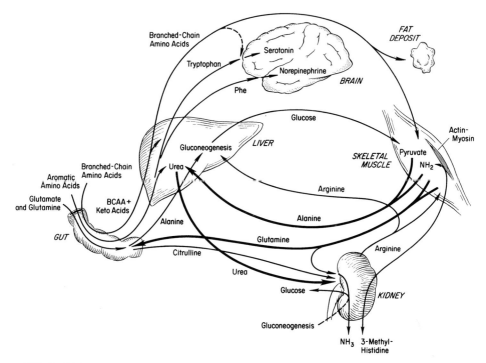

FIG. 6. Major pathways of the transport and fate of various amino acids between and within organs. This representation is an expansion and modification of the original presentation by Munro HN. *Nutritional support of the seriously ill patient.* New York: Academic Press, 1983; 93–105.

permit us to consider, to some extent, how systems such as amino acid transport, urea production, amino acid oxidation, and protein turnover interrelate and serve to achieve a balance between the dietary protein and amino acid supply and the metabolic needs of the organism.

As alluded to earlier, in ureotelic amphibians, turtles, and mammals, including humans, the five urea cycle enzymes, organized within the mitochondrial and cytosolic compartments of the cell, act in concert to form arginine and urea from ornithine, ammonia, and carbon dioxide. This cycle is complete within the normal liver, and only this organ has a major functional capacity for urea synthesis *de novo,* although arginase occurs in most, if not all, extrahepatic tissues. Thus the hepatic urea cycle accounts for a component of *in vivo* arginine metabolism and utilization. However, it has been known for a number of years that urea cycle enzyme activity is present in tissues and organs other than the liver; indeed, the intestine and kidney each function in the extrahepatic synthesis of arginine (see 50 for a review). Based on biochemical and physiological studies, the intestine is capable of synthesizing ornithine and citrulline and the kidney then converts the citrulline to arginine. This endogenous synthesis plays a critical role in the support of the growth needs in the rat (51) and citrulline availability to the kidney appears to be an important factor in determining the rate of extrahepatic arginine synthesis (52). With reference to human nutrition it is of interest to point out that the amount of citrulline converted to arginine by this organ is reduced in patients with chronic renal insufficiency (53). These patients show high plasma citrulline levels, but plasma arginine levels do not differ measurably from those in healthy controls. Hence the nutritional and metabolic significance of the lower synthesis of arginine by the kidney in patients with chronic renal insufficiency is still not clear. To summarize, from these various lines of evidence, Brusilow & Horwich (50) presented a scheme, which is reproduced with slight modification in Fig. 7, showing a metabolic cooperation between the intestine, kidney, and liver in relation to the formation and fate of arginine. Because the skeletal musculature is an important endogenous source of plasma glutamine, which serves as a major precursor for ornithine and citrulline synthesis by the intestine, this organ has been included in this picture of the interorgan aspects of arginine metabolism (Fig. 7). Therefore, it would not be unreasonable to propose that the integrity of the interorgan and metabolic

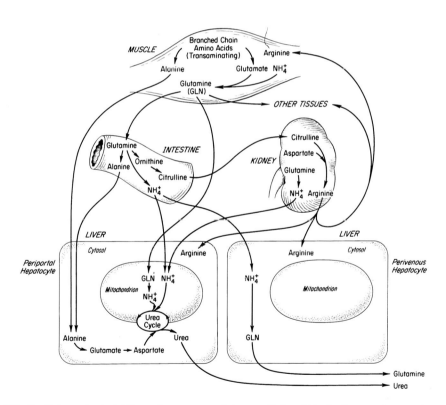

FIG. 7. Scheme to depict the major organs, cellular specificity within the liver, and some of the interactions among amino acids involved in arginine metabolism *in vivo*. [Modified from Brusilow SW, & Horwich AL (50).]

interrelations depicted in Fig. 7 would be affected, or compromised, by various pathophysiological states and that they could change with the developmental stage of individuals.

Based on this scheme (Fig. 7), it might be evident that tissue arginine needs could be met in various ways: (a) from dietary arginine that escapes metabolism in the liver; (b) via the *de novo* synthesis of arginine that can occur in the liver and intestine because of the presence of ornithine transcarbamylase; (c) from conversion of citrulline, which occurs in kidney and brain; and (d) by taking up circulatory endogenous arginine release by the kidneys. An important question for the nutritional biochemist, therefore, is the contribution made by these various sources for meeting tissue arginine requirements and the quantitative extent to which dietary arginine might substitute for, or downregulate, the supply of arginine arising via the metabolic conversions that are depicted here (Fig. 7). Additionally, the impact of altered rates of arginine utilization on arginine homeostasis and the significance of dietary arginine intake should be considered, as we further review the integration of the major systems *in vivo*.

To explore these various issues and questions, we have used L-[di-^{15}N-guanidino,5,5^{2}H]arginine, L-[^{13}C-guanidino]arginine, and [5,5-^{2}H]citrulline as tracers to determine plasma arginine and citrulline fluxes. Although our results are still limited, Table 5 presents some of our findings for plasma arginine and citrulline fluxes in healthy young adult men, studied during the postabsorptive state. We have also estimated the conversion of plasma citrulline to arginine under conditions of an adequate (arginine-rich) or arginine-free intake.

A number of potentially important observations can be made from the results given in Table 5, as follows: first, the measured plasma flux of arginine is much lower than might be expected for the whole body flux of the guanidino moiety for the amino acid. We estimate that the sum of the whole body (plasma?) arginine flux contributed

TABLE 5. *Plasma citrulline and arginine fluxes and conversion of citrulline to arginine (Qc → a) with arginine-rich and arginine-free intakes in young adults during the postabsorptive state[a]*

Diet	Citrulline	Arginine	Qc → a
Arginine-rich			
Mean	10.6[b]	60.0	5.6
SD	3.2	7.3	1.5
Arginine-free			
Mean	10.2	47.0[c]	5.2
SD	2.8	7.4	1.3

From Castillo L, et al. (68).
[a] Diets given for 7 days before tracer studies.
[b] μmol/kg/h ($n = 5$).
[c] Significantly ($p < 0.05$) lower than arginine-rich.

by protein breakdown and the turnover of the guanidino moiety during arginine formation and urea production, via the hepatic urea cycle, amounts to ≥350 μmol/kg/h for subjects consuming an adequate diet (see, e.g., 54). This is far higher than the flux shown here of about 50–60 μmol/kg/h. The likely explanation for this difference is that the intravenous arginine tracer fails to exchange extensively with the arginine pool that is continually being formed and hydrolyzed via the hepatic urea cycle enzymes. Indeed, biochemical evidence now reveals that there is a tight channeling of intermediates between the interacting enzymes of the urea cycle; Cheung et al. (55) have observed a channeling of urea cycle intermediates at each of the three cytoplasmic reactions of the pathway (argininosuccinate synthetase; argininosuccinate lyase; arginase), and it is suggested that these enzymes are spatially organized in relation to the mitochondrial membrane. The channeling of arginine from acylase to arginase is extremely tight, while that of argininosuccinate between synthetase and lyase is somewhat less tight (55). Furthermore, external (intracellular) ornithine also is tightly channeled to ornithine transcarbamylase, which is located within the mitochondrial matrix (56). Additionally, there is a channeling of arginyl-tRNA in the liver cell (57), with the suggestion that one arginyl-tRNA pool is used for protein synthesis and another is involved in the arginylation of proteins, which serves as a signal for ubiquitin-dependent protein degradation.

This intracellular compartmentation of arginine metabolism within the liver, as revealed by these various biochemical studies and further exposed by our tracer studies in humans, is of considerable functional significance; it evidently allows a maintenance of arginine homeostasis while permitting formation of significant quantities of urea without draining the cytosolic arginine compartment required for protein synthesis in the liver or without depleting the plasma pool of arginine, which is used to meet the amino acid needs in extrahepatic tissues and organs. Without this metabolic and structural compartmentation, the daily requirement for arginine, furnished either by way of the diet or by way of endogenous synthesis, would presumably be exceedingly high. There is in fact now extensive evidence that cellular metabolism is largely spatially organized (as for arginine, as discussed above), and this is on a scale much smaller than that of the well-known organelles, with the benefits being efficient substrate transfer, protection of labile intermediates, and improved coordination (58–61).

In addition to this intracellular compartmentation of arginine metabolism, there is also a functional specialization of different hepatocyte populations and regions within the liver (62,63). Thus urea synthesis occurs largely in the periportal compartment, which is thought to provide an effective means for adjusting ammonia flux into either urea or glutamine formation, according to the metabolic needs of the organism (60). Hence there is a distinct interorgan, intraorgan, and intracellular compartmentation of arginine metabolism, adding both complexity and detail to the regulation and control of amino acid metabolism *in vivo*, which would not be readily evident from *in vitro* and cellular studies.

Second, it will be noted from Table 5 that when subjects consumed an arginine-free diet for 7 days the citrulline flux and conversion of citrulline to arginine were

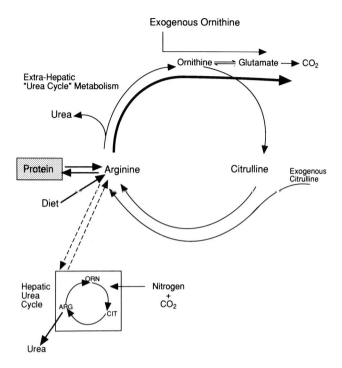

FIG. 8. Schematic view of the relationships between the metabolism of extrahepatic urea cycle arginine and its precursors, ornithine and citrulline. It is proposed that arginine homeostasis is achieved via alterations in the dietary intake level of arginine and the rate of arginine catabolism, via ornithine and glutamate formation. The net synthesis of arginine, via conversion of ornithine to citrulline (in intestine) and of the citrulline to arginine (in kidney), is assumed to be less influenced by a regulatory control. [From Castillo L, et al. Proc Natl Acad Sci USA 1993; 90: 7749–7753.]

unaffected, indicating that the net rate of *de novo* arginine synthesis is not altered by acute changes in arginine intake. On the basis of these and our additional findings (64) we have developed a scheme that helps to explain, in metabolic/regulatory terms, why arginine is required (conditionally indispensable) under various pathological conditions. Specifically, we now propose (see Fig. 8) that whole body arginine homeostasis is achieved, in healthy adults, principally via a regulation in the rate of arginine degradation rather than via changes in the net rate of new arginine formation (from ornithine and citrulline, and especially in extrahepatic tissues). Further, this proposition leads to a prediction that those conditions that greatly increase the metabolic demand for arginine, such as major trauma or during marked catch-up growth (65–67), may precipitate a deficiency of arginine unless there is an adequate or increased supply of arginine, or citrulline, from exogenous (enteral or parenteral) sources.

Additionally, we would predict that in comparison with citrulline and arginine, exogenous ornithine would be a relatively poor substrate for maintaining body arginine balance, and there is evidence in favor of this suggestion (see 68). It seems that a generous supply of ornithine leads to its preferential conversion to glutamate rather than to citrulline and subsequent formation of arginine.

In summary, in this section we have made particular reference to studies of arginine metabolism, as an example, to emphasize *in vivo* aspects of the biochemistry and physiology of amino acid metabolism, especially the importance of metabolic interactions and cooperation among and within organs and the compartmentation of metabolism within cells. Also, with respect to our recent arginine tracer studies in human subjects mentioned above, it has now become possible to propose a metabolically based hypothesis that explains why arginine is a conditionally indispensable amino acid (65–67). Regrettably, further detailed discussion of the metabolism of the other conditionally indispensable and dispensable amino acids (Table 1) cannot be made here, but some further aspects of this topic are covered elsewhere in this volume.

Protein Turnover: Quantitative Aspects and Nutritional Implications

While the mechanisms of protein synthesis and degradation have been discussed briefly above, mention is given here to the *in vivo* rates of these processes, with a comment on their nutritional implications to make the point that protein synthesis rates are high in the newborn and, per unit of body weight, decline with growth and development (Table 6) (30). Three important issues can be raised from the data shown here. First, the higher rate of protein synthesis in the young individual, as compared with the adult, is related not only to the *net* protein deposition associated with growth but also to a high rate of total protein turnover (synthesis and breakdown). Hence protein *synthesis* in the premature infant is about twice as high as in the preschool child and approximately three or four times as high as in the adult. In parallel with these synthesis rates, the rates of organ and tissue protein *breakdown* are also considerably higher in the infant than in the adult. Second, at all ages, the rates of whole body protein *synthesis* and breakdown are considerably greater than intakes of dietary protein necessary to meet maintenance or for support of growth. It follows,

TABLE 6. *Comparison of whole body protein synthesis rates with dietary protein allowances at different ages*

Age group	Protein synthesis (A)[a]	Protein allowance (B)[a]	Ratio A/B
Infant (premature)	11.3, 14	~3.0	4.3
Newborn	6.7	1.85	3.6
Child (15 months)	6.3	1.3	4.8
Child (2–8 yr)	3.9	1.1	3.5
Adolescent (~13 yr)	~5.0	1.0	~5.0
Young adult (~20 yr)	~4.0	0.75	~5.3
Elderly (~70 yr)	~3.5	0.75	4.7

From Young VR, et al. *Acta Pediatr Scand Suppl* 1992; 375: 5–24.
[a] Protein, g/kg/d.

therefore, that there is an extensive reutilization within the body of the amino acids entering tissue pools during the course of protein breakdown. Third, it will be noted that there is a parallelism between age- and development-associated changes in protein turnover and the dietary protein needs. This implies that when rates of synthesis and breakdown of body proteins change in response to various stimuli, such as infection and trauma, the dietary requirement for nitrogen and specific amino acids will be affected.

The potential importance of protein synthesis and turnover as determinants of the dietary requirement for protein and for indispensable amino acids has been discussed previously (30). Also, the oxidation of amino acids is sensitive to alterations in host and dietary factors, as discussed earlier, and we have argued that those factors regulating the rate of amino acid oxidation are key determinants of the quantitative requirement for specific indispensable amino acids (30).

Energy Dependency of Protein and Amino Acid Metabolism

When considering some of the more fundamental aspects of the major systems described above, mention was made that they involved the utilization of metabolic energy. Furthermore, protein and amino acid metabolism are intimately connected with, and affected by, the major energy-yielding substrates and the energy status of the body (69,70). Therefore, in this final section we highlight some of these relationships, including their biochemical basis and quantitative features.

Qualitative Aspects

In Table 7 a list is provided of some of the major processes that require energy, in the form of high energy phosphate bonds (ATP or GTP). These are intimately involved in the utilization of nitrogen and amino acids, in the formation of polypeptides and their further assembly within or outside the cell, and ultimately in the degradation of proteins and the catabolism of their constituent amino acids (69,70).

Briefly, in addition to the well-recognized role of these high energy intermediates in the initiation and elongation of polypeptide synthesis, work in the past five or so years has exposed and further clarified the mechanisms responsible for the regulation of cellular processes, such as protein folding and aggregation (71), intracellular traffic of newly synthesized proteins (72), and transport of proteins across cellular and intracellular membranes (73,74). Most of these involve the participation of energy in the form of ATP or GTP or their derivatives (cAMP; cGMP) through a series of complex protein–protein interactions and second messenger cascades (75,76).

Quantitative Aspects

While the processes mentioned above, as well as others, are energy-dependent, it remains uncertain how much food energy is required to drive most of them *in vivo*.

TABLE 7. *Some energy-dependent processes associated with protein turnover and amino acid homeostasis*

Protein turnover
  Formation of initiation complex
  Peptide bond synthesis
  Protein degradation
    Ubiquitin-dependent
    Ubiquitin-independent
    Autophagic degradation (sequestration, lysosomal proton pump)
RNA turnover
  rRNA; tRNA
  Pre-mRNA splicing (spliceosome) and mRNA
Amino acid transport
Regulation and integrity
  Reversible phosphorylation; enzymes, factors
  GTP-GDP exchange proteins (signal transduction)
  Second messengers (phosphatidyl inositol system)
  Ion pumps and channels
  ATP-dependent heat shock proteins (folding)
  Protein translocation
Nitrogen metabolism
  Glutamate/glutamine cycle
  Glucose-alanine cycle
  Urea synthesis

From Young VR, et al. *Energy metabolism: tissue determinants and cellular corollaries.* New York: Raven Press, 1992; 439–66.

This problem has been pointed out by Waterlow & Millward (77). Hence although we now appreciate better the molecular and cellular events and processes that underlie the interactions between protein and energy metabolism, there remains a considerable challenge to develop the necessary techniques and approaches for quantifying the energy required by these processes *in vivo*. It is abundantly clear that the energy costs of protein turnover exceed the utilization of the ATP and GTP used for activation of amino acids and for the formation of the peptide bonds during the elongation phase of mRNA translation. Earlier estimates of the energy costs of protein turnover (78), therefore, probably underestimate the functional significance of protein turnover in whole body energetics, as has also been emphasized by Hawkins (79).

Four additional points are made here. First, while it has often been assumed that 4 mol of high energy phosphate bond is required for peptide bond formation (e.g., 77,78,80) (2 ATP for amino acid activation and coupling to tRNA and 2 GTP for peptide bond synthesis), it now appears that *three* molecules of GTP are required for each elongation step of protein synthesis (81). Also, the cost of amino acid transport has been taken to be equivalent to 1 mol of ATP per mole of amino acid incorporated into protein (78,80) but if the γ-glutamyl moiety of glutathione participates in the transport of amino acids (cystine and certain other neutral amino acids) then 3 mol of ATP per mole of amino acid would be expended in the operation (or one turn) of the γ-glutamyl cycle (82).

Second, there remains a great deal of uncertainty concerning the energy cost of protein degradation. The complex process of degradation is now beginning to be worked out and it is known, as mentioned above, that this proceeds via a series of lysosomal and non-lysosomal mechanisms and that inhibitors of energy metabolism can arrest protein degradation (83). For example, it appears that at least 14% of ATP utilization in reticulocytes is involved in intracellular protein degradation (84). As mentioned earlier, several mechanisms can account for this energy expenditure, including (a) ubiquitin conjugation; (b) a non-lysosomal protease, which apparently requires two ATPs for the cleavage of each peptide bond (85,86); and (c) the ATP requirement for entry of proteins into lysosomes and to drive the proton pumps involved in maintaining a pH gradient (87). Clearly, energy is transformed and utilized during protein turnover but exactly how much, it is still not possible to state.

Third, we need to understand the quantitative and functional aspects of the link between protein turnover and ion pumps, particularly the Na^+/K^+-ATPase complex. A major function of this pump is to convert the chemical energy from the hydrolysis of ATP into a gradient for Na^+ and K^+; these gradients are used as free-energy sources for a number of processes, such as (a) formation of the membrane potential, (b) cell volume regulation, and (c) transport of glucose and amino acids into cells against concentration gradients (88). The contribution of Na^+/K^+-ATPase activity to cellular and organ energy metabolism has been reviewed recently by Park et al. (89) and they conclude that it accounts for 9–45%, 16–51%, and 17–61% of the oxygen utilization in the skeletal musculature, liver, and gastrointestinal tract, respectively, under a variety of physiological and nutritional conditions. A study by Qian et al. (90) is of interest here, since these investigators showed that the A system amino acid transporter and the mRNA for the α_1 subunit of the Na^+/K^+-ATPase are coordinately controlled by regulatory gene R1. Furthermore, as already implied above, the Na^+ electrochemical gradient, established by the Na^+/K^+-ATPase, across the cell membrane serves as the driving force for the A system of amino acid transport. From these various observations it is reasonable to suggest that the energy needs of ion pumping should be taken into account in a quantitative consideration of the energy

TABLE 8. *Whole body protein turnover in relation to resting metabolic rate (MR) in adults of mammalian and avian species*

Species	Weight (kg)	Protein turnover (A) (g/kg/d)	MR (B) (kJ/kg/d)	Ratio B/A
Mouse	0.04	43.5	760	11
Rat	0.35	22.0	364	17
Rabbit	3.6	9.2	192	20
Wallaby (Parma)	4.2	7.5	163	21
Sheep	63	5.6	96	17
Man	70	4.6	107	23
Cow	575	3.0	60	20
Birds (chickens)	1.4	27	439	16

From summary of literature by Young VR, et al. (70), where original references are given.

cost of protein and amino acid turnover. The problem, of course, is the proportionate extent to which the energetics of ion movement ought to be assigned as the primary cost of protein turnover.

Fourth, *in vivo* and whole body aspects of protein–energy relationships have been reviewed (91) and discussed recently elsewhere (69,70,92). Hence we shall not review these in detail, but from the relationship between resting metabolic rate and whole body protein turnover (Table 8) it can be estimated that about 15–20 kJ (4–5 kcal) of basal energy expenditure is associated with each gram of protein synthesis (92,93). On this basis peptide bond formation alone appears to account for approximately 20% of basal metabolism. It follows, then, that major changes in body protein turnover, due for example in response to severe infection or major trauma (93,94), could not only affect the protein and amino acid requirements but also the status of body energy expenditure and requirements.

SUMMARY AND CONCLUSIONS

We have attempted to present an overview of selected aspects of the biochemistry and physiology of protein and amino acid metabolism, giving particular attention to their nutritional significance. The major systems responsible for the maintenance of body protein and amino acid homeostasis and function have been described in brief biochemical detail; these are protein synthesis, protein degradation, amino acid oxidation and urea production, the synthesis of those amino acids (dispensable; conditionally indispensable) for which there is an enzymatic capacity, and amino acid transport. *In vivo* features of these systems were also considered in an effort to convey both the value and the limitations of the knowledge derived from the more mechanistic or reductionist levels of biological enquiry. Nevertheless, details of the molecular and cellular processes that are involved in these various systems are expanding our appreciation for the functional significance of changes in the dietary level of protein and/or its constituent amino acids. Potentially, this will lead to improved estimates of nutritional requirements and enhanced diagnostic tools for evaluation of nutritional status at various stages of life and under differing physiological and disease states.

ACKNOWLEDGMENT

The authors' studies were supported by NIH grants DK 15856 and DK 42101.

REFERENCES

1. Considine DM, Considine GC. *Amino acids. Van Nostrand Reinhold encyclopedia of chemistry*, 4th ed. New York: Van Nostrand Reinhold, 1984; 50–57.
2. Lee BJ, Worland JN, Davis JN, et al. Identification of a selenocysteyl-tRNA^{ser} in mammalian cells that recognizes the nonsense codon, UGA. *J Biol Chem* 1989; 264: 9724–7.

3. Rose WC, Haines WJ, Warner DJ. The amino acid requirements of man. V. The role of lysine, arginine and tryptophan. *J Biol Chem* 1954; 206: 421–30.
4. FAO/WHO/UNU. *Energy and protein requirements. Report of a joint FAO/WHO/UNU Expert Consultation.* WHO Technical Report Series No 724. Geneva: World Health Organization, 1985; 206 pages.
5. Moncada S, Higgs E, eds. *Nitric oxide from L-arginine: a bioregulatory system.* Amsterdam: Excerpta Medica, 1990.
6. Morimoto RI. Cells in stress: transcriptional activation of heat shock genes. *Science* 1993; 259: 1409–10.
7. Rhoads RE. Regulation of eukaryotic protein synthesis by initiation factors. *J Biol Chem* 1993; 268: 3017–20.
8. Pain VM. Initiation of protein synthesis in mammalian cells. *Biochem J* 1986; 235: 625–37.
9. Moldave K. Eukaryotic protein synthesis. *Annu Rev Biochem* 1985; 54: 1109–49.
10. Melefors O, Hentze MW. Translational regulation by mRNA/protein interactions in eukaryotic cells: ferritin and beyond. *Bio-Essays* 1993; 15: 88–90.
11. Haele DJ, Rouault T, Harfford JB, Kennedy MC, Blondin GA, Beirnert Klausner D. Cellular regulation of the iron-responsive binding protein: disassembly of the culiane iron–sulfur cluster results in high affinity RNA binding. *Proc Natl Acad Sci USA* 1992; 89: 11735–9.
12. deJong FA, Schreiber G. Messenger RNA levels of plasma proteins in rat liver during protein depletion and refeeding. *J Nutr* 1987; 117: 1795–800.
13. Reeds PJ, Davis TA. Hormonal regulation of muscle protein synthesis and degradation. In: Buttery PJ, Boorman KN, Lindsay DB, eds. *Control of fat and lean deposition.* Oxford: Butterworth-Heinemann, 1992; 1–26.
14. Sugden PH, Fuller SJ. Regulation of protein turnover in skeletal and cardiac muscle. *Biochem J* 1991; 273: 21–37.
15. Millward DJ, Rivers JPW. The need for indispensable amino acids: the concept of the anabolic drive. *Diabetes Metab Rev* 1989; 5: 191–211.
16. Dice F. Molecular determinants of protein half-lives in eukaryotic cells. *FASEB J* 1987; 1: 349–57.
17. Rechsteiner M. Natural substrates of the ubiquitin proteolytic pathway. *Cell* 1991; 66: 615–8.
18. Stadtman ER. Protein oxidation and aging. *Science* 1992; 257: 1220–4.
19. Beynon RJ, Bond JS. Catabolism of intracellular protein: molecular aspects. *Am J Physiol* 1986; 251: C141–52.
20. Goldberg AL. The mechanism and functions of ATP-dependent proteases in bacterial and animal cells. *Eur J Biochem* 1992; 203: 9–23.
21. Goldberg AL, Rock KL. Proteolysis, proteasomes and antigen presentation. *Nature* 1992; 357: 375–9.
22. Rechsteiner M, Hoffman L, Dubeil W. The multicatalytic and 26S proteases. *J Biol Chem* 1993; 268: 6065–8.
23. Mortimore GE, Poso AR. Intracellular protein catabolism and its control during nutrient deprivation and supply. *Annu Rev Nutr* 1987; 7: 539–64.
24. Young VR, Marchini JS. Mechanisms and nutritional significance of metabolic responses to altered intakes of protein and amino acids, with reference to nutritional adaptation in humans. *Am J Clin Nutr* 1990; 51: 570–89.
25. Youngman LD, Part J-YK, Ames BN. Protein oxidation associated with aging is reduced by restriction of protein or calories. *Proc Natl Acad Sci USA* 1992; 89: 9112–6.
26. Harper AE. Enzymatic basis for adaptive changes in amino acid metabolism. In: Taylor TG, Jenkins NK, eds. *Proceedings of the XIII international congress of nutrition.* London: John Libbey, 1986; 409–94.
27. Munro HN. Free amino acid pools and their role in regulation. In: Munro HN, ed. *Mammalian protein metabolism,* Vol IV. New York: Academic Press, 1970; 299–386.
28. Krebs HA. Some aspects of the regulation of fuel supply in omnivorous animals. In: Weber G, ed. *Advances in enzyme regulation,* Vol 10. Elmsford, NY: Pergamon Press, 1972; 397–420.
29. Waterlow JC, Fern EB. Free amino acid pools and their regulation. In: Waterlow JC, Stephen JML, eds. *Nitrogen metabolism in man.* London: Applied Science Publishers, 1981; 1–16.
30. Young VR, Meredith C, Hoerr R, *et al.* Amino acid kinetics in relation to protein and amino acid requirements: the primary importance of amino acid oxidation. In: Garrow JS, Halliday D, eds. *Energy and substrate metabolism in man.* London: John Libbey, 1985; 119–33.
31. Jungas RL, Halperin ML, Brosnan JT. Quantitative analysis of amino acid oxidation and related gluconeogenesis in humans. *Physiol Rev* 1992; 72: 419–48.

32. Harper AE, Miller RH, Block KP. Branched-chain amino acid metabolism. *Annu Rev Nutr* 1984; 4: 409–54.
33. Cohen PP. Regulation of the ornithine-urea cycle enzymes. In: Waterlow JC, Stephen JML, eds. *Nitrogen metabolism in man*. London: Applied Science Publishers, 1981; 215–28.
34. Souba WW. Interorgan ammonia metabolism in health and disease: a surgeon's view. *J Parent Enteral Nutr* 1987; 11: 569–79.
35. Meijer AJ, Lamers WH, Chamuleau RAFM. Nitrogen metabolism and ornithine cycle function. *Physiol Rev* 1990; 70: 701–48.
36. Schimke RT. Adaptive characteristics of urea cycle enzymes in the rat. *J Biol Chem* 1962; 237: 459–68.
37. Morris SM. Regulation of enzymes of urea and arginine synthesis. *Annu Rev Nutr* 1992; 12: 81–101.
38. Das TK, Waterlow JC. The rate of adaptation of urea cycle enzymes, aminotransferases and glutamic dehydrogenase, to changes in dietary protein intake. *Br J Nutr* 1974; 32: 353–73.
39. Jackson AA. Critique of protein–energy interactions *in vivo*: urea kinetics. In: Scrimshaw NS, Schurch B, eds. *Protein–energy interactions*. Lausanne: IDECG, Nestlé Foundation, 1992; 63–79.
40. Mitch WE, Lietman PS, Walser M. Effects of oral neomycin and kanamycin in chronic uremic patients. I. Urea metabolism. *Kidney Int* 1977; 11: 116–23.
41. Mitch WE, Walser M. Effects of oral neomycin and kanamycin in chronic uremic patients. II. Nitrogen balance. *Kidney Int* 1977; 11: 123–7.
42. Souba WW, Pacitti AJ. How amino acids get into cells: mechanisms, models, menus, and mediators. *J Parent Enteral Nutr* 1992; 16: 569–78.
43. Biolo G, Chinkes D, Zhang X-J, Wolfe RR. A new model to determine *in vivo* the relationship between amino acid transmembrane transport and protein kinetics in muscle. *J Parent Enteral Nutr* 1992; 16: 305–15.
44. Shotwell MA, Kilberg MS, Oxender DL. The regulation of neutral amino acid transport in mammalian cells. *Biochim Biophys Acta* 1983; 737: 267–84.
45. Christensen HN. Role of amino acid transport and countertransport in nutrition and metabolism. *Physiol Rev* 1990; 70: 43–77.
46. Collarini EJ, Oxender DL. Mechanism of transport of amino acids across membranes. *Annu Rev Nutr* 1987; 7: 75–90.
47. Munro HN. Metabolic basis of nutritional care in liver and kidney disease. In: Winter RW, Green HL, eds. *Nutritional support of the seriously ill patient*. New York: Academic Press, 1983; 93–105.
48. Bier DM. Intrinsically difficult problems: the kinetics of body proteins and amino acids in man. *Diabetes Metab Rev* 1989; 5: 111–32.
49. Christensen HN. Interorgan nutrition: introductory comments from the chain. *Fed Proc* 1986; 45: 2165–6.
50. Brusilow SW, Horwich AL. Urea cycle enzymes. In: Scriver CR, Beaudet AL, Sly WS, Valle D, eds. *The metabolic basis of inherited disease*. New York: McGraw-Hill, 1989; 629–63.
51. Hoogenrood N, Totino N, Elmer H, et al. Inhibition of intestinal citrulline synthesis causes severe growth retardation in rats. *Am J Physiol* 1985; 249: G792–9.
52. Dhanakoti SN, Brosnan JT, Herzberg GR, Brosnan ME. Renal arginine synthesis: studies *in vitro* and *in vivo*. *Am J Physiol* 1990; 259: E437–42.
53. Tizianello A, DeFerrari G, Garibotto G, et al. Renal metabolism of amino acids and ammonia in subjects with normal renal function and in patients with chronic renal insufficiency. *J Clin Invest* 1980; 65: 1162–73.
54. Matthews DE, Downey RS. Measurement of urea kinetics in humans: a validation of stable isotope tracer methods. *Am J Physiol* 1984; 246: E519–27.
55. Cheung C-W, Cohen NS, Raijman L. Channeling of urea cycle intermediates *in situ* in permeabilized hepatocytes. *J Biol Chem* 1989; 264: 4038–44.
56. Cohen NS, Cheung C-W, Raijman L. Channeling of extramitochondrial ornithine to matrix ornithine transcarbamylase. *J Biol Chem* 1987; 262: 203–8.
57. Sivram P, Deutscher MP. Existence of two forms of rat liver arginyl-tRNA for protein synthesis. *Proc Natl Acad Sci USA* 1990; 87: 3665–9.
58. Srere PA. Complexes of sequential metabolic enzymes. *Annu Rev Biochem* 1987; 56: 89–124.
59. Srere PA, Ovadi J. Enzyme–enzyme interactions and their metabolic role. *FEBS Lett* 1990; 268: 360–4.
60. Savageau MA. Metabolic channeling: implications for regulation of metabolism and for quantitative description of reactions *in vivo*. *J Theor Biol* 1991; 152: 85–92.

61. Ovadi J, Srere PA. Channel your energies. *Trends Biochem Sci* 1992; 17: 445–7.
62. Jungermann K, Katz N. Functional specialization of different hepatocyte populations. *Physiol Rev* 1989; 69: 708–64.
63. Haussinger D. Nitrogen metabolism in liver: structural and functional organization and physiological relevance. *Biochem J* 1990; 267: 281–90.
64. Castillo L, Ajami A, Branch S, et al. Plasma arginine kinetics in adult man: response to an arginine-free diet. *Metabolism* 1993 (in press).
65. Zieve L. Conditional deficiencies of ornithine or arginine. *J Am Coll Nutr* 1986; 5: 167–76.
66. Visek WJ. Arginine and disease states. *J Nutr* 1985; 115: 532–41.
67. Visek WJ. Arginine needs, physiological state and usual diets: a reevaluation. *J Nutr* 1986; 116: 36–46.
68. Castillo L, Chapman TE, Sánchez M, et al. Plasma arginine and citrulline kinetics in adults given adequate and arginine-free diets. *Proc Natl Acad Sci USA* 1993; 90: 7749–53.
69. Young VR, Yu Y-M, Fukagawa NK. Protein and energy interactions throughout life: metabolic basis and nutritional implications. *Acta Pediatr Scand Suppl* 1992; 375: 5–24.
70. Young VR, Yu Y-M, Fukagawa NK. Whole body energy and nitrogen (protein) relationships. In: Kinney JM, Tucker HG, eds. *Energy metabolism: tissue determinants and cellular corollaries*. New York: Raven Press, 1992; 139–60.
71. Langer T, Lu C, Echols H, Flanagan J, Hayer MK, Hartl VU. Successive action of Dnak, DnaJ and GroEL along the pathway of chaperone-mediated protein folding. *Nature* 1992; 356: 683–9.
72. Lingappa YR. Intracellular traffic of newly synthesized proteins: current understanding and future prospects. *J Clin Invest* 1992; 83: 739–51.
73. Rapoport TA. Transport of proteins across the endoplasmic reticulian membrane. *Science* 1992; 15: 355–8.
74. Rothman JE, Orci L. Molecular dissection of the secretory pathway. *Nature* 1992; 355: 409–15.
75. Bourne HR, Sanders DA, McCormick F. The GTPase super family: conserved structure and molecular mechanism. *Nature* 1991; 349: 117–27.
76. Simon MI, Strathmann MP, Gautam N. Diversity of G proteins in signal transduction. *Science* 1991; 252: 802–8.
77. Waterlow JC, Millward DJ. Energy cost of turnover of protein and other cellular constituents. In: Weiser W, Gnaiger E, eds. *Energy transformations in cells and organisms*. Stuttgart: George Thieme Verlag, 1990; 277–82.
78. Millward DJ, Garlick PJ, Reeds PJ. The energy cost of growth. *Proc Nutr Soc* 1976; 35: 339–49.
79. Hawkins AJS. Protein turnover: a functional appraisal. *Funct Ecol* 1991; 5: 222–33.
80. Reeds PJ, Fuller MF, Nicholson BA. Metabolic basis of energy expenditure with particular reference to protein. In: Garrow JS, Halliday D, eds. *Substrate and energy metabolism in man*. London: John Libbey, 1985; 46–56.
81. Weijland A, Parmeygiani A. Toward a model for the interaction between elongation factor Tu and the ribosome. *Science* 1993; 259: 1311–4.
82. Meister A, Larsson A. Glutathione synthetase deficiency and other disorders of the γ-glutamyl cycle. In: Scriver CR, Beaudet AL, Sly WS, Valle D, eds. *The metabolic basis of inherited disease*, Vol 1. New York: McGraw-Hill, 1989; 855–68.
83. Goldberg AL, St John AC. Intracellular protein degradation in mammalian and bacterial cells. Part 2. *Annu Rev Biochem* 1976; 45: 747–803.
84. Siems W, Dubiel W, Dumdey R, et al. Accounting for the ATP-consuming processes in rabbit reticulocytes. *Eur J Biochem* 1984; 139: 101–7.
85. Goldberg AL, Menon AS, Goff S, Chin DT. Molecular determinants of degradation of proteins. *Biochem Soc Trans* 1987; 15: 809–11.
86. Menon AS, Waxman L, Goldberg AL. The energy utilized in protein breakdown by the ATP-dependent protease (L_{2a}) from *Escherichia coli*. *J Biol Chem* 1987; 262: 722–6.
87. Dean RT, Barrett AJ. Lysosomes. In: Campbell PN, Aldridge WN, eds. *Essays in biochemistry*. London: Academic Press, 1976; 1–40.
88. Skou JC. The energy coupled exchange for Na^+ and for K^+ across the cell membrane: the Na^+, K^+-pump. *FEBS Lett* 1990; 268: 314–24.
89. Park HS, Kelley JM, Millian LP. Energetics and cell membranes. In: Kinney JM, Tucker HN, eds. *Energy metabolism: tissue determinants and cellular corollaries*. New York: Raven Press, 1992; 411–35.
90. Qian N-X, Pastor-Anglada M, Englesberg E. Evidence for coordinate regulation of the A system for amino acid transport and the mRNA for the α_1 subunit of the Na^+, K^+-ATPase gene in Chinese hamster ovary cells. *Proc Natl Acad Sci USA* 1991; 88: 3416–20.

91. Kinney JM, Tucker HG, eds. *Energy metabolism: tissue determinants and cellular corollaries*. New York: Raven Press, 1992.
92. Young VR, Yu Y-M, Fukagawa NK. Energy and protein turnover. In: Kinney JM, Tucker HG, eds. *Energy metabolism: tissue determinants and cellular corollaries*. New York: Raven Press, 1992; 439–66.
93. Waterlow JC. Protein turnover with special reference to man. *Q J Exp Physiol* 1984; 69: 409–38.
94. Young VR. Tracer studies of amino acid kinetics: a basis for improving nutritional therapy. In: Tanaka T, Okada A, eds. *Nutritional support in organ failure*. Amsterdam: Excerpta Medica, 1990; 3–34.

DISCUSSION FOLLOWING THE PRESENTATION OF DR. YOUNG*

Dr. Garlick: I should like to reemphasize a couple of the points. The first one is the concept of essentiality. I think we often forget just how many processes amino acids are required for, apart from the obvious ones concerned with growth and maintenance. Take arginine, for example. Apart from being a necessary part of the urea cycle and being required for nitric oxide synthesis, another extremely important function is polyamine synthesis. Arginine is the main precursor of polyamines. Glycine is another amino acid which takes part in a considerable number of metabolic processes, being a precursor of creatine, heme, and nucleic acids for a start. It is not like tyrosine, for example, which is mainly used for synthesis of hormones such as thyroid hormone, epinephrine, and so on. You would expect the requirement for such amino acids to be small and perhaps not significant on the nutritional scale. However, amino acids like arginine and glycine are needed in large quantities.

The other point I would like to emphasize is this problem of compartmentation. It is something we tend to ignore because it makes life difficult, particularly when trying to assess rates of protein metabolism using trace isotopes. Compartmentation makes it difficult to look at what is going on in the whole body because it prevents there being a homogeneous pool of free amino acids and a homogeneous pool of protein-bound amino acids. In studies on infants you are not allowed to start sampling intracellular pools or to look at portal amino acids *vs* systemic amino acids. But obviously we have to take account of compartmentation since every tissue will have its own compartments and we have to take note of these. This is exactly the reason why I said during my presentation that we have to be cautious about the precision of the methods we use. We do our best, we make what measurements we can, and we are always taking steps forward. But even now we know that many of the measurements we make must have errors and we must always be trying to identify and evaluate these errors.

Dr. Rassin: One of the things that came out of the brain work was that different compartments could be defined by using different precursors. You have very different metabolism if you use glutamate as a precursor as opposed to glucose in the Krebs cycle, for example. I often wonder whether the differences one sees with the various stable isotope amino acid precursors reflect the compartmentation of their metabolism and whether this could be used for designing more complex models of how those compounds are metabolized.

Dr. Garlick: Yes, this is exactly so. It was done a long time ago: Eddy Fern and I did experiments using double labeling of serine and glycine to demonstrate the synthesis of serine from glycine, and *vice versa,* and illustrate the problem of compartmentation. I do agree with you that the differences in results between using a lysine infusion and a leucine infusion to

---

* Dr. Young was unable to attend the symposium and his paper was read by Dr. Peter Garlick.

measure whole body protein turnover are due to differences in amino acid compartmentation. I think these differences could be used to investigate compartmentation.

Dr. Räihä: Vernon Young's paper is of course very theoretical but I think it brings up some important points that have clinical implications:

1. He has clearly shown that the process of protein oxidation acts as a regulator, such that at low protein intakes you decrease oxidation and at high protein intakes you increase not only synthesis but also breakdown.

2. When we use urea as a marker of excessive or inadequate protein intake we should keep in mind that the very low birthweight infant may not be able to react with an increased urea production as well as an older infant, due to low enzyme capacity.

3. As a child gets older both protein synthesis and protein breakdown decrease. This of course will affect the protein requirement. In the preterm infant, who has rapid protein synthesis, there is also rapid breakdown, and this means a higher protein requirement. As the baby gets older both protein synthesis and protein breakdown decrease, and the protein requirement diminishes.

4. Finally, there is the question of infection. In infected preterm infants, even before the appearance of clinical symptoms, we can see signs of protein breakdown evidenced by increased urea production.

Dr. Young's paper, although theoretical, emphasizes some very important clinical points. It is desirable that we should read it and gain a better understanding of this entire field.

Dr. Garlick: I should like to comment on the point you just made about a faster rate of protein turnover in some way increasing the requirement for protein. The requirement really is for the net deposition. There is nothing in the process of protein synthesis or degradation *per se* that loses amino acids. The processes of protein synthesis and protein degradation leave the amino acids intact and they can then be recycled or used again for another purpose. If, in order to maintain a faster rate of protein turnover, you maintain higher concentrations of amino acids, people tend to think that this will result in greater losses simply because the concentrations are higher. However, it is not intrinsic to the process of turnover that they should be lost. It is to do with the way in which the amino acid degradation steps are regulated.

Dr. Tracey: I am struck by how much we don't know! We refer to processes and pathways and to end products, but if you were to try to develop a drug to improve general nutrition you would first need to identify a target enzyme and from that enzyme you would develop a drug that would up- or downregulate the enzyme. With regard to protein synthesis, as Dr. Rassin pointed out, it has now become very important to look at how the gene is being regulated in skeletal muscle but that is still veiled in mystery. I don't think we know any of the enzymes that are directly responsible for maintaining structural proteins in skeletal muscle or how they work.

Dr. Rassin: The whole area of up- and downregulation of these genes is important. People who are doing research in aging have been trying to understand for example why protein malnutrition results in longevity, which has been well known for a long time, particularly in rats—you can make rats live much longer by starving them. I have been trying to understand what happens to the genetic regulation of metabolism in such cases and it is clear that the amount of protein fed over a long period of time changes many of the genes that are involved in longevity and perhaps some that may be involved in cancer development. Dr. Rey has done some very nice work on when the enzyme activities in the urea cycle appear, but we don't know much about what the DNA is doing and how the genetic regulation of that appearance takes place and what stimulates those enzymes to start becoming active in early development.

Isotopic Methods for Studying Protein Turnover

Peter J. Garlick and Margaret A. McNurlan

*Department of Surgery, State University of New York,
Stony Brook, New York 11794-8191, USA, and The Rowett Research Institute,
Aberdeen AB2 9SB, Scotland UK*

The maintenance of body protein mass depends on the dynamic equilibrium between the opposing processes of protein synthesis and protein degradation. During growth in the young, protein synthesis exceeds degradation, whereas during malnutrition or illness degradation predominates, resulting in the net loss of body protein. An understanding of the nutritional and pathological factors regulating body protein balance therefore requires methods for the accurate assessment of rates of protein synthesis and degradation. These methods are based on the use of isotopically labeled amino acids and generally employ stable isotopes of carbon (^{13}C), nitrogen (^{15}N), or hydrogen (deuterium, ^2H). The aim of this chapter is to review briefly the methods available for measuring protein synthesis and degradation with isotopes in the human, emphasizing particularly some of those which either are, or might possibly become, suited to studies in infants or which have recently undergone significant developments.

WHOLE BODY PROTEIN TURNOVER

The large majority of studies of human protein turnover, particularly those in infants, have measured rates of whole body protein synthesis and degradation, because this can be done relatively non-invasively using samples of blood or urine. The basic approach involves the administration of a labeled amino acid, followed by collection of blood or urine samples over the next few hours, for measurement of the isotopic enrichment in a precursor or product of body protein metabolism. The analysis then requires a metabolic model of body protein metabolism which will fit the data obtained and will permit the rates of protein synthesis and degradation to be computed. The model shown in Fig. 1 is the one which has been most widely used for this purpose (1). In this the body protein metabolism is divided between two compartments, the free amino acid pool (metabolic pool) and the protein pool. All other subcompartments are ignored, such as the distinctions between individual tissues or between

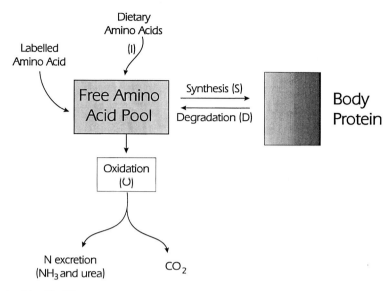

FIG. 1. Simplified illustration of whole body protein metabolism. This two-pool model is frequently used to calculate rates of whole body protein turnover from data obtained by labeling with ^{13}C- or ^{15}N-labeled amino acids. For details, see the text.

plasma and intracellular free pools. The isotope is introduced into the free pool, either intravenously or intragastrically (orally). The distinctions between the various approaches then depend on exactly how the isotope is given (for example, which isotope, in which amino acid, and by what route). However, in all cases the mathematical analysis relies on the same basic principles (1), which are most easily illustrated in the case when the label is given by a constant infusion. The first assumption is of a steady state for the unlabeled species in the free pool, leading to the equation

$$Q \text{ (flux)} = S \text{ (synthesis)} + O \text{ (oxidation)} = D \text{ (degradation)} + I \text{ (intake)} \quad [1]$$

where the flux is the total turnover rate of the free pool. The second assumption is of a steady state in the amount of isotope in the free pool, which occurs during an infusion of isotope when the isotopic enrichment has risen to a constant (plateau) level (A_{max}).

$$i \text{ (rate of isotope infusion)} = Q \times A_{max} \quad [2]$$

Solution of these two equations using the experimentally obtained values for A_{max}, O, and I then allows the rates of protein synthesis and degradation to be calculated. The details, however, depend on the particular isotope used.

Nitrogen-15 and End Product Methods

Historically, the first isotope to be used for measurements in humans was ^{15}N (see ref. 1). Early methods of analysis employed compartmental models with several pools and proved difficult to use (1). However, the method of Picou and Taylor-Roberts (2) used the simple model shown in Fig. 1 and became widely adopted. The label was given as [^{15}N]glycine, which was administered as a continuous intragastric infusion. The labeling of the free amino acid pool was determined by measuring the enrichment of urinary urea, an end product of nitrogen metabolism. It was assumed that urea labeling would represent the average enrichment of the free amino acid pool, as this pool is used to make both protein and urea. Because the label from glycine was distributed among other amino acids, the label was also taken to be representative of total nitrogen rather than only glycine. After about 2 days of infusion the enrichment of urinary urea rose to a plateau value, from which the flux of nitrogen (Q in equation 2) could be calculated. Equation 1 was then used to calculate D, as the rate of nitrogen intake from the food was known. Moreover, S could also be calculated from equation 1, since the rate of oxidation of amino acids could be assumed to be equal to the rate of total nitrogen excretion in the urine.

As this method was totally non-invasive, it was very well suited to measurements in infants. However, a number of modifications have been made to improve both the accuracy and convenience of the technique (see 1,3). The rate of rise to plateau in urinary urea is very slow, sometimes taking as long as 2–3 days, even in preterm infants (e.g., 4). During this time it is inevitable that there will be some recycling of the label from protein back into the free amino acid pool, giving rise to a "plateau" enrichment which is too high (1,5). A more rapid rise to plateau can be achieved by priming (4,6,7), or by choosing a different end product of nitrogen metabolism, urinary ammonia, on the assumption that ammonia, like urea, is synthesized from the same pool of free amino acids as body protein (1,3). Comparisons of urea and ammonia have shown that in general they give different rates of protein turnover (e.g., 3,8), but usually the responses to treatments have been similar (e.g., 3,9), giving some degree of confidence that the detection of physiological changes in metabolism will not be dependent on the end product chosen for measurement.

A second modification involves the use of a single dose of isotope rather than a constant infusion. This is especially convenient and practical for use in infants because it allows the subject to be free from any restraint by infusion lines during the period of measurement. Rates can also be computed using either ammonia (3) or urea (8) as end product. All that is needed is a complete urine collection over 9–12 h, or less in infants, and a blood sample at the beginning and end, to assess the amount of ^{15}N retained in the body urea pool, if urea is used (8).

These methods have been discussed extensively (1,10) and have been used in a variety of nutritional and metabolic studies in adults (e.g., 11,12) and in infants (e.g., 4,13,14). Although in general there is good agreement between the different approaches with ^{15}N and with other isotopic labels (1,10), there are lingering doubts. These mainly concern the metabolism of the ^{15}N derived from a single amino acid such as glycine and its compartmentation within the body. It is assumed that the precursors of urea and ammonia synthesis are the same as those for whole body protein synthesis. However, urea is synthesized in the liver, whereas ammonia is made in the kidney. This spatial heterogeneity has been shown to result in different

TABLE 1. *Rates of nitrogen flux measured in a single subject showing variations resulting from which ^{15}N-labeled amino acid was used as tracer, whether ammonia or urea was used as end product, and whether tracer administration was by the intravenous route*

	Nitrogen flux (g N/12 h)			
	Intravenous		Oral	
^{15}N tracer	Ammonia	Urea	Ammonia	Urea
Alanine	13.3	24.2	15.9	16.9
Aspartate	17.5	22.3	46.2	17.0
Glutamate	28.9	30.7	46.9	20.5
Glutamine	11.6	22.9	18.3	15.3
Glycine	23.7	32.3	28.1	26.2
Leucine	27.3	47.1	27.8	56.6
Lysine	212.8	92.6	191.0	62.5

Modified from Fern EB, et al. *Clin Sci* 1985; 68: 271–282.

rates of protein turnover when calculated from urea and ammonia. Moreover, this difference varies when the isotope is given by the oral or intravenous route (8,10). An additional level of heterogeneity arises because the isotope, although transferred from [^{15}N]glycine to many other amino acids, is not distributed evenly among them (15).

The use of ^{15}N-labeled amino acids other than glycine has been studied in a number of laboratories. An example is shown in Table 1, which shows that for most labeled amino acids, urea and ammonia do not give rise to the same rates of nitrogen flux, and also that the difference depends very strongly on whether the label was given orally or intravenously. Although in general the average of the rates given by urea and ammonia [termed the *end product average* (8)] was similar with oral or intravenous administration, large differences between the individual amino acids persisted. Lysine in particular gave very high rates of nitrogen flux, whereas the metabolic nitrogen carriers, alanine and glutamine, gave very low rates (Table 1). In this regard, the choice of glycine for the majority of studies was fortunate, as this amino acid appears to give values for nitrogen flux that are in the middle of the range, at least in adults. In neonates [^{15}N]glycine gives very high rates of turnover, which is believed might result from the diversion of available glycine into other metabolic pathways (17). To avoid the arbitrary nature of the choice of tracer when a single labeled amino acid is given, there has been a move toward the use of uniformly labeled proteins, for both adults (e.g., 16,18) and infants (13,19). Although this approach seems likely to circumvent some of the problems of precursor heterogeneity, it is not entirely successful. For example, Fern *et al.* (16) showed that both wheat protein and yeast protein gave different rates with urea and ammonia. Moreover, the rate from yeast was much higher than that from wheat or [^{15}N]glycine, possibly because of its high content of lysine (Table 1).

Overall, despite some uncertainty about the exact metabolism of ^{15}N in the metabolic nitrogen pool, there may be compelling reasons for choosing this label when studies are performed in infants (e.g., in developing countries) (20,21). The ability to perform studies with oral administration of label, followed only by a period of urine collection, permits measurements that might otherwise be impossible. In addition, the measurement can be repeated at intervals of 2–3 days (after the background enrichment has declined to an acceptably low value), making longitudinal studies feasible (e.g., 9).

Carbon-13 and Precursor Methods

Carbon isotopes can be used to label an amino acid with limited metabolic pathways and then used to trace the metabolism of that single amino acid. This permits rates of whole body protein turnover to be derived, if the amino acid composition of body protein is known (22,23). A good example is the infusion of [1-^{13}C]leucine when the enrichment of [1-^{13}C]leucine in the plasma rises to a plateau in about 5 h (less in infants or if a priming dose is given), from which the flux of leucine can be calculated from equation 2. Application of equation 1 to calculate the rate of protein synthesis requires a knowledge of the rate of leucine oxidation, and this is conventionally measured by assessing the rate of production of labeled CO_2 in the breath.

The advantages of this method are that the metabolism of leucine is very simple, as it is essential and has basically only one pathway of metabolism other than by protein turnover, its transamination to α-ketoisocaproate (KIC) and subsequent decarboxylation to give $^{13}CO_2$. Thus all of the complexities of metabolism encountered with ^{15}N are avoided. Because the rise to plateau is rapid, it can be used to assess acute changes in protein turnover, such as those associated with feeding (5,9,24,25). However, the method is limited by metabolic compartmentation in a way analogous to ^{15}N. The intracellular enrichment is lower than the plasma, with each tissue being affected differently. The precursor enrichment that is required to solve equation 2 is the average enrichment at the site of protein synthesis in all the tissues, and this is unknown. The enrichment of KIC is usually similar to that of the free leucine in skeletal muscle and has been used in place of the plasma leucine enrichment for calculations of protein turnover (26). However, not all muscles are the same and muscle contributes only about 25% of whole body protein turnover (27). Whereas KIC might be the correct precursor for oxidation, as it is part of the oxidative pathway, it is at best an approximation for the precursor of protein synthesis. The problem of compartmentation is particularly apparent in relation to the splanchnic tissues, which receive the food-derived amino acids before they reach the systemic blood and peripheral tissues by what is known as the first-pass effect, whereas an intravenously infused label reaches the peripheral tissues first. The impact of feeding and fasting on this has been studied by comparing intravenous with intragastric infusions (28).

Infusion of [1-^{13}C]-labeled amino acids has been very widely used for studying the physiological and nutritional control of protein turnover in adults, and to a lesser extent in infants (e.g., 14,25,29). However, a disadvantage of the technique, particularly in relation to studies in infants, is that it is relatively invasive. In many experimental situations intravenous infusions are unlikely to be ethically approved and blood samples might similarly be difficult. Moreover, accurate measurement of breath CO_2 production may be troublesome, particularly when infants are ventilated. However, certain modifications to the basic technique have been made to facilitate such measurements. When intravenous infusion is impossible, oral administration can be used, preferably by nasogastric tube. Similarly, if blood samples cannot be taken, two alternative strategies can be adopted. First, there is good evidence that urinary leucine has a similar enrichment to plasma leucine (30). The attainment of plateau labeling is likely to be slower with urine, but if the infusion is prolonged and the bladder is emptied sufficiently often to wash out the unlabeled urine, protein turnover measurements by this approach are feasible. Second, it is possible to use a combination of the end product and precursor methods, in which the flux rate is calculated from the relationship between the proportion of the infused label expired in the breath and the rate of production of urinary nitrogen (1,31).

When direct measurement of leucine oxidation by CO_2 collection is impractical, it is possible to determine it indirectly from the production of urinary nitrogen, since most of urinary nitrogen is derived directly or indirectly from the oxidation of amino acids (30,32). For this approach to be accurate, it is necessary to collect urine over a precisely timed period coinciding with the period of [^{13}C]leucine infusion. Specially constructed collecting devices (e.g., 32) have enabled accurate collections to be made over quite short times (e.g., 12 h), by starting and ending the collection period at the point when urine is passed and by assuming complete bladder emptying each time. In addition, it is necessary to allow for changes in the body urea nitrogen pool by taking a blood sample at the beginning and end of the collection. Calculation of protein turnover rates then requires that rates in terms of leucine (whole body flux) and nitrogen (urinary excretion) be combined, and this is not straightforward, because the leucine contents of the dietary and body proteins are usually different (33). The dietary nitrogen intake can be converted to leucine intake, allowing the rate of whole body protein degradation (as leucine) to be calculated from the leucine flux by equation 2. Then the leucine retained in body protein can be calculated from the nitrogen balance if a value for the leucine content of body protein is known. Finally, the rate of protein synthesis (as leucine) is calculated as protein degradation plus leucine retention.

For controlled studies in a metabolic facility or hospital ward, the [1-^{13}C]leucine infusion technique is probably the method of choice. Because plateau enrichment can be reached much more rapidly than with [^{15}N]glycine, studies can be performed more quickly, enabling rapid changes in protein turnover rates to be investigated, such as the changes occurring during the first few hours of hormone infusion [e.g., insulin (34)]. The method and its modifications have been used quite extensively in preterm infants to study the effects of diet composition (e.g., 14,29,35). Moreover, because the residual enrichment falls quite quickly after a study is complete, repeat measurements at daily intervals are possible. For example, Fig. 2 illustrates the changes in whole body protein synthesis, degradation and retention in low birthweight neonates during the first few days of life. This shows the progression of increase in

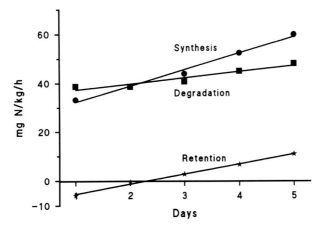

FIG. 2. Rates of protein turnover in low birthweight neonates, as total parenteral nutrition was introduced stepwise over 5 days.

synthesis, and to a lesser extent degradation, as the intake of total parenteral nutrition was increased and the retention of protein changed from negative to positive.

PROTEIN TURNOVER IN INDIVIDUAL TISSUES

Measurements of rates of whole body protein turnover are relatively non-invasive and are a significant advance over the nitrogen balance method used previously. They have enabled much useful information to be obtained on the nutritional and physiological regulation of protein metabolism. However, whole body measurements are limited because they provide no information on the tissues responsible for observed changes. The need for this information is illustrated by studies of tissue protein synthesis in experimental animals, which until relatively recently have been the sole source of information on individual tissues. Table 2 shows the responses of liver and muscle protein synthesis in young rats to a variety of nutritional and pathological treatments. The results show that although rates of protein synthesis in these two tissues change in the same direction during nutrient deprivation, liver and muscle respond differently to various disease states. Whole body measurements in the latter groups of animals would not have revealed these differences.

Until now there appear to have been no studies of individual tissues in human infants, although there have been a number of measurements in adults. There are two basic approaches. The first is to perform an arteriovenous balance study across an organ or limb, such as the forearm, while infusing a labeled amino acid such as [1-^{13}C]leucine (e.g., 36). This has the theoretical capacity to measure both protein synthesis and degradation, but might be of limited value for use in children, because of the need for both arterial and venous sampling lines. The second approach, which can be used to determine the rate of protein synthesis, is to administer a labeled

TABLE 2. *Effects of various nutritional and pathological states on muscle and liver protein synthesis in young rats or mice*

Treatment	Protein synthesis (% of control)	
	Liver	Muscle
Starvation (2 days)	71[a]	47[b]
Protein-free diet (9 days)	70[c]	27[d]
Diabetes	51[c]	41[e]
Malaria infection[e]	66	59
Turpentine infection[f]	116	66[b]
Interleukin-1β infection	145	68
Cancer[g]	126	56

Data from: [a] Ref. 54. [d] Ref. 57. [f] Ref. 52.
[b] Ref. 55. [e] Ref. 51. [g] Ref. 53.
[c] Ref. 56.

amino acid and then to measure directly the amount of label incorporated into tissue protein by taking a biopsy sample. The main limitation has been the ability to obtain a large enough biopsy sample of the tissue of interest. In adults, percutaneous muscle biopsy or open biopsy during surgery provide enough sample (about 100–200 mg) for measurement of the enrichment of ^{13}C-labeled amino acids as $^{13}CO_2$ by gas isotope ratio mass spectrometry (27,37), but in infants this size of sample would usually not be possible. However, a new development of gas chromatography/mass spectrometry (GC/MS) has enabled [$^2H_5$]phenylalanine to be measured accurately at the very low enrichments found in protein, using very small tissue samples [about 5 mg (38)]. In addition, the recently developed GC combustion isotope ratio mass spectrometer can also measure ^{13}C enrichments in much smaller samples than previously was possible (39). It seems likely, therefore, that studies of tissue protein synthesis will soon become possible in children. The methods available for making these measurements will now be discussed.

Constant Infusion of Labeled Amino Acids

During constant intravenous infusion of a labeled amino acid, a plateau enrichment in the free amino acid enrichment in plasma is reached, as described above for whole body protein turnover measurement. Rates of protein synthesis in individual tissues can then be determined by measuring the enrichment of the amino acid in protein at the end of the infusion (40–42). The calculation requires, in addition to the enrichment in protein, the average enrichment of the free amino acid precursor of protein synthesis during the incorporation period.

Constant infusion of [1-^{13}C]leucine has been used to study the effects of nutrition, hormones, and a variety of disease states on muscle protein synthesis in human adults (see 43). The advantage of the infusion is that it ensures that the free amino acid enrichment in both plasma and tissues is nearly constant, therefore requiring only a small number of measurements to define the average precursor enrichment accurately. However, a difficulty arises because the enrichments in the plasma and intracellular pools are different (1). Studies in cultured cells and isolated tissues, and *in vivo*, have shown that, in general, neither the plasma nor the intracellular free amino acid pools have the same enrichment as the aminoacyl-tRNA, which is the direct precursor of protein synthesis (see 1,44). This occurs because the tRNA can be charged with amino acid channeled from a variety of labeled or unlabeled sources (see Fig. 3). Direct measurement of the enrichment of aminoacyl-tRNA is not feasible for routine measurements of protein synthesis *in vivo*. Furthermore, there is evidence that the relationships between its enrichment and those of the measurable pools in the plasma and tissue are not constant under different metabolic conditions (see 1,44).

Two approaches have been taken in an attempt to minimize this problem of precursor measurement. For measurement of liver protein synthesis and of plasma proteins synthesized in the liver (e.g., albumin), the enrichment in the apoB protein of very low density lipoprotein in the plasma rises very rapidly to the same as that of its own precursor, because of its very fast turnover rate (e.g., 45). This enrichment can then be used as the precursor for other liver synthesized proteins. For skeletal muscle

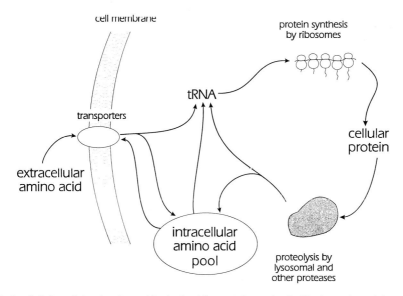

FIG. 3. Cellular origin of amino acids destined for protein synthesis. The isotopic enrichment of the tRNA during isotopic labeling experiments depends on relative charging from each of three possible sources: (a) direct from the membrane transporter at an enrichment similar to that in the plasma; (b) from the mixed intracellular pool, with a lower enrichment than that in the plasma; and (c) directly recycled from proteolysis, with zero enrichment. If (a) predominates, the enrichment of aminoacyl-tRNA will be higher than that of the intracellular pool; if (c) predominates, the aminoacyl-tRNA enrichment will be lower than that of the intracellular pool. In practice, the relative proportions of these three pathways are likely to vary in different metabolic and nutritional states.

a more indirect assessment has been used. KIC is believed to be synthesized from leucine, mainly in skeletal muscle, and is then secreted into the plasma. Its enrichment has therefore been taken as a reflection of the intracellular enrichment for studies on skeletal muscle (see 43). Although measurement of tRNA in a single muscle during surgery provided confirmation for this (46), it seems unlikely that KIC can reflect the enrichment in all the muscles of the body, as the relationship between the enrichments of KIC and plasma or tissue free leucine varies appreciably (see 44).

The Flooding Method

Measurements on cultured cells and isolated tissues and *in vivo* have shown that if the labeled amino acid is presented not as a tracer amount but together with a large amount of unlabeled amino acid, the isotopic enrichment in the various extracellular and intracellular compartments becomes nearly the same (see 44). Accurate measurement of the precursor enrichment is therefore facilitated. In other respects, the method is similar to constant infusion, in that the rate of protein synthesis is calculated from the average enrichment in the plasma and the enrichment in protein at the end of the incorporation period (see Fig. 4). The additional advantages are that the period of measurement is shorter (1–1.5 h *vs* 4–6 h for infusion), and there is no requirement for a metabolic steady state, both of which make the method particularly suitable for use in the clinical environment. A potential disadvantage is the possibility that

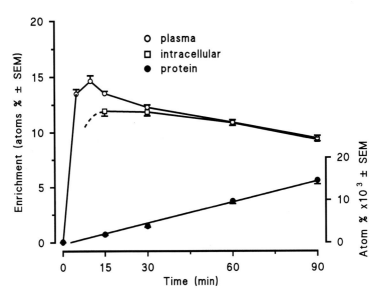

FIG. 4. Enrichments of phenylalanine in plasma and muscle free pools (*left axis*) and in muscle protein (*right axis*) after intravenous injection of 45 mg/kg of [^{2}H$_5$]phenylalanine into healthy adults.

TABLE 3. *Rates of protein synthesis measured by flooding with [1-^{13}]leucine or [^{2}H$_5$]phenylalanine in a range of adult human tissues*

	Protein synthesis %/day ± SEM
Skeletal muscle[a]	1.95 ± 0.12
Heart (atrium)[b]	5.2 ± 0.8
Lymphocytes[c]	6.2 ± 0.4
Colon (mucosa)[d]	9.4 ± 1.2
Liver[d]	20.7 ± 1.9
Esophagus[e]	29.5 ± 2.0
Stomach (antrum)[e]	42.9 ± 5.0
Stomach (fundus)[e]	44.5 ± 10.3

Data from: [a] Ref. 58.
[b] Ref. 59.
[c] Ref. 60.
[d] Ref. 37.
[e] Park KGM, McNurlan MA, Garlick PJ, unpublished data.

the large amount of amino acid given might alter the rate of protein synthesis. However, despite many studies designed to detect such changes, none have been demonstrated convincingly (see 44).

This approach has been used to examine the responses of muscle protein synthesis in adult humans to feeding and starvation, leading to the conclusion that the sensitivity to nutrition is much lower than that reported from studies in rats (see Table 2). However, the rat studies shown in Table 2 were performed in young growing animals. By contrast, adult rats have been shown to be unresponsive to nutritional changes (47), thus resembling the human adult. Studies of human infants are now needed to detect whether these discrepancies reflect differences due to species or to growth.

The flooding method is particularly well suited to measurements on tissues sampled at surgery, and this has enabled rates of protein synthesis in a wide range of human tissues to be evaluated in normal and disease states (Table 3). Studies of this type can include a period of treatment prior to measurement, and can therefore be used to investigate the responses to such as nutrient intake (e.g., 48,49). By use of GC/MS to measure the enrichment of [^{2}H$_5$]phenylalanine in very small tissue samples (38), this approach appears to have considerable potential for gaining insight into the regulation of protein metabolism in infants. It is also possible to measure the rate of synthesis of plasma proteins by the flooding technique (50). This is particularly well suited to measurements in infants, as it can be achieved with only small samples of plasma.

CONCLUSION

A variety of methods for measuring protein turnover with stable isotopes are available for use in human beings, and most of these can be used to study infants and children. For whole body protein turnover, methods using either [^{15}N]glycine or [1-^{13}C]leucine can give valuable information, and the choice of technique will depend on the exact conditions under which the measurement is to be made and the equipment available. In particular, adaptations of the constant infusion of [1-^{13}C]leucine method, so that measurements can be made without blood sampling or breath CO_2 collection, have facilitated studies of preterm neonates. It seems likely, however, that the limitations of the whole body approach will compel future research efforts to address the challenge of measuring protein turnover in individual tissues of infants, as has occurred in adults. New developments in mass spectrometry have enabled measurements of the amino acid enrichment in protein of very small tissue samples to be made. Use of these techniques can make studies of protein synthesis in tissues of infants attainable, and lead further toward the goal of understanding the factors that control protein metabolism during growth.

REFERENCES

1. Waterlow JC, Garlick PJ, Millward DJ. *Protein turnover in mammalian tissues and in the whole body.* Amsterdam: North-Holland, 1978.
2. Picou D, Taylor-Roberts T. The measurement of total protein synthesis and catabolism and nitrogen turnover in infants in different nutritional states and receiving different amounts of dietary protein. *Clin Sci* 1969; 36: 283–96.
3. Waterlow JC, Golden MHN, Garlick PJ. Protein turnover in man measured with ^{15}N: comparison of end products and dose regimes. *Am J Physiol* 1978; 235: E165–74.
4. van Lingen RA, van Gouldever JB, Luijendijk IHT, Wattimena JLD, Sauer PJJ. Effects of early amino acid administration during total parenteral nutrition on protein metabolism in pre-term infants. *Clin Sci* 1992; 82: 199–203.
5. Melville S, McNurlan MA, McHardy KC et al. The role of degradation in the acute control of protein balance in adult man: failure of feeding to stimulate protein synthesis as assessed by L-[1-^{13}C]leucine infusion. *Metabolism* 1989; 38: 248–55.
6. Jeevanandam M, Brennan MF, Horowitz GD, et al. Tracer priming in human protein turnover studies with [^{15}N]glycine. *Biochem Med* 1985; 34: 214–25.
7. Jackson AA, Persaud C, Badaloo V, deBenoist B. Whole-body protein turnover in man determined in three hours with oral or intravenous ^{15}N-glycine and enrichment in urinary ammonia. *Hum Nutr Clin Nutr* 1987; 41C: 263–76.
8. Fern EB, Garlick PJ, McNurlan MA, Waterlow JC. The excretion of isotope in urea and ammonia for estimating protein turnover in man with [^{15}N]glycine. *Clin Sci* 1981; 61: 217–28.
9. Garlick PJ, Clugston GA, Waterlow JC. Influence of low-energy diets on whole-body protein turnover in obese subjects. *Am J Physiol* 1980; 238: E235–44.
10. Fern EB, Garlick PJ, Waterlow JC. The concept of the single body pool of metabolic nitrogen in determining the rate of whole body nitrogen turnover. *Hum Nutr Clin Nutr* 1985; 39C: 85–99.
11. McNurlan MA, McHardy KC, Broom J, et al. The effect of indomethacin on the response of protein synthesis to feeding in rats and man. *Clin Sci* 1987; 73: 69–75.
12. Velasco N, Long CL, Nelson KM, Blakemore WS. Whole-body protein kinetics in elective surgical patients receiving peptide or amino acid solutions. *Nutrition* 1991; 7: 28–32.
13. Stack T, Reeds PJ, Preston T, Hay S, Lloyd DJ, Aggett PJ. ^{15}N tracer studies of protein metabolism in low birth weight preterm infants: a comparison of ^{15}N glycine and ^{15}N yeast protein hydrolysate and of human milk- and formula-fed babies. *Pediatr Res* 1989; 25: 167–72.
14. Duffy G, Pencharz P. The effects of surgery on the nitrogen metabolism of parenterally fed human neonates. *Pediatr Res* 1986; 20: 32–5.
15. Golden MHN, Jackson AA. Assumptions and errors in the use of ^{15}N-excretion data to estimate whole body protein turnover. In: Waterlow JC, Stephen JML, eds. *Nitrogen metabolism in man.* London: Applied Science Publishers, 1981; 323–44.

16. Fern EB, Garlick PJ, Waterlow JC. Apparent compartmentation of body nitrogen in one human subject: its consequences in measuring the rate of whole-body protein synthesis with ^{15}N. *Clin Sci* 1985; 68: 271–82.
17. Jackson AA, Badaloo AV, Forrester T, Hibbert JM, Persaud C. Urinary excretion of 5-oxoproline (pyroglutamic aciduria) as an index of glycine insufficiency in normal man. *Br J Nutr* 1987; 58: 207–14.
18. McNurlan MA, El-Khoury AE, Milne E, Fern EB. Assessment of protein metabolism in man with ^{15}N-labelled soybean (Abstract). *Proc Nutr Soc* 1991; 50: 46A.
19. Wutzke K, Heine W, Drescher U, Richter I, Plath C. ^{15}N-labelled yeast protein—a valid tracer for calculating whole-body protein parameters in infants: a comparison between ^{15}N-yeast protein and ^{15}N-glycine. *Hum Nutr Clin Nutr* 1983; 37C: 317–27.
20. Tomkins AM, Garlick PJ, Scholfield WN, Waterlow JC. The combined effects of infection and malnutrition on protein metabolism in children. *Clin Sci* 1983; 65: 313–24.
21. Golden MHN, Waterlow JC, Picou D. The relationship between dietary intake, weight change, nitrogen balance, and protein turnover in man. *Am J Clin Nutr* 1977; 30: 1345–8.
22. Waterlow JC. Lysine turnover in man measured by intravenous infusion of L-[U-^{14}C]lysine. *Clin Sci* 1967; 33: 507–15.
23. James WPT, Sender PM, Garlick PJ, Waterlow JC. The choice of label and measurement technique in tracer studies in man. In: *Dynamic studies with radioisotopes in medicine*, Vol 1. Vienna: IAEA, 1974; 461–72.
24. Motil KJ, Matthews DE, Bier DM, Burke JF, Munro HN, Young VR. Whole body leucine and lysine metabolism: response to dietary protein intake in young men. *Am J Physiol* 1981; 240: E712–21.
25. Denne SC, Karn CA, Liechty EA. Leucine kinetics after a brief fast and in response to feeding in premature infants. *Am J Clin Nutr* 1992; 56: 899–904.
26. Matthews DE, Schwartz HP, Young RD, Motil KJ, Young VR. Relationship of plasma leucine and α-ketoisocaproate during a L-[1-^{13}C]leucine infusion in man: a method for measuring human intracellular leucine tracer enrichment. *Metabolism* 1982; 31: 1105–12.
27. Halliday D, Pacy PJ, Cheng KN, Dworzak FD, Gibson JNA, Rennie MJ. Rate of protein synthesis in skeletal muscle of normal man and patients with muscular dystrophy; a reassessment. *Clin Sci* 1988; 74: 237–40.
28. Biolo G, Tessari P, Inchiostro S, *et al*. Leucine and phenylalanine kinetics during mixed meal injection: a multiple tracer approach. *Am J Physiol* 1992; 262: E455–63.
29. Beaufrère B, Putet G, Pachiaudi C, Salle B. Whole body protein turnover measured with ^{13}C-leucine and energy expenditure in preterm infants. *Pediatr Res* 1990; 28: 147–52.
30. de Benoist B, Abdulrazzak Y, Brooke OG, Halliday D, Millward DJ. The measurement of whole body protein turnover in the preterm infant with intragastric infusion of L-[1-^{13}]leucine and sampling of the urinary leucine pool. *Clin Sci* 1984; 66: 155–64.
31. Golden MHN, Waterlow JC. Total protein synthesis in elderly people: a comparison of results with [^{15}N]glycine and [^{14}C]leucine. *Clin Sci* 1977; 53: 277–88.
32. Mitton SG, Calder AG, Garlick PJ. Protein turnover rates in sick, premature neonates during the first few days of life. *Pediatr Res* 1991; 30: 418–22.
33. Mitton SG, Garlick PJ. Changes in protein turnover after the introduction of parenteral nutrition in premature infants: comparison of breast-milk and egg protein based amino acid solutions. *Pediatr Res* 1992; 32: 447–54.
34. Fukagawa NK, Minaker KL, Rowe JW, *et al*. Insulin-mediated reduction of whole body protein breakdown: dose-response effects on leucine metabolism in postabsorptive man. *J Clin Invest* 1985; 76: 2306–11.
35. Pencharz PB, Farri L, Papageorgiou A. The effects of human milk and low-protein formulae on the rates of total body protein turnover and urinary 3-methyl-histidine excretion of preterm infants. *Clin Sci* 1983; 64: 611–6.
36. Cheng KN, Dworzak F, Ford GC, Rennie MJ, Halliday D. Direct determination of leucine metabolism and protein breakdown in humans using L-[1-^{13}C, ^{15}N]-leucine and the forearm model. *Eur J Clin Invest* 1985; 15: 349–54.
37. Heys SD, Park KGM, McNurlan MA, Millar JDB, Eremin O, Garlick PJ. Protein synthesis rates in colon and liver: stimulation by gastrointestinal pathologies. *Gut* 1992; 33: 976–81.
38. Calder AG, Anderson SE, Grant I, McNurlan MA, Garlick PJ. The determination of low d_5-phenylalanine enrichment (0.002–0.09 atom percent excess), after conversion to phenylethylamine, in relation to protein turnover studies by gas chromatography/mass spectrometry. *Rapid Commun Mass Spectrom* 1992; 6: 421–4.

39. Yarasheski KE, Smith K, Rennie MJ, Bier DM. Measurement of muscle protein fractional synthetic rate by capillary gas chromatography/combustion isotope ratio mass spectrometry. *Biol Mass Spectrom* 1992; 21: 486–90.
40. Waterlow JC, Stephen JML. The effect of low protein diets on the turnover rates of serum, liver and muscle proteins in the rat measured by continuous infusion of L-(^{14}C)lysine. *Clin Sci* 1968; 35: 287–305.
41. Garlick PJ. Turnover rate of muscle protein measured by constant intravenous infusion of ^{14}C-glycine. *Nature* 1969; 223: 61–2.
42. Garlick PJ, Millward DJ, James WPT. The diurnal response of muscle and liver protein synthesis *in vivo* in meal-fed rats. *Biochem J* 1973; 136: 935–46.
43. Smith K, Rennie MJ. Protein turnover and amino acid metabolism in human skeletal muscle. In: Harris JB, Turnbull DM, eds. *Muscle metabolism: Ballière's clinical endocrinology and metabolism*. London: Ballière Tindall, 1990; 461–98.
44. Garlick PJ, McNurlan MA, Essén P, Wernerman J. Measurement of tissue protein synthesis rates *in vivo*: a critical analysis of contrasting methods. *Am J Physiol* (in press).
45. Parhofer KG, Barrett PHR, Bier DM, Schonfeld G. Determination of kinetic parameters of apolipoprotein B metabolism using amino acids labeled with stable isotopes. *J Lipid Res* 1991; 32: 1311–23.
46. Watt PW, Lindsay Y, Scrimegeour CM, *et al*. Isolation of aminoacyl-tRNA and its labeling with stable-isotope tracers: use in studies of human tissue protein synthesis. *Proc Natl Acad Sci USA* 1991; 88: 5892–6.
47. Baillie AGS, Garlick PJ. Attenuated response of muscle protein synthesis to fasting and insulin in adult female rats. *Am J Physiol* 1992; 262: E1–5.
48. Heys SD, Park KGM, McNurlan MA, *et al*. Stimulation of protein synthesis in human tumours by parenteral nutrition: evidence for modulation of tumour growth. *Br J Surg* 1991; 78: 483–7.
49. Park KGM, Heys SD, Blessing K, Kelly P, Eremin O, Garlick PJ. The stimulation of human breast cancers by dietary L-arginine. *Clin Sci* 1992; 82: 413–7.
50. Ballmer PE, McNurlan MA, Milne E, *et al*. Measurement of albumin synthesis in humans: a new approach employing stable isotopes. *Am J Physiol* 1990; 259: E797–803.
51. Pain VM, Albertse EC, Garlick PJ. Protein metabolism in skeletal muscle, diaphragm, and heart of diabetic rats. *Am J Physiol* 1983; 245: E604–10.
52. Ballmer PE, McNurlan MA, Southorn BG, Grant I, Garlick PJ. Effects of human recombinant interleukin-1β on protein synthesis in rat tissues compared with a classical acute-phase reaction induced by turpentine. *Biochem J* 1991; 279: 683–88.
53. Pain VM, Randall DP, Garlick PJ. Protein synthesis in liver and skeletal muscle of mice bearing an ascites tumor. *Cancer Res* 1984; 44: 1054–7.
54. McNurlan MA, Tomkins AM, Garlick PJ. The effect of starvation on the rate of protein synthesis in rat liver and small intestine. *Biochem J* 1979; 178: 373–9.
55. Garlick PJ, Fern EB, McNurlan MA. Methods for determining protein turnover. In: Rapoport S, Schewe T, eds. *Processing and turnover of proteins and organelles in the cell*. Oxford: Pergamon Press, 1979; 85–94.
56. McNurlan MA, Garlick PJ. Protein synthesis in liver and small intestine in protein deprivation and diabetes. *Am J Physiol* 1981; 241: E238–45.
57. Garlick PJ, Millward DJ, James WPT, Waterlow JC. The effect of protein deprivation and starvation on the rate of protein synthesis in tissues of the rat. *Biochim Biophys Acta* 1975; 414: 71–84.
58. Garlick PJ, Wernerman J, McNurlan MA, *et al*. Measurement of the rate of protein synthesis in muscle of postabsorptive young men by injection of a 'flooding dose' of [1-^{13}C]leucine. *Clin Sci* 1989; 77: 329–36.
59. Park KGM, Heys SD, McKenzie J, Burns J, Eremin O, Garlick PJ. Comparison of protein synthesis rates in skeletal and heart muscle. *J R Coll Surg Edinb* 1990; 35: 121.
60. Park KGM, Heys SD, Eremin O, Garlick PJ. Lymphocyte protein synthesis: an *in vivo* measure of activation. (Abstract). *Proc Nutr Soc* 1992; 51: 100A.

DISCUSSION FOLLOWING THE PRESENTATION OF DR. GARLICK

Dr. Uauy: In the last 20 years we have learned a lot about manipulations of diet and energy. One of the questions that comes to mind is: What are the functional applications of various levels of protein turnover? Is it better for turnover to be high, low, or intermediate? I raise this question because of the issue of amino acid requirements and how much are needed to enhance turnover. What is your view?

Dr. Garlick: I think the evidence points to the fact that the more protein in the diet, the higher the rate of protein turnover. There seems to be no obvious explanation for this. It may be that it is equivalent to a substrate cycle where the more rapidly the proteins turn over the more adaptable is the organism and the greater the ability to recycle amino acids to the new sets of proteins that are required in changing circumstances. I think that if you extend this concept, then you can argue that for an animal—a wild animal (don't consider humans for the moment)—then the advantage is always in terms of being more adaptable. So the faster it can adapt, the better it is going to survive illness, injury, or rapid change in requirements. Thus a more generous supply of protein in the diet is obviously an advantage in allowing increased protein turnover rates. If you extend this concept to human beings, however, you might come into conflict with the animal data; there is evidence, for example, that the higher the protein in the diet, the sooner animals die. In the wild, long-term survival is relatively unimportant, so the high rate of protein turnover is definitely advantageous. But diseases such as tumors, which do not matter to the animal in the wild, may increase with aging when there is a higher protein turnover. Such diseases are of obvious importance to humans. So perhaps we have to think in terms of the potential advantages and disadvantages of these increases in protein turnover. But I think they do occur and are probably there in order to improve adaptability.

Dr. Rassin: When you flood the transporter with a particular amino acid you are inhibiting the access of other amino acids into the intracellular free pool. Is that going to affect protein synthesis negatively because you are downregulating transport precursors?

Dr. Garlick: This is something which has concerned everyone from the time of the first experiments in the early 1950s. Nobody has ever detected any change. We experimented in rats a great deal, looking at the incorporation of other labeled amino acids in the presence of the flooding amino acids. We found that occasionally one amino acid will perhaps decrease but another will increase incorporation, which we assume is just due to differences in transport. Obviously the competition with transport will affect the penetration of the label to the site of protein synthesis and may do it either positively or negatively, but we have no evidence that protein synthesis itself is altered.

We have done similar experiments in humans to see if protein synthesis is altered by flooding. We can't show any effect of a single amino acid, leucine, given as a flood on the incorporation of another amino acid, phenylalanine, also given as a flood. The reason for looking at this, as I am sure you are aware, is that leucine has a reputation for being able to stimulate muscle protein synthesis, while phenylalanine has been shown not to affect the rate of protein synthesis. Also, valine gives exactly the same rate of protein synthesis as leucine or phenylalanine when given by flooding dose. I don't see how these amino acids can all be having the same effect on protein synthesis. I think the conclusion in the end is that they are probably not having any effect at all.

Furthermore, in human muscle, if you look at polysome profiles after giving a large flood of leucine, you don't get any change at all in the profile in the next 90 minutes or so after the flood, again suggesting that there is not really any change in muscle protein synthesis. I think on balance the evidence is that there is no change. However, each method is a compromise. With a method such as continuous infusion of a tracer amount of a labeled amino acid there are potential problems of interpretation from not being able to measure the precursor enrichment accurately, whereas with the flooding dose there remains the possibility that you may alter the rate you are trying to measure. One must be aware of these possibilities, but these are the only methods we have and we have to work with them.

Dr. Rassin: Most of what you are seeing is probably the response of muscle protein synthesis. My concern is more what goes on with tissues where the individual amino acid is much more regulated because of the blood-brain barrier and competition for transport into the central nervous system. I know that this is not a large mass of tissue, but I wonder if you are really seeing the final number; muscle may not always be of primary importance in terms of function.

Dr. Garlick: No, remember we are actually measuring individual tissues here, so we are measuring specifically muscle, or in other cases we may be measuring specifically liver, or specifically colon. We have not studied the brain much, but the results we obtained with labeled phenylalanine in the brain are similar to what you get if you use continuous infusion to measure brain protein synthesis. Dunlop did these studies many years ago with different amino acids and got very similar rates of protein synthesis with each of them (1).

Dr. Marini: I read a paper last year in the *American Journal of Clinical Nutrition* (2) in which it was shown in neonates that there were different rates of turnover for three amino acids (leucine, phenylalanine, and glycine). From the clinical point of view, if we have to make a judgment about the relationship between protein synthesis and growth, which amino acid should one study?

Dr. Garlick: Each amino acid has its own problem of measurement or interpretation. One particular problem is that the site of protein synthesis is not the plasma. With the carbon labeling method, what I called a "precursor method," we use the enrichment of the amino acid in plasma to calculate the rate of protein synthesis when we should be using the enrichment at the site of protein synthesis. Of course, that enrichment is very indeterminate because the whole body contains all sorts of cells, in all the tissues in the body, and presumably they all differ. If you go back to the work of Vernon Young's group with [^{13}C]leucine (3), you find that when they also used lysine as a labeled amino acid all the rates they got were about 50% lower than they were with leucine. Compare this with Simon's work on pigs (4), in which he showed that the intracellular specific activity was much lower for lysine relative to the plasma than it was for leucine. Calculations of whole body protein turnover from the plasma leucine specific activity were much higher than those from plasma lysine, whereas rates calculated on the basis of intracellular specific activities were very much more in agreement. These are very important aspects of this kind of study, and unfortunately at the level of the whole body they are things we really can't do a great deal about.

My own personal philosophy now is where possible to try to look at individual tissues, because then you can ignore some of the complexities of the whole body model. You can focus on getting one answer only, which is a rate of protein synthesis in an individual tissue; there is then a good chance that it will be more accurate.

Dr. Marini: There is one unusual amino acid, threonine. Do you have any data concerning label in threonine and the metabolism of threonine? I am asking this because we are now using whey formulas and we find a rather high level of threonine in plasma compared to the other formulas. I don't know if this can have any influence from the metabolic point of view.

Dr. Garlick: I have very limited personal experience of the effects of changing single amino acids. However Sally Mitton studied whole body protein turnover in preterm infants over the first 5 days of life and she found that a change in the balance of amino acids in the intravenous feeding solutions had no effect on turnover (5). Any effect on turnover relating to single amino acids probably depends most on whether that amino acid is limiting; I suspect that a person or an animal has quite a wide range of ability to adapt to individual amino acid changes as long as the limiting one is still there in adequate amount. The rest of them can be present in higher amounts, and unless they go enormously high they won't have much influence.

Dr. Priolisi: By using [^{13}C]leucine as a precursor method, when you reach the isotopic enrichment plateau is it possible to calculate the fraction of the plasma leucine that is derived from protein catabolism, and if this is possible, is there some kind of association between this variable and the synthesis and breakdown of protein?

Dr. Garlick: It can be calculated by a simple manipulation of the data that you have, but I can't immediately see any reason why the absolute rate of protein degradation should not be more useful.

Dr. Priolisi: In other words, if the protein degradation is increased, the fraction of the plasma pool that is derived from body protein should be higher too?

Dr. Garlick: Yes it would be, but I don't see why that should have any functional implications.

Dr. Priolisi: There are functional implications. For example, instead of a solution of amino acids we used albumin as an intravenous protein source in a pilot study. When we look at the amino acid concentrations, we think that some of the plasma pool of amino acids in these extremely low birth weight infants was coming from the very rapid degradation of albumin. So I want to know if the contribution to the plasma pool is high when protein degradation is high.

Dr. Garlick: The problem is to know where the site of degradation is. For most proteins degradation is intracellular and not within the plasma compartment, although for albumin it might be in the plasma. How much of the protein that is degraded intracellularly appears in the plasma is not something I find it very easy to envisage as an entity in itself.

Dr. Priolisi: Is it not measurable?

Dr. Garlick: It is measurable as a number, but I can't see how it would have any bearing on what I think of as physiological regulation.

Dr. Guesry: Is the increased degradation of protein that occurs on refeeding after protein–energy malnutrition due to the fact that when you increase synthesis you also increase degradation, or is it because you are making a comparison with a period of very low degradation during fasting or malnutrition, when the body tries to conserve all possible protein?

Dr. Garlick: The thing that affects protein turnover in addition to whether the individual is well nourished or malnourished is the immediate intake of protein at the time of study. For example, in the adult, as the dietary protein intake increases, the rates of protein synthesis and degradation in the fasted state tend to rise; in the fed state, however, they are affected by the fact that during feeding the response to food intake, in the adult at least, is mostly to depress protein degradation. So you have a general increase in rates of synthesis and degradation, and imposed on that is an effect of the immediate intake of the diet, which is to depress proteolysis. The result of this is quite a complex pattern of changes as you increase the amount of protein in the diet.

I don't know exactly how this process is influenced by malnutrition, but your interesting suggestion that the low rate of protein degradation during malnutrition will be stimulated by feeding a normal diet is consistent with what is known of the effects of feeding and malnutrition.

Dr. Heine: There is apparently a lot of confusion concerning the choice of the optimal tracer substance. According to Dennis Bier and others, the tracer which is used for studies should, if we use ^{15}N-labeled substances, represent the total amino nitrogen of the protein pool and the non-protein pool. We know that [^{15}N]glycine is poorly transaminated and that the amino group is poorly transferred to other amino acids with the exception of serine. On the other hand, glycine is particularly deaminated in the liver and kidney, and is involved in many other metabolic processes besides protein turnover. So our group is of the opinion that it would be better to use tracer substances that are uniformly labeled when running studies with ^{15}N.

You mentioned the data given by Edward Fern and the poor results he has obtained when using these proteins. Couldn't this be methodologic error, due to poor absorption of the tracer substances?

Dr. Garlick: This does highlight one of the difficulties of the oral administration of amino acids and of using labeled proteins. Labeled proteins have to be given orally unless you can get them passed for human use, which is unlikely in the case of a protein hydrolysate. You then have to assume they are completely absorbed, which you are never quite sure about. I think the simplest way of advancing this question is to refer it to Dr. Fern.

Dr. Fern: The absorption was about 70%, but we made allowances for that in the calculation. But I have subsequently done other work with other proteins and if you vary the protein, you vary the rate, but not as much as you do with single amino acids.

Dr. Tracey: You told us about the importance of the urea pool and how difficult it is to measure because of the long half-life of urea. But since in protein catabolism and perhaps in starvation urea forms 80% of the nitrogen lost, one wonders whether urea labeling studies might be used to assess protein turnover.

Dr. Garlick: If you give labeled urea, then you measure the turnover rate of the urea pool, which is a different phenomenon altogether from protein turnover. The urea pool in a human adult has a half-life of, say, 10 hours, in a rat it is 6 hours. These values are related to the rate of catabolism of amino acids, the rate of nitrogen excretion, in other words, rather than protein turnover.

Dr. Schöch: We have compared data from premature infants, from sick infants, and from cancer patients, all data that are extremely difficult to interpret because of the many confounding influences, and we have found that the protein degradation rates are astonishingly stable. I am amazed that there is no real theory of macromolecular degradation.

Dr. Räihä: You showed that in adult patients with HIV there is both an increased synthesis rate and an increased degradation rate. Do you have any information about infants? We have seen that some apparently normal infants, who later developed clinical infection, had increased urea excretion. Then, within a few days, the signs of infection appeared. This implies to me that catabolism was increased in these infected infants before the infection was clinically obvious. Do you have any information on whether synthesis and catabolism both increase with infection in neonates, or is it mostly catabolism and may synthesis even decrease? Clinically, I think this is important.

Dr. Garlick: I have no information on this. In our own study we were interested to find that what we called "sick" infants were actually the ones with low protein turnover rates, from which we made the simple conclusion that perhaps this was because of malnutrition rather than because of inflammation, because we would have expected stress in any of its forms to have increased turnover rates. In animals that are infected you get increases in protein turnover in some tissues but not in others. We don't know how this balances out on a whole body basis.

Dr. Marini: There are experimental data that show that if you increase the amount of glutamine in the diet, animals survive better when you provoke septic shock. These may be important data because the total parenteral nutrition we give to preterm babies for 10 days before starting full feeding does not contain sufficient glutamine by these experimental criteria.

There has been much discussion of the need for glutamine in adult parenteral nutrition, but it is just as important for neonates. In particular it is required for intestinal development.

Dr. Garlick: There is an enormous amount of literature about glutamine relating to adults, yet the points that are made about the growth and maintenance of the gut may be even more important in the infant.

Dr. Bremer: Are there data available on the influence of anoxia on protein degradation?

Dr. Garlick: There is some work on perfused tissues, particularly heart, where the effect of hypoxia or anoxia is to reduce synthesis (6).

Dr. Marini: I remember some very early work on growth and synthesis of nucleotides in babies with congenital heart disease which showed that there is a reduction of protein synthesis.

REFERENCES

1. Dunlop DS, van Elden W, Lajtha A. A method for measuring brain protein synthesis rates in young and adult rats. *J Neurochem* 1975; 24: 337–44.
2. Wykes LJ, Ball RO, Menendez CE, Ginther DM, Pencharz PB. Glycin, leucine, and phenylalanine flux in low-birth-weight infants during parenteral and enteral feeding. *Am J Clin Nutr* 1992; 55: 971–975.
3. Motil KJ, Matthews DE, Bier DM, Burke JF, Munro HN, Young VR. Whole body leucine and lysine metabolism: response to dietary protein intake in young men. *Am J Physiol* 1981; 240: E712–21.
4. Simon O, Munchmeyer R, Bergner H, Zebrowska T, Buraczewska L. Estimation of rate of protein synthesis by constant infusion of labelled amino acids in pigs. *Br J Nutr* 1978; 40: 243–52.
5. Mitton SG, Garlick PJ. Changes in protein turnover after the introduction of parenteral nutrition in premature infants: comparison of breast-milk and egg protein based amino acid solutions. *Pediatr Res* 1992; 32: 447–54.
6. Preedy VR, Smith DM, Sugden PH. The effects of 6 hours of hypoxia on protein synthesis in rat tissues *in vivo* and *in vitro*. *Biochem J* 1985; 228: 179–85.

Interrelations between the Degradation Rates of RNA and Protein and the Energy Turnover Rates

Gerhard Schöch and Heinrich Topp

Forschungsinstitut für Kinderernährung Dortmund, Heinstück 11, 44225 Dortmund, Germany

BACKGROUND AND METHODOLOGICAL ASPECTS

The whole body degradation rates of cytoplasmic transfer RNA (tRNA), ribosomal RNA (rRNA), and messenger RNA (mRNA) in mammals can be determined noninvasively using high-performance liquid chromatography, by measuring the urinary excretion of special modified RNA catabolites (ribonucleosides, nucleobases) which are excreted virtually quantitatively (1–4). The method is based on the pathways illustrated in Fig. 1. During the post-transcriptional processing of the RNA precursor molecules to the mature tRNA, rRNA, and mRNA molecules, some nucleosides are modified in highly specific ways. These modified building blocks are released during naturally occurring RNA degradation and cannot be reutilized for *de novo* RNA synthesis. Some modified RNA catabolites have been shown to be excreted virtually quantitatively (Fig. 1) and are therefore suitable as RNA degradation markers (1–4). The degradation rates of tRNA, rRNA, and mRNA can be calculated selectively from the excreted amounts of these catabolites because the average frequencies of their occurrence within the various RNA classes are known (1–4).

In analogy to the post-transcriptional RNA modifications, some proteins are modified post-translationally, also in highly specific ways (Fig. 1). In the course of protein degradation the modified amino acids are released and cannot be reutilized for protein synthesis. In the case of the two modified amino acids 3-methylhistidine (m^3His) and γ-carboxyglutamic acid (Gla), virtually quantitative excretion in urine has been proved in humans and rats (5,6). Therefore, urinary m^3His stemming predominantly from skeletal muscle can be used as an indicator of muscle protein degradation (5,7). Gla occurs in proteins such as prothrombin, coagulation factors VII, IX, and X, plasma proteins C, S, and Z, osteocalcin, and matrix Gla-protein (8). Urinary Gla can be used as a common degradation marker of these proteins.

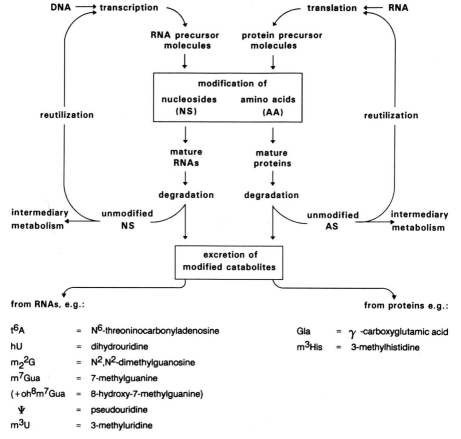

FIG. 1. Molecular basis underlying the noninvasive determinations of the whole body degradation rates of tRNA, rRNA, and mRNA as well as of muscle protein and Gla-containing proteins.

RESULTS

Figure 2 shows the average whole body degradation rates of tRNA, rRNA, and mRNA and the average excretion rates of Gla determined by us in variously sized mammalian species as functions of the respective body weights. The average muscle protein degradation rates determined by us in humans and rats of different ages are given in Fig. 2 as function of their respective muscle masses. Furthermore, published whole body protein turnover rates in variously sized mammalian species determined by the stable isotope technique (9) as well as values of the metabolic activity [resting metabolic rate (RMR)] of the various mammals are given as functions of their respective body weights. The values of the RMR were calculated on the basis of the body weights using an empirical formula (9): RMR (kJ/d) $\triangleq$ 240 × kg body weight$^{0.74}$. The virtually parallel courses of the curves in Fig. 2 indicate that the variables shown are highly correlated (1–4,10,11).

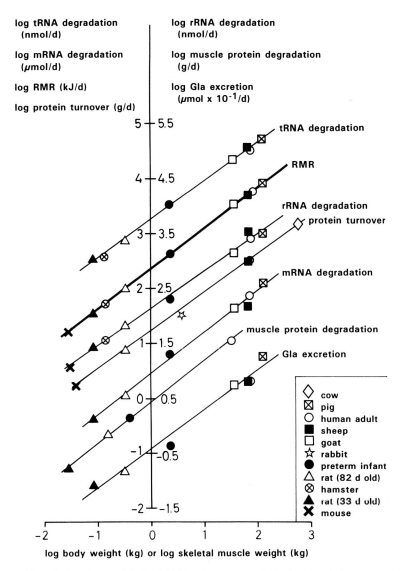

FIG. 2. Compilation of own data (2–4,10,11) of average whole body degradation rates of tRNA, rRNA, mRNA, and muscle protein as well as of average urinary excretion rates of Gla (γ-carboxyglutamic acid) found in variously sized mammalian species. The data of the whole body protein turnover rates as well as the formula for calculating the resting metabolic rates (RMR) in the various species on the basis of body weights (see the text) were taken from literature (9). All data are given in a double-logarithmic form as functions of the respective average body weights. As growth-related changes in body composition affect relative muscle mass, muscle protein degradation rates are given as a function of muscle masses.

SPECULATION

From these findings we have hypothesized that there is a causal relationship between the body-size-related energy turnover (i.e., the oxygen consumption) and the degradation rates of RNA and proteins. It is tempting to speculate that oxygen radicals were generated in proportion to the oxygen consumed as a consequence of the energy turnover. These radicals would attack RNA and protein macromolecules at random and initiate their degradation.

ACKNOWLEDGMENTS

This work was supported by the Bundesministerium für Gesundheit, by the Ministerium für Wissenschaft und Forschung des Landes Nordrhein-Westfalen, and by the Deutsche Forschungsgemeinschaft.

REFERENCES

1. Schöch G, Sander G, Topp H, Heller-Schöch G. Modified nucleosides and nucleobases in urine and serum as selective markers for the whole-body turnover of tRNA, rRNA and mRNA-cap: future prospects and impact. In: Gehrke CW, Kuo KC, eds. *Chromatography and modification of nucleosides*. Part C. Amsterdam: Elsevier, 1990; C389–441.
2. Schöch G, Topp H, Held A, *et al.* Interrelation between whole-body turnover rates of RNA and protein. *Eur J Clin Nutr* 1990; 44: 647–58.
3. Topp H, Duden R, Schöch G. 5,6-Dihydrouridine: a marker ribonucleoside for determining whole body degradation rates of transfer RNA in man and rats. *Clin Chim Acta* 1993; 218: 73–82.
4. Topp H, Kikillus K, Heller-Schöch G, Schöch G. The whole-body degradation rates of cytoplasmic mRNA in mammals can be determined by measuring the urinary excretion of the modified RNA catabolites 7-methylguanine and 8-hydroxy-7-methylguanine. *Biol Chem Hoppe Seyler* 1991; 372: 770.
5. Young VR, Munro HN. N^τ-methylhistidine (3-methylhistidine) and muscle protein turnover: an overview. *Fed Proc* 1978; 37: 2291–2300.
6. Shah DV, Tews JK, Harper AE, Suttie JW. Metabolism and transport of γ-carboxyglutamic acid. *Biochim Biophys Acta* 1978; 539: 209–17.
7. Sjölin J, Stjernström H, Henneberg S, *et al.* Splanchnic and peripheral release of 3-methylhistidine in relation to its urinary excretion in human infection. *Metabolism* 1989; 38: 23–9.
8. Vermeer C. γ-Carboxyglutamate-containing proteins and the vitamin K-dependent carboxylase. *Biochem J* 1990; 266: 625–36.
9. Waterlow JC. Protein turnover with special reference to man. *Q J Exp Physiol* 1984; 69: 409–38.
10. Topp H, Gu X, Iontcheva V, Schöch G. The excreted amounts of the modified protein catabolite γ-carboxyglutamic acid in urine correlate with the basal metabolic rates in mammals of various sizes. *Biol Chem Hoppe Seyler* 1991; 372: 769.
11. Topp H, Schöch G. The degradation rates of muscle protein determined via urinary m^{3}His correlate with the resting metabolic rates in human adults, preterm infants and rats. *Amino Acids* 1993; 5: 424.

Digestibility and Absorption of Protein in Infants

Bo Lönnerdal

Department of Nutrition, University of California, Davis, California 95616-8669, USA

Proteins in breast milk and infant formulas are often assumed to be well utilized by infants and to cover their amino acid requirements. While this is likely to be largely correct and while insufficient amino acid availability is unlikely, some aspects require further scrutiny. Several proteins in human milk are not well digested and can escape digestion, making a discussion of protein requirements of breast-fed infants necessary. In addition, processing and protein composition of formulas will affect protein digestibility, necessitating a reevaluation of the optimum protein intake of formula-fed infants. The increased use of metabolic indices in infants to evaluate the adequacy of different diets will also demand a better understanding of the physiological events in the gut that precede the amino acid levels in plasma.

DIGESTIBILITY OF HUMAN MILK PROTEINS

The ability of human milk proteins to serve as a source of amino acids ("nutritional protein") to the breast-fed infant has been an area of controversy. While it was earlier believed that human milk protein was very well digested and utilized, it has recently been argued that some of the proteins in human milk that resist digestion should not be included when calculating the amounts of protein and amino acids provided to the breast-fed infant (1). The truth is somewhere in between—not all proteins in human milk are well digested; on the other hand, even if a proportion of these proteins escape digestion and are found in the stool, this fraction is quantitatively minor (2).

Secretory IgA (SIgA) was detected in the feces of breast-fed infants over 20 years ago (3). This particular form of IgA, which is present in high concentrations in breast milk, particularly in colostrum, consists of two molecules of IgA linked together with two other molecules, the secretory component and the J-chain (3). This molecular arrangement renders SIgA uniquely stable against degradation by proteolytic enzymes (4). However, it should be emphasized that there is not absolute protection; with time, SIgA will ultimately be degraded by digestive enzymes. The extent of this degradation (i.e., gastric acid secretion and pH, pepsin and pancreatic enzyme output, and transit time) will be dependent on the maturity of the infant.

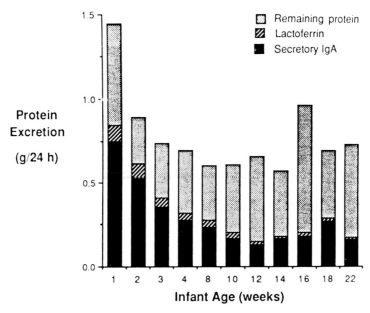

FIG. 1. Contribution of lactoferrin and secretory IgA to total protein excretion of term breast-fed infants. [From Davidson LA, & Lönnerdal B *Acta Paediatr Scand* 1974; 63: 588–594.]

A study of the persistence of human milk proteins in exclusively breast-fed infants showed that three proteins could be detected in the stool in significant quantities: secretory IgA, lactoferrin, and α_1-antitrypsin (5,6). The proportion of intact milk proteins found in the feces varied with the age of the infant; about 10% of the total protein intake of the breast-fed infant was found during the early neonatal period (0–1 month of age), while only 3% was found at 4 months of age (Fig. 1). That the excreted proteins originate from human milk and are not secretory products from bile, pancreatic fluid, or intestinal cells is evident from the fact that only negligible quantities of these proteins are found in the feces collected from infants fed formulas based on cow's milk (7). Some early studies also describe lysozyme in the stool of infants and children (8); however, the origin of this lysozyme was not known.

The presence of intact lactoferrin in the stool of infants has been used to suggest a physiological role for lactoferrin in the gastrointestinal tract of breast-fed infants (9). It has been proposed that lactoferrin facilitates iron absorption (10), inhibits the growth of intestinal pathogens (11), and stimulates the growth of the intestinal mucosa (12), all of which require the presence of intact lactoferrin within the gut lumen. *In vitro* experiments have shown that lactoferrin has an unusual stability against proteolytic enzymes; however, at low pH, lactoferrin will eventually be degraded (13).

The survival of some breast milk proteins may not only be explained by the stability of these proteins against proteolytic attack, but also by the presence of protease inhibitors in human milk that may inhibit some of the proteolytic capacity of the infant (6,14). In particular, the finding of considerable quantities of α_1-antitrypsin in breast milk and the stool of breast-fed infants (6) suggests that this protein may inhibit part of the trypsin activity in the gut. As much as 0.3–0.6 mg of α_1-antitrypsin per milliliter was found in early human milk, and even if concentrations were lower in mature milk, the protein was still detectable after 3–4 months of lactation. The proportion of α_1-antitrypsin in the stool decreased during lactation: during the first weeks 50–60% was found intact, while 20–30% was found at 3–4 months of age. *In vitro* digestion experiments showed that α_1-antitrypsin on its own was digested, while α_1-antitrypsin together with trypsin resisted digestion by pancreatic enzymes. Other protease inhibitors have also been detected in breast milk, such as α_1-antichymotrypsin and antielastase (14); however, the concentrations of these inhibitors are considerably lower than that of α_1-antitrypsin and their quantitative role may be less significant.

The digestibility of human milk proteins has also been studied in preterm infants (15,16). It is obvious that the premature infant's capacity to digest milk proteins is more limited than that of the term infant: not only does a larger percentage of proteins resist proteolysis, but other milk proteins, such as lysozyme and serum albumin, are also found in their stools (Table 1). When the infants were fed a combination of preterm milk and formula, a higher proportion of lactoferrin (13% *vs* 5%) and lysozyme (18% *vs* 3%) was found in the feces (16). Thus the presence of other proteins in the gut of the preterm infant may limit proteolysis of breast milk proteins. It should be emphasized that the immunological techniques used to detect and quantitate human milk proteins also recognize larger fragments (epitopes) of these molecules. Chromatographic separation of soluble proteins from the feces showed that part of the proteins is present in intact form, but that partial proteolysis creates some larger peptides (Fig. 2). It is possible, however, that some of these larger fragments still exert some biological activity; for example, the intestinal lactoferrin receptor will bind larger fragments of human lactoferrin (17).

TABLE 1. *Intake and fecal excretion of human whey proteins*[a]

	PTM group		PTM + formula group	
	Intake	Excretion (%)[b]	Intake	Excretion (%)[b]
α-Lactalbumin	425 ± 104	0 (0)	171 ± 32[b]	0 (0)
Serum albumin	102 ± 67	0.56 ± 0.13 (0.5)	57.21	0.34 ± 0.6 (0.6)
Lactoferrin	492 ± 217	24 ± 26 (5)	262 ± 80	33 ± 22 (13)
Lysozyme	34 ± 10	1.1 ± 1.2 (3)	11 ± 8	2.0 ± 1.5 (18)
Secretory IgA	460 ± 383	110 ± 150 (24)	138 ± 83	38 ± 58 (27)

Adapted from Donovan SM, et al. Am J Dis Child 1989; 143: 1485–1491.
[a] Values represent mean ± SD mg/kg/d.
[b] Calculated as a percentage of intake.

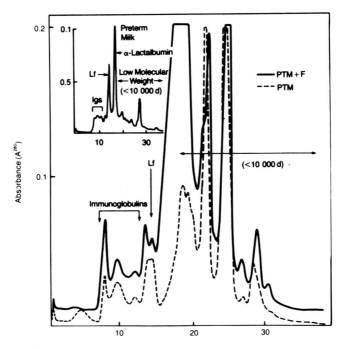

FIG. 2. Gel filtration chromatography of soluble fecal extracts from infants fed preterm milk (PTM) or PTM and formula (PTM + F) on Superose 12. Inset shows chromatographic separation of preterm human milk. [From Donovan SM, et al. Am J Dis Child 1989; 143: 1485–1491.]

DIGESTIBILITY OF FORMULA PROTEINS

Infant formulas normally contain higher levels of protein than human milk (18). The rationale for these higher levels of protein has been that they will assure the infant of an adequate supply of amino acids from a protein source with assumed lower digestibility. However, the digestibility of heat-treated milk proteins should be relatively high and it has been suggested that the commonly used protein level in formula of 15–20 g/liter may be unnecessarily high (19,20). This notion was supported by observations that metabolic indices, such as plasma amino acids and blood urea nitrogen, are quite different in formula-fed infants than in breast-fed infants.

If the protein level of infant formula is lowered, the protein and amino acid intake of the formula-fed infant will by definition become closer to the requirement of the infant, at the same time narrowing the "safety" margin. In this situation, the protein quality of formula becomes more critical. All formulas are exposed to heat treatments which vary with regard to both temperature and duration. There has been very limited consideration of the significance of using different heat treatments for production of liquid formulas ("in-can" sterilization, >110°C, extended time periods) and powdered formulas (spray-drying, 60°C, brief exposure). More recently, ultrahigh temperature treatment (UHT, >130°C, 3–5 s) has become more common for production of

TABLE 2. *Gastric acid and pepsin output at different ages*

	Acid (mEq/h)	Pepsin (mg/h)
Infant (1 day)	13.2	0.18
Infant (3–8 days)	0.06	0.21
Infant (10–11 days)	0.12	0.46
Infant (14–17 days)	0.19	0.88
Infant (67–110 days)	0.47	1.34
Children (4–8 yr)	4.9	18.5
Adults	13.1	41.9

Adapted from Agunov M, et al. *Am J Dig Dis* 1969; 14: 400–414.

ready-to-feed formula. Besides the heat treatment, the protein composition (casein-predominant *vs* whey-predominant) of the formula may be expected to affect protein digestibility (21).

The digestibility of milk proteins can be assessed by *in vitro* methods or *in vivo*, using experimental animals or human infants. The *in vitro* methods usually employ a combination of proteolytic enzymes such as pepsin, trypsin, or chymotrypsin, or, as in some studies, gastric and/or duodenal aspirates from infants in order to mimic more closely the conditions of the infant's gastrointestinal tract. An important consideration in all these methods is the time factor; with time, most proteins will be digested; however, passage through the gastrointestinal tract in infants is rapid and the time of exposure to enzymatic digestion is short. In addition, gastric acid secretion is relatively low (Table 2), making the stomach pH comparatively high and limiting pepsin activity (22). Development of the exocrine pancreas is not complete at birth (23), making the activities of several pancreatic enzymes lower than normal (24,25).

We have assessed the digestibility of milk proteins in infant formulas that had been exposed to various types of heat treatment (26,27). For the identical product, UHT-treated formula showed a significantly higher protein digestibility than formulas that were powdered (spray-dried) or sterilized (84% *vs* 72%). When comparing products that are present on the market in both powdered and sterilized form, the powdered products consistently showed higher protein digestibility, although the difference was less pronounced than that found for the UHT products (Table 3). Thus the more intense the heat treatment, the lower the protein digestibility. As an index of the heat treatment that the formulas had been exposed to, we analyzed available lysine in the product. Powdered products were consistently found to have significantly higher levels of available lysine, although the magnitude of the difference was not large. To explore the cause of the different digestibilities found, we subjected the formulas to various pH exposures, centrifugation, gel filtration, and gel electrophoresis. By using these techniques, we were able to study protein solubility and the distribution of protein between the fat, the soluble fraction, and the insoluble fraction. We found protein–protein and protein–lipid interactions in all formulas tested and these were more pronounced in the sterilized products than in powdered and UHT

TABLE 3. Digestibility of proteins in infant formulas

Formula	Protein (mg/ml)	Protein digestibility (%)
Pre-Aptamil (powder)	17.61	72.0
Pre-Aptamil (sterilized)	16.03	72.8
Pre-Aptamil (UHT)	19.26	83.9
Enfamil (powder)	14.24	81.9
Enfamil (sterilized)	15.92	71.7
S.M.A. (powder)	16.11	76.6
S.M.A. (sterilized)	14.29	72.5
Similac (powder)	14.55	81.3
Similac (sterilized)	16.45	74.5
Baby Semp (powder)	14.32	80.6
Allomin (UHT)	13.97	69.2

Adapted from Rudloff S, & Lönnerdal B *J Pediatr Gastroenterol Nutr* 1992; 15: 25–33.

products. These results support the hypothesis by Rowley & Richardson (28) that the composition of the lipid–fluid phase interface in milk is determined by the temperature and the duration of heat exposure used during processing. In addition, they stated that if the pH reaches the isoelectric point of the protein (e.g., pH 4.6 for bovine caseins), proteins are less charged and more hydrophobic and can therefore interact with uncharged lipids. We found that enhanced protein–lipid interactions occurred in the pH range 4.0–5.0, which is similar to the pH in the stomach of young infants (29,30). In particular, casein and β-lactoglobulin in sterilized products was found associated with lipids at this pH range. This was further documented by gel filtration and gel electrophoresis, showing that few proteins were soluble at this pH and also that casein and β-lactoglobulin were associated with the lipid fraction. Since heat treatment opens up disulfide bridges within proteins such as β-lactoglobulin and κ-casein, reactive sulfhydryl groups may now form covalent bonds with other components. It has been suggested that the formation of such bonds is responsible for the lower protein digestibility. Thus the new disulfide bonds formed and the modifications of some amino acid residues (e.g., lysine) by Maillard reactions may limit the accessibility of proteolytic enzymes and consequently, protein digestibility.

Individual milk proteins have different capacities to resist proteolysis. Jakobsson *et al.* assessed the digestibility of some human milk and bovine milk proteins *in vitro* by using either duodenal juice from infants (31) or trypsin and elastase (32). They found that human lactoferrin was very slowly digested, while bovine casein was degraded rapidly. Interestingly, human α-lactalbumin was found to be more slowly digested than bovine α-lactalbumin, even if their structure and composition are very similar.

The suckling rat pup model has also been used to assess protein digestibility *in vivo* (33; Lönnerdal *et al.*, unpublished). At this developmental stage, the rat pup's capacity to digest proteins has not yet been fully developed and transit through the

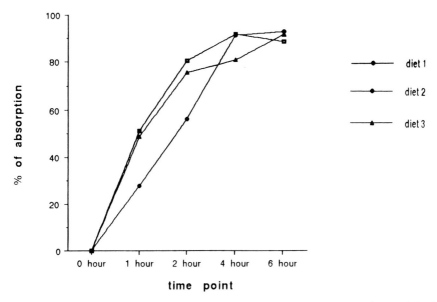

FIG. 3. *In vivo* digestibility of infant diets intubated into the stomach of suckling rat pups. Animals were killed at different time points after intubation and stomach and intestine were perfused with saline (diet 1, UHT; diet 2, powdered; diet 3, sterilized). (Unpublished data.)

gastrointestinal tract is rapid, limiting the time for digestion to occur. We therefore believe that this is a good model for the human infant and, at the least, it allows us to compare the digestibility of proteins in formulas that have been processed in various ways. It should be noted, however, that the older rat (weanling or adult) has a very efficient digestive "machinery"—differences among various formulas are rarely observed and digestibility is always very high (34). In our model (Lönnerdal et al., unpublished) diets are intubated into the stomach of fasted pups and the animals are killed at various time points (1–4 h) after the dose has been given. The stomach and the small intestine are perfused separately and the perfusates are then separated into soluble and insoluble proteins by centrifugation. Nitrogen and nonprotein nitrogen are analyzed in all fractions. By this method we found that proteins in UHT-treated and powdered products were most rapidly digested, while sterilized formulas were more slowly digested (Fig. 3). Protein digestibility of soy formulas was also found to be slow. Although most formula proteins were digested with time, our results suggest that when transit time is short, as in newborn infants, slow digestibility may result in lower amino acid availability to the infant.

IMPLICATIONS OF DIFFERENCES IN PROTEIN DIGESTIBILITY IN INFANT NUTRITION

It is obvious that a lower digestibility of proteins in infant formula will affect the amount of amino acids and smaller peptides available for absorption by the infant.

Whether the differences in digestibility found for formulas that have been exposed to different heat treatments are large enough to have a significant quantitive effect in infants remains to be studied. However, it appears prudent to attempt to optimize protein utilization from infant formula, as undigested fragments may affect utilization of other nutrients (see below). In addition, knowledge about protein digestibility is important when deciding on the appropriate level of protein to use in formula. For example, fasting levels of plasma amino acids are often used to evaluate protein and amino acid adequacy from different formulas in clinical studies in infants. In such studies, the type of heat treatment is rarely presented, only the brand name. Thus data may be derived from sterilized or powdered products and the results may be as dependent on the degree of heat treatment as on the protein level or composition being studied. An important illustration of this is given by the recent study of Räihä et al. (personal communication), who gave premature infants either formulas exposed to different heat treatments or human milk. Plasma amino acid profiles were taken at hourly intervals and protein intake was normalized between the diets. In infants fed human milk, plasma amino acids rose quickly after a meal and then decreased rapidly and reached the baseline after about 2 h. The plasma amino pattern of infants fed UHT-treated formula was similar to that of breast-fed infants, while the peak and decline in amino acids were somewhat slower in infants fed powdered formula. More dramatically, however, infants fed sterilized formula had a much slower increase in plasma amino acids, and the levels had not returned to baseline even 4 hours after feeding. This observation has two important implications: (a) protein digestibility from formula in infants is affected by the degree of heat treatment; and (b) the slower protein digestibility of sterilized formula makes the fasting plasma amino acid pattern an unreliable diagnostic tool for assessing protein quality.

Lower protein digestibility causes a higher percentage of larger peptides and proteins to persist longer in the small intestine. Since some of these proteins/peptides can bind minerals and trace elements, they may limit the uptake of such essential nutrients, either at the site of their maximum absorption (often the duodenum), or within the entire intestine if they ultimately leave for the colon. We have shown both in human studies (35) and in experimental animals (36) that casein-predominant formulas result in lower zinc bioavailability than whey-predominant formulas, presumedly because of less complete digestion of casein. However, it should also be recognized that some peptides formed during digestion of casein, the so-called casein phosphopeptides, may have a positive effect on calcium absorption (37,38). Furthermore, we have shown in our suckling rat pup model that trace element bioavailability was highest for UHT-treated and powdered products, but lower from sterilized products (Lönnerdal et al., unpublished). Thus protein composition and protein digestibility can affect mineral and trace element bioavailability from formulas.

REFERENCES

1. Räihä NCR. Nutritional proteins in milk and the protein requirement of normal infants. *Pediatrics* 1985; 75: 136–41.

2. Hambraeus L, Fransson GB, Lönnerdal B. Nutritional availability of breast milk proteins. *Lancet* 1984; 11: 167–8.
3. Hanson LÅ, Winberg J. Breast-milk and defense against infection in the newborn. *Arch Dis Child* 1972; 47: 845–8.
4. Lindh E. Increased resistance of immunoglobulin A dimers to proteolytic degradation after binding of secretory component. *J Immunol* 1975; 114: 284–6.
5. Davidson LA, Lönnerdal B. Persistence of human milk proteins in the breast-fed infant. *Acta Paediatr Scand* 1987; 76: 733–40.
6. Davidson LA, Lönnerdal B. Fecal α_1-antitrypsin in breast-fed infants is derived from human milk and is not indicative of enteric protein loss. *Acta Paediatr Scand* 1990; 79: 137–41.
7. Prentice A, Ewing G, Roberts SB, et al. The nutritional role of breast milk IgA and lactoferrin. *Acta Paediatr Scand* 1987; 76: 592–8.
8. Haneberg B, Finne P. Lysozyme in feces from infants and children. *Acta Paediatr Scand* 1974; 63: 588–94.
9. Spik G, Brunet B, Mazurier-Dehaine C, Fontaine G, Montreuil J. Characterization and properties of the human lactotransferrins extracted from the feces of newborn infants. *Acta Paediatr Scand* 1982; 71: 979–85.
10. Lönnerdal B. Iron in breast milk. In: Stekel A, ed. *Iron nutrition in infancy and childhood.* Nestlé Nutrition Workshop Series, Vol 4. New York: Raven Press. 1984; 95–118.
11. Bullen JJ, Rogers HJ, Leigh L. Iron-binding proteins in milk and resistance to *Escherichia coli* infection in infants. *BMJ* 1972; 1: 69–75.
12. Nichols B, McKee KS, Henry JF, Putman M. Human lactoferrin stimulates thymidine incorporation into DNA of rat crypt cells. *Pediatr Res* 1987; 21: 563–7.
13. Brock JH, Arzabe F, Lampreave F, Piñeiro A. The effect of trypsin on bovine transferrin and lactoferrin. *Biochim Biophys Acta* 1976; 446: 214–25.
14. Lindberg T, Ohlsson K, Weström B. Protease inhibitors and their relation to protease activity in human milk. *Pediatr Res* 1982; 16: 479–83.
15. Schanler RJ, Goldblum RM, Garza C, Goldman AS. Enhanced fecal excretion of selected immune factors in very low birth weight infants fed fortified human milk. *Pediatr Res* 1986; 20: 711–5.
16. Donovan SM, Atkinson SA, Whyte RK, Lönnerdal B. Partition of nitrogen intake and excretion in low birth-weight infants. *Am J Dis Child* 1989; 143: 1485–91.
17. Davidson LA, Lönnerdal B. Fe-saturation and proteolysis of human lactoferrin: effect on brush-border receptor-mediated uptake of Fe and Mn. *Am J Physiol* 1989; 257: G930–4.
18. Donovan SM, Lönnerdal B. Non-protein nitrogen and true protein in infant formulas. *Acta Paediatr Scand* 1989; 78: 497–504.
19. Lönnerdal B, Zetterström R. Protein content of infant formula: how much and from what age? *Acta Paediatr Scand* 1988; 77: 321–5.
20. Räihä NCR. Milk protein quantity and quality and protein requirements during development. *Acta Pediatr* 1989; 36: 347–68.
21. Rudloff S, Lönnerdal B. Solubility and digestibility of milk proteins in infant formulas exposed to different heat treatments. *J Pediatr Gastroenterol Nutr* 1992; 15: 25–33.
22. Agunov M, Yamaguchi N, Lopez R, Luhby AL, Glass GB. Correlative study of hydrochloric acid, pepsin, and IF secretion in newborn and infants. *Am J Dig Dis* 1969; 14: 400–14.
23. Lebenthal E, Lee PC. Development of functional response in human exocrine pancreas. *Pediatrics* 1980; 66: 556–60.
24. Lindberg T. Proteolytic activity in duodenal juice in infants, children and adults. *Acta Paediatr Scand* 1974; 63: 805–8.
25. Antonowicz I, Lebenthal E. Developmental pattern of small intestinal enterokinase and disaccharidase activities in the human fetus. *Gastroenterology* 1977; 72: 1299–303.
26. Rudloff S, Lönnerdal B. Effect of processing of infant formulas on calcium retention in suckling rhesus monkeys (Abstract). *FASEB J* 1990; 4: A521.
27. Rudloff S, Lönnerdal B. Calcium and zinc retention from protein hydrolysate formulas in suckling rhesus monkeys. *Am J Dis Child* 1992; 146: 588–91.
28. Rowley BO, Richardson T. Protein–lipid interactions in concentrated infant formula. *J Dairy Sci* 1985; 68: 3180–8.
29. Sondheimer JM, Clark DA, Gervaise EP. Continous gastric pH measurement in young and older

healthy preterm infants receiving formula and clear liquid feedings. *J Pediatr Gastroenterol Nutr* 1985; 4: 352–5.
30. Adamson I, Esangbedo A, Okolo AA, Omene JA. Pepsin and its multiple forms in early life. *Biol Neonate* 1988; 53: 267–73.
31. Jakobsson I, Lindberg T, Benediktsson B. *In vitro* digestion of cow's milk protein by duodenal juice from infants with various gastrointestinal disorders. *J Pediatr Gastroenterol Nutr* 1982; 1: 183–91.
32. Jakobsson I, Borulf S, Lindberg T, Benediktsson B. Partial hydrolysis of cow's milk proteins by human trypsins and elastases *in vitro*. *J Pediatr Gastroenterol Nutr* 1983; 2: 613–6.
33. Britton JR, Koldovský O. Luminal digestion of human milk proteins in suckling and weaning rats. *Nutr Res* 1987; 7: 1041–9.
34. Harris D, Burns RA, Ali R. Evaluation of infant formula protein quality: comparison of *in vitro* with *in vivo* methods. *J Assoc Off Anal Chem* 1988; 71: 353–7.
35. Lönnerdal B, Cederblad A, Davidsson L, Sandström B. The effect of individual components of soy formula and cow's milk formula on zinc bioavailability. *Am J Clin Nutr* 1984; 40: 1064–70.
36. Lönnerdal B, Bell JG, Hendrickx AG, Burns RA, Keen CL. Effect of phytate removal on zinc absorption from soy formula. *Am J Clin Nutr* 1988; 48: 1301–6.
37. Lee YS, Noguchi LT, Naito H. Intestinal absorption of calcium in rats given diets containing casein or amino acid mixture: the role of casein phosphopeptides. *Br J Nutr* 1983; 49: 67–76.
38. Sato R, Noguchi T, Naito H. Casein phosphopeptide (CPP) enhances calcium absorption from the ligated segment in rat small intestine. *J Nutr Sci Vitaminol* 1986; 32: 67–76.

DISCUSSION FOLLOWING THE PRESENTATION OF DR. LÖNNERDAL

Dr. Räihä: I think this is the beginning of a very important new area concerning the manufacture of infant formulas. What Dr. Lönnerdal has shown *in vitro* and partly in his rat studies, and which we have been able to confirm in the premature infant *in vivo*, is that human milk has a very different pattern from formula when you look at the speed of absorption of amino acids and also at the height of the postprandial plasma amino acid concentration curve. We don't really know what this means. We can only say that some formulas are more like human milk and others are less like it when you look at these variables. However, I think these findings have clinical and technical implications. For example, the speed at which the amino acids increase in the plasma and then return to baseline may have an effect on release of gastric hormones, insulin, and so on, and this of course can affect the baby. Also, you have to be very precise about the time of sampling. We have compared studies in premature infants given either human milk or formula in the same quantities. If you take preprandial samples, usually about 2½–3 hours after the last feed, it is always clear that the infants fed human milk have lower plasma amino acid levels than those of infants fed formula. I think the explanation for this may be that in the formula-fed infants the amino acids have not yet come down to baseline. We must remember this when we refeed the baby, because we may be pushing up the levels higher and higher without allowing them time to return to baseline. So I think there are many clinical implications related to these findings.

Dr. Guesry: The smoother postprandial amino acid curve with casein-predominant than with adapted whey formulas is in keeping with work by Jacques Senterre showing that gastric emptying was much slower with casein formulas than with adapted whey formulas. My second comment is for our colleagues from developing countries who may follow a routine of pasteurizing their powdered formulas in the autoclave, perhaps for as long as 1 hour at high temperature. This is likely to harm the protein, which was originally of good quality and well absorbed.

I have a question relating to the quantitative assessment of protein. It used to be assumed when calculating the protein content of human milk that the functional proteins, such as secretory IgA, lactoferrin, and lysozyme, were not digested, and only the true nutritional protein was taken into account. However, you have now shown us that 85–90% of these functional proteins are indeed digested and could be used for nutrition. I think it is time that we reassessed the quantity of protein in human milk so that we can judge the right quantity for infant formula.

Dr. Räihä: No one has ever claimed that the protective proteins are not absorbed at all. But even if we were to assume that only between 3% and 5% of the total proteins in human milk are not absorbed, we would still find that the nutritional protein concentration in human milk is very low compared to all the formulas on the market today. Even if you take into account the amount of nonabsorbed whey protein, many formulas today still have almost twice as much nutritionally available protein as human milk.

Dr. Rassin: I wonder if another way to interpret the curves of the total amino acids is that the amino acid concentrations are regulating the feeding patterns of the infant, because you get much more frequent feeding in breast-fed infants than in formula-fed infants. Perhaps what really happens is that you see different responses to those amino acids.

Dr. Räihä: The study that we showed was based on one bolus feed.

Dr. Rassin: I can understand that, but in a normal situation your breast-fed baby is going to feed maybe every 2 or 2½ hours, while frequently in the nursery we have no problem getting formula-fed babies to go for several hours between feeds. What I am saying is that I wonder whether these amino acids are playing a role in the regulation of these feeding patterns through an influence on hormone release and brain response.

Dr. Axelsson: Dr. Lönnerdal, what was the composition of the cereal-based formula you used that was poorly digested? Was it based only on cereal proteins?

Dr. Lönnerdal: It was based on a mixture of rice and barley. We have also looked at mixtures of cereals and milk which are quite commonly used, and their digestibility is not much better. Digestibility is slower for such formulas and this reemphasizes that we should not start to feed cereals to infants at too young an age.

Dr. Heine: Is anything known about the enzymes that split off the carbohydrate side chain of glycoproteins and food proteins such as casein? What enzyme is it?

Dr. Lönnerdal: Some of the brush border enzymes can have this role. The digestibility of the protein is important because the activity of some of the carbohydrate metabolizing enzymes may be dependent on how much of the polypeptide has been split to allow access to the branch structures.

Dr. Heine: You mentioned the loss of bioavailability of lysine during heat treatment. Are there any other amino acids that lose bioavailability under these conditions? What about tryptophan, for instance?

Dr. Lönnerdal: This has been reported in the literature; serine and threonine can also be affected but I am not aware of any very good methods for easily detecting these effects.

Dr. Räihä: Hydrolyzed and partially hydrolyzed formulas are becoming more and more popular, and are sometimes used in premature infants. According to your data, these should be absorbed very differently and probably much faster, so there may be marked physiological effects.

Dr. Lönnerdal: It is quite likely that these formulas may have marked effects on hormonal responses following rapid surges in substrate levels in the blood, but I have not seen much work on this.

Dr. Marini: I would like to comment on this. We have studied gastrointestinal hormones in full-term babies fed with hydrolyzed formulas. In general, the responses were much the same as with human milk, but we found differences in enteroglucagon and pancreatic polypeptide inhibitors. It is easy to explain this because they don't need pancreatic hydrolysis. In another study we looked at gastric function and found that basal and maximum acid output with hydrolyzed formula was more similar to that found with human milk than with artificial formulas.

Dr. Guesry: If we consider that gastric emptying is an index of digestibility, then in the curve which starts with low digestibility for raw cow's milk and proceeds to the best digestibility or best gastric emptying for human milk, hydrolyzed formula is closest to human milk, with a gastric emptying time of less than 2 h.

Dr. Räihä: You showed the difference between human milk and human milk to which formula was added. We often supplement human milk with bovine proteins when feeding premature infants. Do you know of any studies looking at protein digestibility in such cases?

Dr. Lönnerdal: We have looked at some of these human milk fortifiers and they are remarkably insoluble. They are much more insoluble than milk formulas and I have the feeling that the processing of some of these cow's milk proteins is quite different. When you prepare a whey protein milk powder there are many other substances present which help to stabilize the mixture during spray drying. In the case of milk fortifiers, consisting only of proteins, the proteins become very hydrophobic. If you centrifuge fortified human milk you find a pellet at the bottom consisting of the fortifier proteins. The various commercial brands that we have investigated have all been more or less totally insoluble.

Dr. Pettifor: We are becoming increasingly aware of the role of the large bowel in carbohydrate absorption. Is there any evidence that bacterial flora play a role in protein digestibility?

Dr. Uauy: This is an interesting question and has particular relevance to urea. Urea is secreted into the upper gut, and flora definitely play a role in urea recycling; there is also a possibility that non-absorbed proteins are modified by the intestinal flora. I know that Alan Jackson is currently looking at the effect of the gut flora in producing some of the essential amino acids and we shall probably have interesting surprises relating to the role of the flora and the colon modifying the amino acid supply. In the ruminants the role of the flora in determining the amino acid supply is well established.

Dr. Heine: We have given ^{15}N-labeled protein into the colon of babies with colostomies and we observed 80% nitrogen absorption. Since the colon has no proteolytic enzymes, we must assume that the bacterial flora are splitting this protein down to the amino acid label to allow its absorption. We have run the same experiments with bacterial nitrogen and found the same absorption rate.

Dr. Lönnerdal: Were those studies done in preterm infants?

Dr. Heine: There were preterm infants among the subjects.

Dr. Räihä: How much protein are we talking about? Is it a substantial amount or of only minor importance?

Dr. Heine: The amounts absorbed in our studies were of the order of grams rather than milligrams. This might be important in the malnourished infant when the protein supply is low. The mode of absorption from the colon is not quite clear. It might be that some protein is digested within macrophages and other cells, but we can't explain the high absorption rate in any other way than that there is digestion by the intestinal flora.

Dr. Marini: Dr. Lönnerdal, you used 10 to 14-day-old rats for your study. Wouldn't it have been better to have used neonatal guinea pigs? It is said that neonatal guinea pigs are more similar to human beings.

Dr. Lönnerdal: It is quite possible that it would have been better. However, the rat produces very large litters and is a good laboratory animal for screening tests even if it may not be the ideal model. When we see relevant differences we can move on to studies in primates or in premature infants.

Dr. Ali Dhansay: I would like to ask a question related to lipid protein interactions. There are situations when one uses medium-chain triglycerides as an additive to formula feeding. Have you any suggestion as to when and how much one should add with respect to digestibility of proteins in the formula itself?

Dr. Lönnerdal: My feeling is that it would not matter much for this phenomenon whether there are medium-chain triglycerides or long-chain fatty acids present; you would most likely see the same type of hydrophobic interaction regardless of chain length.

International Recommendations on Protein Intakes in Infancy: Some Points for Discussion

Brian A. Wharton

Old Rectory, Belbroughton, Worcestershire, DY9 9TF, United Kingdom

BACKGROUND

Populations

The definitive international report on protein requirements is the one published by the WHO in 1985 (1). It forms the basis of many other recommendations, including the American ones published in 1988 (2) and the recent British *Dietary Reference Values* (3). These reports also include energy requirements, which are obviously closely related to protein requirements.

Foods

In addition, there are other recommendations and in some cases legislation concerning the composition of infant formula, follow-on milk, and weaning foods. Since these foods are consumed by infants to meet their nutritional requirements, the reports are in effect recommendations concerning the infants' nutritional intake, or at least part of it. A few reports are "international" (e.g., Codex Alimentarius of WIIO/FAO, reports of the European Society of Pediatric Gastroenterology and Nutrition, and reports of the Committee on Food of the European Community). There is a plethora of national reports concerning infant foods. References will be given to individual reports at relevant points in the paper.

There is little point in doing a line-by-line comparison of all these reports, whether for populations or foods. Instead, I shall discuss five areas where there is room for debate.

NUMERATOR

For numerical recommendations a numerator and a denominator are involved (e.g., protein per person per day). In this section the numerator is considered.

Multiples of Nitrogen

The major problem with the numerator is its expression as weight or as a weight of nitrogen contained within the food. The use of nitrogen arises from the chemical analysis of foods, but it extends into other nutritional considerations when assessing balance data (e.g., amount of nitrogen in feces). In parenteral and enteral nutrition it has become the custom to express protein, peptides, and amino acids as nitrogen.

The conversion of weight of nitrogen to weight of protein uses a conventional multiple of 6.25. For milk the more appropriate factor 6.38 is often used. For weaning foods containing cereal protein only, the appropriate figure is 5.7–6.0, depending on the cereal. In practice, the content of protein stated on a label of a weaning food is likely to be calculated from food composition tables rather than from direct chemical analyses, and these tables will have used multiples of analyzed nitrogen appropriate for the source of protein.

Non-protein Nitrogen

The multiple of nitrogen method does, of course, count non-protein nitrogen as protein and where this is amino nitrogen it is likely to be available for synthesis of protein by the baby. The problem of how to account for the non-protein nitrogen of breast milk is well known. In the British national "pooled sample" of breast milk (4), total nitrogen was 210 mg per 100 ml, of which non-protein non-amino-acid nitrogen was 46 mg. If total nitrogen found in the milk is multiplied by 6.38, a figure of 1.3 g of protein per 100 ml is obtained, but when non-protein non-amino-acid nitrogen is excluded, the figure is 1.05 g.

It is not clear, however, that non-protein nitrogen compounds, particularly urea, should be excluded. Urea reaching the colon either directly from the food or via diffusion from the blood is hydrolyzed by the gut flora releasing ammonia, which travels via the portal vein to the liver for production of non-essential amino acids. The concept of urea cycling in protein economy, particularly during rapid catch-up growth, has been developed by Jackson (5). We should beware of assuming that non-protein nitrogen is of no nutritional significance. To regard the protein and amino acid nitrogen content of breast milk as true protein and to use this as a basis for determining protein requirements may lead to an underestimate because it ignores urea. Heine has explored the bioavailability of urea nitrogen in breast milk (6).

Further caution is necessary if we are tempted to deduct from this "true protein" the immunological proteins such as lactoferrin and secretory IgA (on the basis that they are not absorbed and are not nutritionally available), to give a final nutritionally available protein content in breast milk that is perhaps as low as 0.8 g per 100 ml. Two studies have found that relatively small amounts of the secretory IgA and lactoferrin ingested in breast milk appear in the feces of infants a few weeks old (7,8). This suggests that large amounts of these protective proteins are broken down and so become nutritionally available higher in the gut. Our own observations were that only 1% of estimated lactoferrin intake appeared in the feces of breast-fed babies aged 2 weeks; the figure for bovine lactoferrin was around 0.5% (9).

In summary, neither non-protein nitrogen nor the "immunological proteins" should be discarded when assessing the amount of protein available to and required by an infant.

DENOMINATOR

Individuals and Their Weight

Recommendations on protein intake in children, which are part of national or international recommendations for all ages, use either an individual or his weight as the denominator; that is, the "safe level," "advisable intake," or "reference nutrient level" is expressed as grams per person at a particular age or as grams per kilogram of body weight.

The WHO figures from which the others follow are based on the factorial method, or at least a modification of it, since the determination of obligatory nitrogen loss on very low protein diets is not relevant for a rapidly growing infant. A typical calculation for 0–4 months quoted in the WHO report (1) is an increase in tissue protein of 3.5 g per day plus the equivalent of 0.5 g each day lost from desquamating skin and sweat. At this age average retention is 45% of intake. Therefore, an intake of 8.9 g is necessary to retain 4 g; that is, for an average weight at this age of 5.25 kg, an intake of 1.7 g/kg is needed. Some factor may then be applied to allow for safety to give an advisable or safe intake of around 1.9 g/kg.

Volume, Weight, and Energy of Food

In recommendations for food, values expressed as grams per infant or grams per kilogram of infant clearly will not do. For infant formulas the choices of denominator are weight of powder, volume of feed, or energy content of the feed. Few people will think in terms of the weight of powder in a formula (about 13 g per 100 ml in most), and the volume fed is the obvious measurement. Furthermore, if nutrients are expressed per unit volume, this stresses the role of water as an essential nutrient, which is vital when considering the renal solute load of a diet and its effect on the concentration of urine and on water balance. This type of reasoning led to British recommendations on the composition of infant formulas being expressed per unit volume (10).

However, volume is not a suitable denominator for a weaning food; the consistency of a cereal and sugar gruel, whether homemade or provided by a manufacturer, could vary considerably. In addition, as weaning progresses renal function, including concentrating ability, increases, so the renal solute load of a diet becomes less important unless there are very abnormal losses of water from lungs, skin, or gut. Of the possible denominators, therefore, only energy can be applied to food for both suckling and weanling infants. Apart from this pragmatic reason there are also physiological reasons for the choice of energy. It recognizes the interrelationships between the requirements of protein and energy and the relatively large proportion of both energy and protein used for growth in the rapidly growing infant. Many international reports therefore express protein per unit energy, both for infant formulas [ESPGAN (11), Codex Alimentarius (12), European Community (13)] and for weaning foods (14).

Protein/Energy Ratio in Physiological States

Since there are physiological as well as pragmatic reasons for expressing protein content of foods in terms of energy, should this concept be extended to the recommendations for individuals? In other words, should the reference, safe, or advisable intakes of protein by groups of people be expressed not per person or per kilogram of body weight, but per 100 kcal of food ingested? Fomon (15) has argued in favor of this method of calculation and we have also used it as a way of expressing advisable intakes of many nutrients for children of various ages (see Table 1).

Where recommendations for protein and energy are expressed separately, then calculation of the ratio from these figures is, in theory at least, invalid since the protein figure is chosen to meet the requirements of 95% of the population (that is, more than most people need) while the energy figure is the observed average intake of the population (that is, less than what 50% of the population consumes). The mathematics are complex since the ratio involves two factors, each with its own variance. The concept and statistical manipulations are discussed in some detail by Beaton & Swiss (17).

Table 2 shows the protein/energy ratios calculated in this way for infants. The ratios are all above the ratio found in breast milk. It is probably unwise to adopt the very low protein/energy ratio in breast milk as a minimum safe standard, since the nitrogen present is utilized with unusual efficiency [an argument developed by Millward (19)]. For older children the "safe" protein/energy ratios calculated this way are well below those seen in the normal diet of Western children. In my opinion there should be some caution in accepting very low ratios of 1.7 and 1.5 g per 100 kcal as "safe" in a community until there have been extensive observations of children living on such diets. Therefore, in our recommendations shown in Table 1 (16), we adopted higher protein/energy ratios (as consumed by normal children) than the international recommendations imply.

Protein/Energy Ratios in Weaning Diets

Consideration of the protein/energy ratio can also help in unscrambling the vexing question of the composition of milk or formula feeds given during weaning.

Figure 1 shows the mathematical relationships of protein/energy ratios in milks, formulas, and solid foods used during weaning to achieve the advisable intake of not less than 1.7 g per 100 kcal. We have described the mathematical derivation of these

TABLE 1. *Advisable intake of nutrients per unit energy*

	Age of child		
Nutrient/100 kcal	0–11 months	1 year–1 year 11 months	2–10 years
Protein (g)	2.25–3.0	2.25–3.0	2.25–3.5
N_2 (g)	0.36–0.48	0.36–0.48	0.4–0.56
Non-protein Kcal N_2 ratio	160:1–238:1	193:1–228:1	194:1–175:1
Fat (g)	3.3–6.5	3.3–5.5	3.3–4.2
C8:0 % total fat	Max. 20%		
C10:0 % total fat			
C12:0 % total fat	Max. 15%		
C14:0 % total fat	Max. 15%		
Linoleic acid (g)	0.3–1.2	0.3–1.2	0.3–1.2
Total carbohydrate (CHO) (g)	7–14	10–15	12–15
Lactose	Min. 3.5		
Sucrose	<20% total CHO		
Fructose	Not usually added		
Precooked starch and/or gelatinized starch	<30% total CHO		
Minerals			
Iron (mg)	0.5–1.5	0.6–1.2	0.6–1.2
Calcium (mg)	Min. 45	Min. 45	Min. 35
Phosphorus (mg)	25–90	50–100	50–100
Sodium (mmol)	1.0–2.6	1.0–3.0	1.0–3.0
Potassium (mmol)	1.6–3.8	1.6–5.7	1.6–5.7
Chloride (mmol)	1.4–3.5	1.4–3.5	1.4–4.0
Magnesium (mg)	5–15	10–20	10–20
Copper (μg)	60–120	Min. 100	Min. 100
Zinc (mg)	0.3–1.0	0.5–1.0	0.5–1.0
Manganese (μg)	2–8	Min. 100	Min. 100
Iodide (μg)	Min. 5	Min. 5	Min. 5
Chromium (μg)	3–10	3–10	3–10
Molybdenum (μg)	5–15	5–15	5–15
Selenium (μg)	3–10	3–10	3–10
Fluoride (mg)	0.02–0.1	0.04–0.15	0.04–0.15
Vitamins			
Vitamin A (μg RE)	60–180	25–45	20–45
Vitamin D (μg cholecalciferol)	1–2	0.8–1.0	0.6–1.0
Vitamin E (mg)	Min. 0.5	Min. 0.4	Min. 0.4
(mg/g PUFA)	Min. 0.5	Min 0.4	Min. 0.4
Vitamin K (μg)	2.5–15.0	1.5–15.0	1.5–15.0
Vitamin C (mg)	Min. 5	Min. 1.5	Min. 1.5
Thiamine (mg)	0.04–0.09	0.04–0.09	0.04–0.09
Riboflavin (mg)	0.06–0.4	0.1–0.4	0.1–0.4
Nicotinic acid (mg equivalent)	0.250–1.25	0.6–1.25	0.6–1.25
Folic acid (μg)	4–10	8–20	8–20
B-12 (μg)	0.1–0.5	0.15–0.5	0.15–0.5
Pyridoxine (mg)	0.035–0.16	0.08–0.16	0.08–0.16
Biotin (μg)	1.5–5.00	Min. 5	Min. 5
Panthothenic acid (mg)	0.30–0.75	0.20–0.75	0.20–0.75
Energy	110 − (3 × age) × kg body weight		

From Wharton BA, & Clark B *Med Int* 1990; 3375–81.

TABLE 2. *Protein, energy, and protein/energy (Pr:En) ratio in recommendations and observed intakes*

Source	Age (months)	Energy (kcal)	Protein (g/100 kcal)	Pr:En ratio
WHO (1)				(calculated)
kg^{-1}	1–2	116	2.25^a	1.9
4 kg^b		464	9	
kg	3.6	99	1.86	1.9
7 kg		693	13	
kg	6–9	97	1.65	1.7
8 kg		776	13	
kg	9–12	100	1.48	1.5
9 kg		900	13	
				(recommended)
Fomon (15)	0–4			1.9
	4–12			1.7
Wharton & Clark (16)	0–12			2.25–3.0
Intakes from milks/formula (% energy intake)				
Breast milk at 150 ml/kg (100%)	3	105	1.95^a	1.9^c
Infant formula at 150 ml/kg (100%)	3	100	2.25–3.0	2.25–3.0^d
Follow-on milk providing 300 kcal (40%)	8	300	8–16	2.25–4.5^d
Observed intakes in British weanlings (18)				(calculated)
Total intake	6–9	815	27.4	3.4
Intakes from milks/formula		384	13.1	3.4
% total		47	48	
Intake from other foods		431	14.3	3.3
% total		53	52	
Total intake	9–12	928	34.4	3.7
Intake from milks/formula		335	14.7	4.4
% total		36	43	
Intake from other foods		593	19.7	3.3
% total		64	57	

^{a} Reference protein for WHO figures; other figures refer to actual protein.
^{b} Typical weight for age.
^{c} Protein at nitrogen × 6.38 (4).
^{d} Pr:En ratio advised by European Community (13).

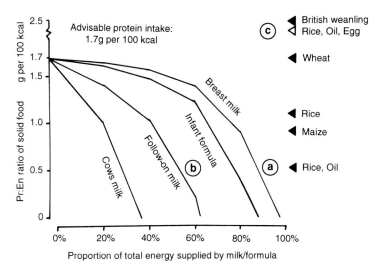

FIG. 1. Protein/energy (PR:En) ratios in milks, infant formulas, and solid foods to achieve advisable protein intake of 1.7 g per 100 kcal in older infants. For example, if an older infant received 40% of his energy intake from a follow-on milk (protein 2.7 g per 100 kcal) and the remainder of his energy intake from rice (protein 1.1 g per 100 kcal), then since the intersecting point is just above the line for a follow-on milk, the advisable protein intake of 1.7 g per 100 kcal would be achieved. Intersecting points below and to the left of the line would not achieve the advisable intake. All Pr:En ratios in the milks, formula, and solid foods have been converted for net protein utilization in children: breast milk 1.0, cow's milk 0.9, wheat 0.49, rice 0.63, maize 0.36. [Modified from Wharton BA, et al. *Acta Paediatr* 1994 (in press).]

relationships previously (20). A brief description is given in the legend to Fig. 1. As an example, a child getting 40% of his energy from an infant formula (2.25 g per 100 ml) would need to consume weaning foods with a protein/energy ratio of 1.5 g per 100 kcal to meet the desirable ratio in the overall diet of 1.7 g per 100 kcal.

The protein/energy ratio in British weanlings' supplementary food (3.3 g per 100 kcal or 2.5 g after corrections for NPU; see Table 2) is well above the 1.7 g "advisable" figure even if these infants consumed no milk of any kind.

If, however, the supplementary food were entirely rice or entirely maize, as, for example, in Asia or Africa, the safety limits are met very much earlier; for example, rice alone is satisfactory only if breast milk is still providing about 70% of the energy intake; if the child receives an infant formula (providing 2.25 g protein per 100 kcal), rice is satisfactory as long as the formula meets 55% of the energy requirement, and so on.

The protein/energy ratio of rice plus oil is much lower, so that if the infant were receiving less than 85% of his energy intake from breast milk, the safe ratio would not be met; a weaning food that provided only 15% of energy intake would be of limited value.

From this unsatisfactory position (point a in Fig. 1), the protein/energy ratio of the diet could be improved in two ways:

1. *By the use of a milk or formula with higher protein/energy ratio.* This child's position moves horizontally from point (a) to, say, point (b). A safe protein/energy ratio is achieved if a follow-on milk accounts for only 55% of the energy intake.
2. *By the use of a weaning food with enhanced protein as well as enhanced energy.* The child's position moves vertically upward from point (a) to (c). This is achieved by using a "double mix" weaning food based on rice and oil plus another protein source, such as egg.

Which of these strategies is better will depend on local factors such as availability and cost of the additional foods. In both instances care is necessary to ensure microbiological safety of the milk and rice gruel used.

Protein/Energy Ratio in Pathological States

The ratio thus deals effectively with safe or advisable intakes in infants and with guidelines or directives for foods designed for infants. A further advantage is that the concept of the protein/energy ratio can be extended to pathological states of nutrition.

Hypercatabolic States

Protein supplementation when there is already an adequate energy intake is often ineffective because the liver and muscles cannot synthesize any more protein from an increased amino acid intake even though abnormal losses of nitrogen are occurring. Even in hypercatabolic syndromes it is probable that the protein/energy ratio should not exceed the upper suggested limit of 3 g per 100 kcal (3.5 g per 100 kcal for 2- to 10-year-olds). One rule of thumb for dealing with the increased requirements following burns is to increase energy intake over "normal" requirements by 20 kcal for every percent of surface area burned (for example, 50% burns would require an extra 1000 kcal), and protein intake by 1 g of protein for every percent of burned area. The same effect is achieved by increasing total energy intake as indicated and maintaining the protein/energy ratio of the intake at 3.0–3.5 g per 100 kcal. In practice, if a suitable enteral feed is chosen (3.0–3.5 g per 100 kcal), the volume of feed can be increased according to the proportion of surface area burned, without individual consideration of protein and energy.

Diets for Catch-up Growth

A further use of the protein/energy ratio is as a check on diets that have been manipulated in some way. It has become a common practice in British hospitals when trying to induce catch-up growth in a malnourished child to give an infant formula to which extra fat and carbohydrate have been added. A brief consideration of protein/energy ratios shows that this maneuver has potential difficulties.

Using Fomon's figures (15,21), deposition of new tissue from birth to 4 months of age on average requires 5.6 kcal/g and the tissue contains 11.4% protein. The following protein/energy ratios are necessary after allowing for varying levels of the efficiency of protein utilization:

100% efficiency = 2.04 g per 100 kcal (0.114 g per 5.8 kcal)

90% efficiency = 2.26 g per 100 kcal

80% efficiency = 2.54 g per 100 kcal

75% efficiency = 2.71 g per 100 kcal

Clearly, the deposition of new tissue will require all essential nutrients as well as energy, and the figures above indicate that when supplementing a standard formula to support faster growth in early infancy, the *supplement* should contain *at least* 2.26 g of protein per 100 kcal and probably more. Use of an energy-only supplement is of limited value.

AMINO ACIDS

The quantity of protein in any recommendation must necessarily be accompanied by some statement on quality. This generally involves the use of a defined reference protein with a stated amino acid composition. For the suckling infant there is no doubt that the reference protein should be breast milk protein. For the weanling infant the choice is less clear, but the FAO reference protein is widely used for amino acid scoring. An additional measure, the protein efficiency ratio, may be used. This is weight gain (g) divided by protein consumed (g) in rats under standardized conditions. The reference protein used is usually casein, and a protein/efficiency ratio of 70% is regarded as suitable (14). Casein may also be used as a reference protein for amino acid scoring.

Questions of protein quality and amino acid content become more important as protein quantity is reduced. This has been most apparent as the protein content of infant formulas has fallen. Cow's milk protein contains less cysteine (about 50%) and less tryptophan (about 65%) than breast milk protein.

Cysteine

The story of cysteine is well known, so I shall deal with it only briefly. The lower intake of cysteine with cow's milk formulas is not accompanied by lower plasma cysteine concentrations in either low birthweight or normal-sized babies (20,22,23), but plasma concentrations may not be a good guide to requirements. An increase in the cow's milk whey fraction in infant formulas increases the cysteine content of the formula, so that when the total protein intake is 2.25 g per 100 kcal, the cysteine intake matches that of breast-fed babies. The FAO protein standard combines methionine and cysteine and should not be used for reference in the first 6 months of life.

Tryptophan

While cysteine has received considerable attention, tryptophan has only recently been examined in any detail, probably because of problems in estimation. In 1990 we published reference values for plasma amino acids in early life using reversed-phase HPLC (23). This method allows satisfactory determination of plasma tryptophan. Normal bottle-fed babies receiving a standard formula (2.25 g protein per 100 kcal) had lower plasma tryptophan concentrations during the first 3 weeks of life than those of breast-fed babies (Fig. 2). These biochemical differences were not associated with clinical or biochemical disadvantages. Growth velocity and other biochemical measures of protein nutrition (plasma albumin, transferrin, urea, amino acid ratios), although differing in detail, were broadly similar to those in breast-fed babies. Nevertheless, tryptophan has many biological functions, such as a role in neurotransmitter metabolism, so we should be wary of casting aside differences in plasma concentrations in the newborn.

It has recently been shown that if low protein formulas (1.3 g per 100 ml, 1.95 g

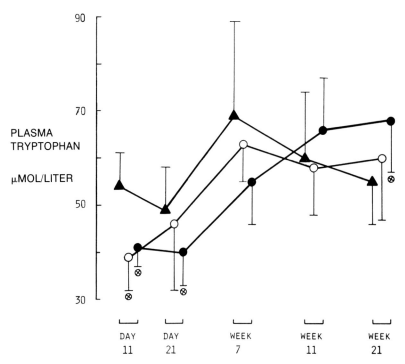

FIG. 2. Plasma tryptophan in normal breast-fed babies (▲) and in bottle-fed babies receiving either a casein-predominant (○) or whey-predominant (●) formula *ad libitum* (protein: 2.25 g per 100 kcal). *Results significantly different from those in breast-fed babies. Original data from Scott *et al.* (23) [Modified from Wharton BA, *et al. Acta Paediatr Scand* 1994 (in press).]

per 100 kcal) are supplemented with tryptophan, plasma tryptophan concentrations are similar to those in breast-fed infants (24,25).

Recommendation

The metabolic importance of cysteine and tryptophan illustrate the importance of amino acid composition in determining protein quality. Recommendations for amino acid requirements at present use little or none of the data available from kinetic studies and estimates of turnover. Denne & Kalhan (26) have used this approach in the newborn and Young & Cortiella (27) in older children. The nutritional properties of whole proteins and peptides are not just the sum of their constituent amino acids. This makes me wary of the recommendation that a low total protein intake is allowable as long as the intake of the constituent amino acids is not less than (say) that of a breast-fed baby.

ARE INTESTINAL EVENTS IMPORTANT?

Consideration of intestinal events in classical determinations of protein quality and requirements is limited. Net absorption of nitrogen is considered only in so far as it contributes to net protein utilization (NPU). In simple terms, where I is intake, F is fecal nitrogen, and U is urinary nitrogen,

$$\text{NPU} = \frac{I - F - U}{I} = \text{retention as a proportion of intake}$$

Net absorption $(I - F)$ is therefore included. It is, however, completely excluded in such indices as the chemical score, which depends solely on a comparison of the amino acid content of the protein with that of a reference protein. Different proteins do, however, have different effects on the gastrointestinal tract, and four examples that have attracted my interest are discussed next.

Net Nutrient Absorption

Table 3 shows the results of two nitrogen balance investigations that compared the quality of proteins when quantity was held constant. In low birthweight babies a whey-predominant protein was better absorbed than a casein-predominant protein. In children recovering from kwashiorkor, soya protein was less well absorbed than whole cow's milk protein with added casein (28,29). Apart from direct effects on nitrogen absorption, however, the type (quality) of the protein affects the absorption of other nutrients. The intestinal effects of soya protein are complex and even in soya protein isolates, the phytate present alters the absorption of phosphorus, iron, and zinc. Fat absorption from a whey-predominant formula is greater than from a

TABLE 3. Nitrogen and fat absorption in children undergoing rapid catch-up growth[a]

Children studied:	Low birthweight babies aged 3 weeks[b]		Toddlers recovering from kwashiorkor[c]	
Quality of protein:	Casein-predominant formula	Whey-predominant formula	Cows' milk plus casein	Soya flour
Fecal weight (g)	11.8	7.8		
Nitrogen Intake (mmol)	35.3	35.5	49.2	51.2
Net absorption[d] (mmol)	31.9	33.1	38.8	31.3
% intake	90	93	79	61
Fat intake (mmol)	27.8[e]	29[e]		
Net absorption[d] (mmol)	20.4	24.7		
% intake	73	85		
Fecal bile acids (μmol)	12.1	8.2		

[a] All values per kg/24 h.
[b] Berger et al. (29).
[c] Rutishauser & Wharton (28).
[d] Net absorption = intake minus fecal excretion
[e] Brown et al. (30).

casein-predominant formula (see Table 3), so the metabolizable energy available to the baby is greater (30).

Other Effects in the Lumen

Casein delays gastric emptying, induces lactobezoar formation, reduces the secretion of pancreatic carboxypeptidases, reduces binding of bile salts, and reduces the bioavailability of divalent cations such as calcium and iron. Fragments produced during digestion can act as gut hormones, inducing intestinal movement, insulin secretion, and so on [see Miller et al. for a review (31)].

Fecal Flora

A modern infant formula has metabolic effects similar to those of breast milk. Babies grow at about the same rate whichever food they receive, and plasma and urinary concentrations of various metabolites are similar in the two groups. This similarity is not found in microbiological observations, however (see Fig. 3). The fecal flora of breast-fed babies are very different from those of formula-fed babies, despite all the modifications aimed at making the composition of cow's milk formula similar to that of breast milk. Bifidobacteria and lactobacilli are the most common organisms in the feces of breast-fed babies, while coliforms, enterococci, and bacteroides predominate in the feces of bottle-fed babies (see Whey, Fig. 3). Casein-predominant formulas cause a change in the fecal flora even further from the breast-fed baby. The addition of bovine lactoferrin also has the same effect (see Whey +

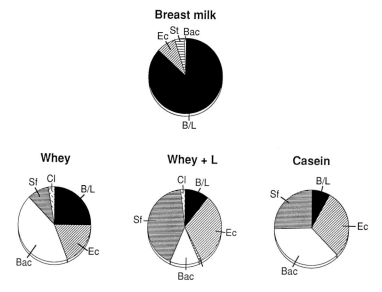

FIG. 3. Fecal flora (mean proportion of total bacterial counts) at 14 days of age in breast-fed babies and in bottle-fed babies receiving either a whey-predominant or a whey plus bovine lactoferrin formula, protein 2.25 g per 100 kcal. B/L, *bifidobacteria* and *lactobacilli;* Ec, *Escherichia coli;* St, *staphylococci;* Bac, *bacteroides;* Sf, *enterococci,* including *Streptococcus faecalis;* Cl, *clostridia.* Original data from Balmer & Wharton (32) and Balmer et al. (33). [Modified from Wharton BA, & Balmer SE *New perspectives in infant nutrition.* Stuttgart: Georg Thiem Verlag, 1992; 89–94.]

L, Fig. 3). It seems that different proteins have different effects on the fecal flora (32–34).

The nutritional consequences of these microbiological differences are not known. There is evidence of effects on the digestion and absorption of fat (35). The hydrolysis of urea and the production of short-chain fatty acids in the colon may also be affected, resulting in changes in both protein and energy metabolism.

Metabolic Factors Affecting Gut Function

Once the whole protein and products of digestion have left the intestinal lumen, the amount and type of peptides and amino acids absorbed may affect gut function. It is well known that breast milk contains more taurine (as a free amino acid, not as part of the protein) than is contained in cow's milk. In 1978 we showed that breast-fed babies preferentially conjugate their bile acids with taurine, whereas those receiving a formula, which in those days contained no added taurine, conjugated predominantly with glycine (36). There have subsequently been several investigations relating diet to changes in bile composition and its effect on intestinal function, for example, that by Jarvenpaa (37).

To summarize, if some present regulations and recommendations are based on the aim of mimicking the effect of breast milk on intermediary metabolism, in the future should we not also aim to mimic its effect on gastrointestinal function and on the microbial flora?

THE REAL WORLD

Recommendations, guidelines, regulations, and directives are all very well, but can the food industry meet them? What are the tolerance limits around a stated concentration on a label? If in the regulation the denominator is volume (or weight of powder), then the numerator (i.e., nitrogen $\times$ 6.38) is the only variable. If the denominator is energy, the variability of the other energy-providing proximates in the food (i.e., fat and carbohydrate) will cause a substantial increase in the variation of the actual content around the value claimed on the label.

If the "regulation" is, say, 1.8–3.0 g per 100 kcal and for a particular product the label claim is, say, 2.2 g, what are the limits of variability around this in individual batches of the product? A variation of, say, +15% would give a range of 1.9–2.5, values still within the regulatory range but somewhat different from those the pediatrician thinks that he has prescribed. In practice, a manufacturer usually aims at a concentration above that claimed on the label. This approach probably has its roots in the fight against diluted and adulterated foods which began in the nineteenth century. There are two other factors that will cause manufacturers to provide excess nutrients in their products. Many nutrients are only given a lower limit in composition recommendations, so label claims will always tend to err on the generous side; some nutrients, such as vitamins, deteriorate on storage, so the original amount in the product has to be higher, to meet the label claim when the product is used many months later.

Whatever the reason, however, the actual amount of protein in a product, as determined by nitrogen $\times$ 6.38, may well be somewhat different from the label claim. Since the amount is usually higher, the nutritional implications are related more to the effects of higher protein intakes than to protein deficiency. The higher than expected intake of casein, for example, may increase gastric emptying time and the buffering capacity within the intestine. The effects on intermediary metabolism will be different; for example, casein-predominant formulas give rise to higher plasma concentrations of aromatic amino acids, while whey-predominant formulas lead to increased threonine. The renal solute load, and hence the obligatory water losses, will be higher with a higher protein intake.

Pediatricians are fond of discussing and studying quite small changes in the protein intake of infants. This is relevant in nutritional terms, but whether it is relevant in the real world, where variations around stated label concentrations could be greater than the fine tuning suggested by the physiological experiments, remains to be shown.

Manufacturers need to improve the understanding of pediatricians and nutritional scientists on these matters. Cook (38) has started this process and it needs to be continued. In the meantime, if a label states, say, that the protein content is 1.5 g per 100 ml, what are the 95% confidence limits after allowing for all variations due to formulation, nutrient stability, analytical methods, and so on?

CONCLUSIONS AND SUMMARY

1. International recommendations concerning protein for normal infants are based primarily on the WHO report of 1985 (1). Five points deserving further attention are discussed.

2. Neither non-protein nitrogen nor the immunological proteins in breast milk should be ignored when assessing the amount of protein available to and required by an infant.

3. There are many advantages, both pragmatic and physiological, in considering protein requirements in relation to energy intake (i.e., as grams per 100 kcal) both in an individual and in a food. This method can be applied to situations as diverse as weaning diets, catch-up growth, and hypercatabolic states.

4. The nutritional properties of whole proteins and peptides are not just the sum of the constituent amino acids. We should be wary of making recommendations that a low total protein intake is acceptable as long as the intake of each of the constituent amino acids is not less than, say, that of a breast-fed baby.

5. Since some recommendations and regulations are based on the effect of breast milk on intermediary metabolism, should we in future also aim to mimic the effects of breast milk on gastrointestinal function and microbial flora?

6. Pediatricians are fond of discussing and studying quite small changes in protein intake of infants. This is relevant in nutritional terms but it is less certain that it is relevant in the real world, where variations around label concentrations could be greater than the fine tuning suggested by physiological experiments.

REFERENCES

1. World Health Organization. *Energy and protein requirements.* WHO Technical Report Series No. 742. Geneva: WHO, 1985; 98–105.
2. National Academy of Sciences. *Recommended dietary allowances,* 10th ed. NAS, Washington, DC, 1989.
3. Department of Health. *Dietary reference values for food energy and nutrients for the United Kingdom.* London: HMSO, 1991; 78–84.
4. Christie AA, Darke SJ, Paul AA, Wharton BA, Widdowson EM. *The composition of mature human milk.* Department of Health & Social Security. Report on Health and Social Subjects No 12. London: HMSO, 1977; 19–20.
5. Jackson AA, Doherty J, de Benoist M-H, Hibbert J, Persaud C. The effect of the level of dietary protein, carbohydrate and fat on urea kinetics in young children during rapid catch up weight gain. *Br J Nutr* 1990; 64: 371–85.
6. Heine W, Tiess M, Wutzke KD. ^{15}N tracer investigations of the physiological availability of urea protein in mothers' milk. *Acta Paediatr Scand* 1986; 75: 439–43.
7. Prentice A, Ewing G, Roberts SB. The nutritional role of breast milk IgA and lactoferrin. *Acta Paediatr Scand* 1987; 76: 592–8.

8. Davidson LA, Lönnerdal B. Persistence of human milk proteins in breast fed infants. *Acta Paediatr Scand* 1987; 76: 733–40.
9. Balmer SE, Scott PH, Wharton BA. Diet and faecal flora in the newborn: lactoferrin. *Arch Dis Child* 1989; 64: 1685–90.
10. Oppe TE, Barltrop D, Belton NR, *et al*. *Artificial feeds for the young infant*. DHSS Report on Health and Social Subjects No 18. London: HMSO, 1980: 1–10.
11. ESPGAN. Recommendations for the composition of an adapted formula. *Acta Paediatr Scand* 1977; 262: 1–20.
12. Codex Alimentarius Commission. *Recommended international standards for foods for infants and children*. FAO/WHO publication CAC/RS 72/74. Rome: FAO, 1976; 3–11.
13. Rey J, Astier-Dumas M, Fernandes J, *et al*. *First report of the Scientific Committee for Food on the essential requirements of infant formulae and follow up milks based on cow's milk proteins*. 14th Series, EUR 8752 EN. Luxembourg: Commission of the European Communities, 1983; 198.
14. Wharton BA, Oppe TE, Astier-Dumas M, *et al*. *First report of the Scientific Committee for Food on the essential requirements for weaning foods*. 24th Series, EUR 131140 EN. Luxembourg: Commission of the European Communities, 1990; 9–38.
15. Fomon SJ. *Infant nutrition*. Philadelphia: WB Saunders, 1974; 139–41.
16. Wharton BA, Clark B. Childhood nutrition. *Med Int* 1990; 3375–81.
17. Beaton GH, Swiss LD. Evaluation of the nutritional quality of food supplies: prediction of desirable or safe protein:calorie ratio. *Am J Clin Nutr* 1974; 27: 485–504.
18. Mills A, Tyler H. *Food and nutrient intakes of British infants aged 6–12 months*. London: HMSO, 1992; 12–44.
19. Millward DJ. Protein requirements of infants. *Am J Clin Nutr* 1989; 50: 406–7.
20. Wharton BA, Balmer SE, Berger HM, Scott PH. Faecal flora, protein nutrition and iron metabolism: the role of milk based formulas. *Acta Paediatr Scand* 1994 (in press).
21. Fomon SJ, Thomas LN, Filer LJ, Ziegler EE, Leonard MT. Food consumption and growth of infants fed milk based formulas. *Acta Paediatr Scand* 1971; 223: 1–36.
22. Scott PH, Berger HM, Wharton BA. Growth velocity and plasma amino acids in the newborn. *Paediatr Res* 1985; 5: 446–50.
23. Scott PH, Sandham S, Balmer SE, Wharton BA. Diet related reference values for plasma amino acids in newborn measured by reversed-phase HPLC. *Clin Chem* 1990; 36: 1992–7.
24. Hanning RM, Paes B, Atkinson SA. Protein metabolism and growth of term infants in response to a reduced protein, 40:60 whey casein formula with added tryptophan. *Am J Clin Nutr* 1992; 56: 1004–11.
25. Fazzolari-Nesci A, Domianella D, Sotera V, Räihä NCR. Tryptophan fortification of adapted formula increases trytophan concentrations to levels not different from those found in breast fed infants. *J Pediatr Gastroenterol Nutr* 1992; 14: 456–9.
26. Denne SC, Kalhan SC. Leucine metabolism in human newborns. *Am J Physiol* 1987; 253: E608–15.
27. Young VR, Cortiella J. Protein and amino acid requirements of healthy 6–12 month old infants. In: Heird WC, ed. *Nutritional needs of the six to twelve month old infant*. New York: Raven Press, 1991: 149–74.
28. Rutishauser IHE, Wharton BA. Toasted full-fat soya flour in treatment of kwashiorkor. *Arch Dis Child* 1968; 43: 463–7.
29. Berger HM, Scott PH, Kenward C, Scott P, Wharton BA. Curd and whey proteins in the nutrition of low birth weight babies. *Arch Dis Child* 1979; 54: 98–104.
30. Brown GA, Berger HM, Brueton MJ, Scott PH, Wharton BA. Nonlipid formula components and fat absorption in the low birthweight newborn. *Am J Clin Nutr* 1989; 49: 55–61.
31. Miller MJS, Witherlys A, Clark DA. Casein: a milk protein with diverse biological consequences. *Proc Soc Exp Biol Med* 1990; 195: 143–59.
32. Balmer SE, Wharton BA. Diet and faecal flora in the newborn: breast milk and infant formula. *Arch Dis Child* 1989; 64: 1672–7.
33. Balmer SE, Scott PH, Wharton BA. Diet and faecal flora in the newborn: casein and whey proteins. *Arch Dis Child* 1989; 64: 1678–84.
34. Wharton BA, Balmer SE. Diet and faecal flora in the newborn. In: Renner B, Sawatzki Q, eds. *New perspectives in infant nutrition*. Stuttgart: Georg Thiem Verlag, 1992; 89–94.
35. Combe E, Demarne Y, Guegen L, Ivorec Szylit O, Meslin JC, Sacquet E. Some aspects of the relationships between gastrointestinal flora and host nutrition. *World Rev Nutr Diet* 1976; 24: 1–57.
36. Brueton KJ, Berger HM, Brown GA, Ablitt C, Iynkaran N, Wharton BA. Duodenal bile acid conjugation patterns and dietary sulphur amino acids in the newborn. *Gut* 1978; 19: 95–8.

37. Jarvenpaa A-L. Feeding the low birth weight infant. IV. Fat absorption as a function of diet and duodenal bile acids. *Pediatrics* 1983; 72: 684–9.
38. Cook DA. Nutrient levels in infant formulas: technical considerations. *J Nutr* 1989; 119: 1773–8.

DISCUSSION FOLLOWING THE PRESENTATION OF DR. WHARTON

Dr. Guesry: Concerning formula feeds, Dr. Wharton has asked me to comment on the discrepancy that is often found between what is on the label and what is in the container. Of greatest importance for us is the daily requirement. What is in the can should be more than or equal to what is on the label. This is why, to be on the safe side, we tend to add a little more. The second point is that we work with an unstable raw material. Cow's milk is not constant. In spring there is more fat; during the winter, when the animals receive other types of food, there is less fat. Because of this, the protein/energy ratio is quite variable. We try to measure what is in the raw material and to adapt it, but this is not very easy to do on a day-to-day basis. This explains one type of variability. The law is also different from country to country and we have to take this into account.

Dr. Wharton also suggested that the quality of the formula should be adapted to the quality of the rest of the weaning food, the beikost. That is what we are doing. We have different types of formulation for different situations. For developing countries we tend to have higher levels of protein than for the industrialized countries. In spite of this the protein intake may be very high in industrialized countries, due to the fact that the beikost is rich in protein, and that mixed feeding is often started early, for example in the USA. So the trend now in industrialized countries is for a further decrease in the protein quantity.

Years ago ESPGAN recommended that follow-on formulas in developing countries should cover the daily protein needs of the baby (i.e., around 15 g of protein) in a volume of half a liter. This means that we need to provide about 4.5 g of protein per 100 kcal. Such a high protein concentration may still be necessary in certain developing countries.

Dr. Pettifor: I am interested in your comments about the need to maintain an adequate protein intake in your milk formula whatever formula you use, as you approach the weaning age, particularly in developing countries, where the protein is often relatively poorly available or of low biological value. Related to that is the issue of repeated infections. Does infection change the ratio of protein to energy requirements?

Dr. Wharton: With an infection you will certainly get nitrogen loss. We heard this morning that both synthesis and degradation will be increased but there will be a brisk nitrogen loss even in the mildest infection or even following immunization procedures. So the protein/energy ratio should presumably be increased following such an episode.

Dr. Bremer: Much more is now known about specific functions of individual amino acids. Is there any indication that we should now be deriving new figures for amino acid requirements based on the functions of individual amino acids?

Dr. Wharton: I am not aware of any new work on this other than that of Vernon Young. He has started to argue that we should be using kinetic data as opposed to net balance data in determining requirements of specific amino acids (1).

Dr. Garlick: One has to recognize that requirements derived from kinetic methods are strikingly different from those derived by traditional methods. This discrepancy has not yet been resolved, and until it is I don't think we can really rely on kinetic data.

Dr. Wharton: Could you give us some examples?

Dr. Garlick: From what has been published, the estimates for leucine requirements in adults by the kinetic method appear to be considerably higher than by traditional nitrogen balance methods. The kinetic method measures leucine oxidation on different leucine intakes and attempts to derive a value for the amount of leucine required to maintain leucine balance. Until the reasons for these differences have been investigated thoroughly it is hard to comment on them further.

Dr. Uauy: Kinetic data provide values for specific amino acid requirements. For several amino acids these may be twice the values obtained from nitrogen balance data. I agree that these data are not yet generally accepted, but the point that remains to be resolved is the functional implication of a given level of protein oxidation or amino acid oxidation. I think until this is settled we shall not be able to abandon the nitrogen balance data. Up to 6 months of age the model was human milk. The real problem starts after 6 months, and up to 1 year the only values available were the 1-year data, partly from Fomon and partly from Wang. So the values from 6 months to 1 year are just a direct interpolation of the 6-month value on the 1-year value, and they are mainly derived from short-term nitrogen balances. We now know much more about the limitations of such short-term studies. Probably not more than 10 or 15 infants have been studied on long-term nitrogen balances, and this is the work of Tontiserin in Thailand.

With the information we have today we are going to be more demanding on what we are going to call an acceptable level of protein intake. We are going to be demanding data on immune function and on growth, and not only growth in weight gain but in body composition. In particular, as Waterlow points out, we should be considering the protein requirements for optimal linear growth. I don't think we are prepared to answer this question right now, but it is the question that comes up whenever and wherever stunting occurs. There is quite a strong correlation between linear growth and protein intake, specifically animal protein intake. I am not saying that we have the answer but I think we should probably keep an open mind when considering both the right kind of metabolic studies and the right kind of epidemiologic studies. Nitrogen balances have already given an answer but probably not the final answer.

Dr. Axelsson: I was surprised by the protein content in the supplementary food you showed. Is it homemade or canned baby food? The protein value is much lower than we have in Sweden and lower than the ESPGAN recommendations. I agree with Dr. Guesry that the recommendations for formulas for weaning must be based on the composition of available food supplements.

Dr. Wharton: The weaning foods in Fig. 1 are both homemade (e.g., rice, rice with oil, rice with oil and egg) and commercial products (included in the British weanling diet). In many countries almost all of the weaning food would be rice. In South Africa I think there are many children whose only weaning food will be maize, perhaps with minor changes. This provides a lowish, but not bad, protein/energy ratio, but then you have to make your adjustment for net protein utilization to derive the figures I presented. Foods such as cassava or plantain, which is used in Uganda, have crude protein/energy ratios well below 1 g per 100 kcal and when you apply the adjustment for net protein utilization, the value becomes very low indeed. In Jamaica, historically, the weaning food was sucrose. Most manufactured infant foods available in Europe based on rice have other foods added. Sometimes high-protein foods are added to the rice, such as a milk powder or soya.

Dr. Yamashiro: I understand that protein requirements in infants are higher in developing countries than in highly developed countries. I wonder if babies in their first 3 months still require more protein than those in developed countries.

Dr. Wharton: As far as I know the answer to that is no. The recommendations relate to increased requirements resulting mainly from infections, which are much more prevalent after 3 months than before.

Dr. Uauy: There is only one recommended value for healthy individuals the world over. Of course, there is a long list of environmental stress factors, infection being number one, chronic gastrointestinal problems being number two, and energy deficit being number three, which all affect protein requirements. So as soon as the infant stops being breast fed, or even before because of inadequate weaning foods, he is likely to end up requiring more protein in developing countries, but this is not because he is biologically or genetically different.

Dr. Rassin: I have a couple of technical comments. First, with respect to Olson Schneidemanen's work, I think we sometimes put too much trust in some of the earlier amino acid studies. Some of these studies involved only one infant. You need to be really careful about how you interpret such data; such studies might not even be published today. The second point is about tryptophan. I am concerned about the methodology related to tryptophan measurements. In our experience it is extremely difficult to get accurate tryptophan measurements in blood unless you use very specific HPLC or fluorometric methodologies. You see very different results depending on the methods used.

Dr. Wharton: Our results quoted (2) were determined by reversed-phase HPLC, using a quaternary solvent system, and fluorimetry.

Dr. Marini: I would like to present our experience with growth of babies fed with a hypoallergenic diet. We had a program for very high risk atopic babies, giving breast milk until 1 year, during which time the mother's diet was supervised, or hypoallergenic formula containing 1.5 g of protein per 100 ml. We started solid foods at about 6 months of age, beginning with rice alone because it is hypoallergenic, and graduating to meat products such as turkey, rabbit, or lamb, which are also not very allergenic. The only dairy product during the first year was Parmesan cheese because it consists of partially hydrolyzed protein and is rich in calcium. We calculated the intakes of these elements in comparison with the WHO recommendations and they are very similar. These babies have grown very well. Some of them are now 5 years old and have no problems at all. I believe it is quite possible to achieve good growth in developed countries following the WHO recommendations. Of course, it is not easy to obtain full cooperation from the family, but if there are good medical reasons this can usually be achieved.

Dr. Wharton: Do you think it is possible to get as good growth with protein hydrolysates?

Dr. Marini: Yes, in terms of both stature and body composition.

Dr. Wharton: The question of protein quality and protein hydrolysates is an intriguing and difficult area. I think the standards used should not just rely on amino acid scores, and so on, which is the common approach in legislation relating to infant formulas; it should also include some type of biological assessment, such as the protein efficiency ratio.

Dr. Heine: I should like to add something to the information Dr. Wharton has given us relating to gut microflora. Alan Jackson showed that there is recycling of ammonia from the splitting of urea in the large bowl, and it is certainly possible that intestinal bacteria may produce essential amino acids from the ammonia that is split by bacterial urease. We don't know whether this has any importance and if it has, to what degree. But it is well known that you can harvest bacteria on culture media which contain ammonium salts exclusively and that they are capable of producing essential amino acids in large amounts, whereas if the ammonia is absorbed only non-essential amino acids can be produced. Further work is needed to decide whether this is of importance for nutrition, especially in situations of undernutrition.

Dr. Uauy: I should like to make a point about adaptation to low protein intakes. In the case of children "adaptation" usually means growing less well and being more likely to become infected. In adults it has been very well demonstrated that you can reduce the protein intake to half and if you wait long enough nitrogen balance will be regained. But this is a very artificial situation, in which the subjects need to be protected from environmental stress. So I don't think that we should consider adaptation as being something potentially good. It is a survival state and should not be the basis of any sort of assessment of a reasonable protein requirement.

Dr. Marini: We have been considering protein, but there are of course many other food constituents that are also important for growth. I did a study on growing preterm babies of about 2 kg birthweight who grew rather well on about 2 g of protein per kilogram per day, achieving weight gains of about 16 g/kg/d with good retention. However, when these babies had an inadequate intake of chlorine or potassium, they gained weight at only 13 g/kg/d instead of 16 g/kg/d. So when we consider protein we should not only be considering the amount of protein but the other constituents of the diet as well.

Dr. Wharton: That is absolutely right. We tend to think mainly of protein and energy in relation to limiting nutrients for growth, but there are certainly situations where other factors may be naturally limiting: for example, zinc. I think that very occasionally vitamin D could be a limiting nutrient for growth. We should not be obsessed by the macro elements of the diet.

REFERENCES

1. Young VR, Yu YM, Fukagawa NK. Protein and energy interactions throughout life. *Acta Paediatr Scand* 1991; 373: 5–24.
2. Scott PH, Sandham S, Balmer SE, Wharton BA. Diet related reference values for plasma amino acids in newborn measured by reversed-phase HPLC. *Clin Chem* 1990; 36: 1992–7.

Protein Metabolism During Infancy, edited by
Niels C. R. Räihä. Nestlé Nutrition Workshop
Series, Vol. 33. Nestec Ltd., Vevey/
Raven Press, Ltd., New York © 1994.

Protein Content of Human Milk, from Colostrum to Mature Milk

Niels C. R. Räihä

Department of Pediatrics, University of Lund, Malmö General Hospital, 21401 Malmö, Sweden

COMPARATIVE ASPECTS OF PROTEIN CONTENT IN MAMMALIAN MILKS

By definition, mammals are those animals that suckle their young and they comprise the highest class of vertebrates. Thus lactation is one important distinguishing characteristic of this class of animals. Mammals are born at markedly different stages of maturity and if it is assumed that their nutritive requirements correspond to their physiological maturity, one may reason teleologically that the milk of a given species is best adapted to the nutritional needs of the young of that same species. The nutritional adequacy of mother's milk for the young depends not only on the composition of the milk, but also on the quantity of milk produced. Linzell's comparative studies (1) of milk production in 22 different species indicate a mean daily yield of 0.126 kg milk/kg body weight and a daily energy output of 140 kcal/kg. The milk yield is closely related to the body size of the lactating mother. In the human, however, there seems to be a great individual variation in the milk yield and it may therefore be more common in humans than in animals for the milk production to be insufficient for optimal growth.

Protein in milk is the source of amino acids needed for protein synthesis and growth of the young. The protein content of milk produced by various species ranges from about 1% in humans to about 20% in rabbits. At present the compositions of milks of about 150 different mammalian species are known. For six species in addition to humans (horse, goat, cow, sheep, and reindeer) the data are extremely comprehensive and detailed (Table 1), due to the fact that the milks of all these animals, particularly the cow and goat, have commonly been used for human consumption.

The protein content of milk has been related to the growth rate of the young. Bernhart (2) suggested that there is a direct correlation between milk protein and the time required to double the weight of the young, based on studies of nine different species: humans, horse, cow, goat, pig, sheep, dog, cat, and rabbit. The data range from humans, with 125 days to double birthweight and 7% of energy intake from protein, to the rabbit, with 7 days to double birthweight and nearly 30% of the energy

TABLE 1. *Protein content of milk from some mammals*

Species	Protein (%)
Humans	1.0
Horse	2.5
Goat	2.9
Cow	3.4
Sheep	5.5
Reindeer	11.5

intake from protein. This generalization, however, does not hold true for the arctic and aquatic mammals, in which milk energy is derived mainly from fat and in which the growth of the young is due to a large extent to deposition of fat.

Figure 1 shows that in most mammals the concentration of lactose in milk is inversely proportional to that of fat and protein (3). It is seen that human milk has the

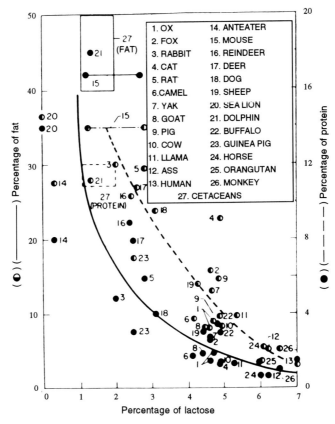

FIG. 1. Concentrations of lactose, fat, and protein in milk from various animals. [From Bernhart FW *Nature* 1961; 191: 358–360.]

highest lactose content and the lowest protein content of all the milks shown in the figure. In this chapter I shall discuss the quantitative and some qualitative aspects of the major proteins in human milk during the various stages of lactation.

TOTAL PROTEIN IN HUMAN MILK DURING PHASES OF LACTATION

Colostrum may be defined as "the thin, yellow, milky fluid secreted by the mammary gland a few days before or after parturition." In general, the milk excreted during the first 3–4 days is called colostrum. *Transitional milk* is that excreted from day 6 to day 15 and *mature milk* is the milk excreted from day 15 (4). The mean concentrations of total protein during these phases of lactation are shown in Fig. 2. Colostrum has a mean protein content of about 20 g/liter, transitional milk about 15 g/liter, and mature milk between 10 and 11 g/liter (5). After the first month of lactation the total nitrogen content decreases somewhat and then fluctuates, following no particular pattern, in the period between 3 and 6 months' postpartum. The changes in concentration of the major macronutrients of human milk are shown in Fig. 3. As the concentration of the two major energy-containing components, fat and lactose, increases, the concentration of protein decreases (5). This leads to a change in the energy distribution (Table 2). In early lactation during the colostral stage, protein is responsible for 17% of the energy content, but by the third week of lactation it accounts for only 7% of the total energy. Thus one may ask from a teleological point of view whether the infant needs more protein during the first days of life than later.

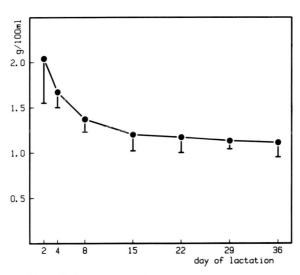

FIG. 2. Change in total milk protein concentration during lactation. Values are means, bars = SD. [From Harzer G, & Bindels JG *New aspects of nutrition in pregnancy, infancy and prematurity.* Amsterdam: Elsevier Scientific Publishers, 1987; 83–94.]

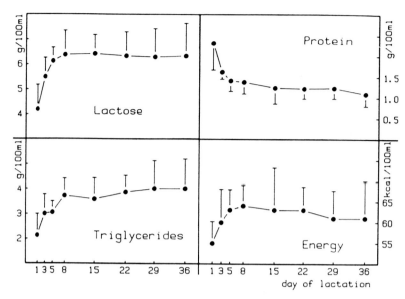

FIG. 3. Lactose, total protein, fat, and energy of human milk during the 5 weeks of lactation. Daily means, bars = SD. [From Harzer G, & Bindels JG *New aspects of nutrition in pregnancy, infancy and prematurity.* Amsterdam: Elsevier Scientific Publishers, 1987; 83–94.]

TABLE 2. *Distribution of energy in human milk*

Stage of lactation	Fat (%)	Protein (%)	Lactose (%)	Total energy (kcal/100 ml)
Colostral	44	17	39	56
Transitional	48	9	43	60
Mature	50	7	43	61

In order to answer this question one must examine the functions of the different milk proteins.

CHANGES IN THE PATTERN OF MAJOR MILK PROTEIN FRACTIONS DURING LACTATION

The concentration of the whey protein secretory IgA decreases sharply during the first days of lactation, whereas lactoferrin shows only a moderate decrease. By contrast, casein, α-lactalbumin, and serum albumin concentrations are more or less constant, as shown in Fig. 4. Thus, according to Harzer & Bindels (5), the whey protein/casein ratio of human milk decreases from 80:20 in colostrum to 55:45 in mature milk. This has recently been confirmed by Kunz & Lönnerdal (6), who quanti-

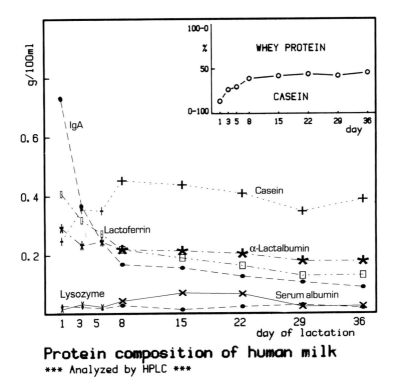

FIG. 4. Human milk protein composition during the first 5 weeks of lactation: casein; α-lactalbumin; lactoferrin; lysozyme; SIgA; serum albumin. [From Harzer G, & Bindels JG (5).]

tated total casein as well as whey proteins in human milk samples during lactation using two independent methods: precipitation at pH 4.3 in the presence of calcium ions followed by Kjeldahl analysis, and polyacrylamide gradient gel electrophoresis (PAGGE) followed by densitometric scanning. Figure 5 shows that casein concentration is low early in lactation and then increases rapidly during the transitional phase to a subsequent slow decrease during late lactation. In early colostrum, concentrations of whey proteins are very high but subsequently decrease significantly. These changes lead to a change in the whey protein/casein ratio during lactation. It was estimated by Kunz and Lönnerdal to be about 90:10 in early lactation, 55:45 in mature milk, and 50:50 late in lactation (6). Thus the decrease which is seen in the total protein content during lactation is due mainly to a decrease in secretory IgA and lactoferrin with lesser changes in the concentrations of the other milk proteins.

NUTRITIONAL VALUE OF PROTEINS IN HUMAN MILK

The nutritional availability of human milk proteins for amino acid metabolism and protein synthesis of the infant has recently been discussed by several investigators: Hambraeus et al. (7), Räihä (8), and Butte et al. (9). The functions of many major

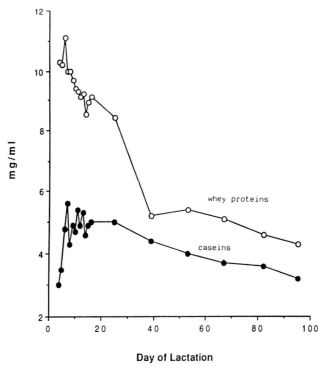

FIG. 5. Whey protein and casein concentration in milk during lactation. [From Kung C, & Lönnerdal B *Acta Paediatr Scand* 1992; 81: 107–112.]

whey proteins, secretory IgA, lactoferrin, and lysozyme are to some extent physiological rather than nutritional. Secretory IgA is the main immunoglobulin of human milk and acts locally in the infant's gut to prevent attachment of microbes to intestinal cells, thereby preventing infection. Lactoferrin is the major iron binding protein of human milk and is believed to prevent microbial proliferation by reducing the availability of iron. Lysozyme attaches to the bacterial cell wall, causing lysis.

These three protective proteins comprise about 30% of the total proteins in mature human milk. Recent studies have shown that these proteins are resistant to low pH and to proteolytic enzymes (10,11), which explains why it was possible for Davidson & Lönnerdal (12) to detect large amounts of both secretory IgA and lactoferrin in the stools of exclusively breast-fed infants. Figures 6 and 7 show that the excretion of these proteins correlates with the concentration in the milk, both of which decrease with the age of the infant. The amount of secretory IgA excreted is high, 60% of intake, in the early weeks of lactation and decreases to 10% of the intake during the later phase of lactation. A similar pattern is found for lactoferrin (Fig. 7).

By calculating the amounts of whey proteins ingested and subsequently excreted it has been estimated that between 3% and 10% of the milk proteins are unavailable to the infants as a nutritional source of amino acids (13,14). The higher percentage

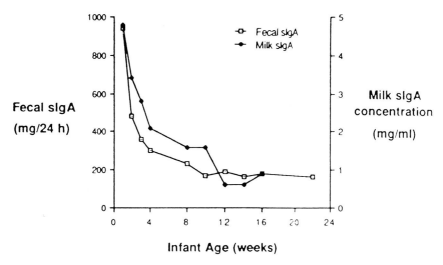

FIG. 6. Comparison of the change in fecal excretion and milk concentration of secretory IgA with infant's age. [From Davidson LA, et al. Protein and non-protein nitrogen in human milk. Boca Raton, FL: CRC Press, 1989; 161–172.]

of excretion is found in early lactation when the milk content of secretory IgA and lactoferrin is greater. The non-protein nitrogen (NPN) fraction of human milk comprises 20–25% of the total nitrogen. Of this, urea forms some 50%, but its concentration fluctuates with the state of lactation. The NPN fraction of human milk also contains some peptides and free amino acids, of which taurine is the most prominent. The utilization of urea nitrogen for *de novo* synthesis of amino acids and body proteins

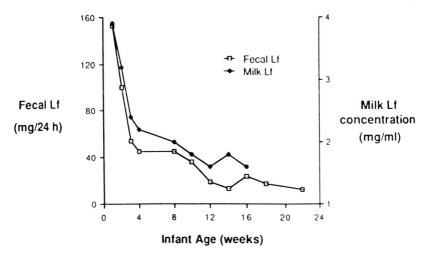

FIG. 7. Comparison of the change in fecal excretion and milk concentration of lactoferrin with infant's age. [From Davidson LA, et al. (11).]

TABLE 3. *Theoretical estimation of the nutritional value of proteins in human milk*

	g/liter
Total protein (N × 6.38)	11.6
Non-protein N (25%) (2% utilized for protein synthesis)	2.8
True protein	8.8
Non-nutritional proteins (3%)	0.3
Nutritional proteins	8.5

in the normal newborn infant has been much debated and the data are conflicting (15). Fomon (16) has estimated that an average of 13% of the ingested urea could be available for endogenous synthesis of amino acids in term infants. Most of this synthesis may be carried out by intestinal bacteria (17). The total nutritional contribution of the protective whey proteins and the NPN components in human milk for the normal infant is still not fully understood and needs further elucidation. Thus the minimum nutritionally available protein in mature human milk may be as low as 8.5 g/liter (8), as shown in Table 3. From a nutritional point of view, these results also imply that the whey/casein ratio of the nutritionally available proteins of mature human milk may be 50:50 or perhaps even slightly casein predominant and thus different from that in the total milk proteins.

One of the methods used to estimate protein requirements of infants is based on measurements of protein intake from breast milk in healthy breast-fed infants maintaining satisfactory growth (18). The true nutritionally available protein content of human milk must be considered when assessing protein requirements of infants. On the basis of the facts presented it is obvious that the requirement may be considerably less than recommended previously.

AMINO ACID CONTENT OF NUTRITIONALLY AVAILABLE PROTEINS OF HUMAN MILK

Harzer & Bindels (5) have studied the amino acid profile of the major human milk proteins. Figure 8 shows the amino acid concentrations in the various human milk proteins, secretory IgA, lactoferrin, α-lactalbumin, and casein. In comparison to the other milk proteins, secretory IgA is rich in threonine and has more valine than α-lactalbumin. Casein is rich in tyrosine and has a low content of tryptophan and cystine. Due to the change in the nutritional availability of secretory IgA, the threonine and valine available *in vivo* may be less than what would be predicted on the basis of the amino acid profile of the hydrolyzed total protein of human milk.

The quantities of amino acids ingested by healthy breast-fed infants have been estimated on the basis of milk consumption and the amino acid concentrations of human milk samples (19). Since not all human milk proteins are fully hydrolyzed it is clear that these estimations have been too high and qualitatively misleading. Harzer & Bindels (5) have measured the nutritionally available amino acid composition of

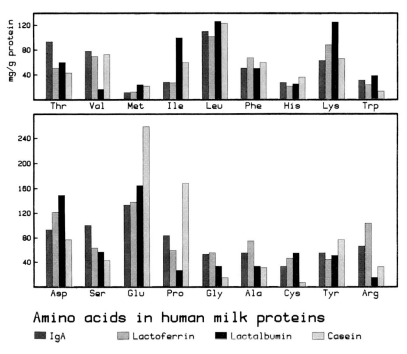

FIG. 8. Amino acid concentrations in various human milk proteins. [From Harzer G, & Bindels JG *New aspects of nutrition in pregnancy, infancy and prematurity.* Amsterdam: Elsevier Scientific Publishers, 1987; 83–94.]

human milk at 2, 8, and 36 days of lactation (Fig. 9). Their data show that although the amino acid content of human milk decreases considerably from day 2 to day 36 of lactation, the pattern of nutritionally available amino acids remains essentially the same.

Thus it is difficult to design infant formulas on the basis of the amino acid composition of human milk. It is essential to study the composition of the nutritionally available protein amino acids *in vivo*. By simply changing the quantity or the whey/casein ratio of bovine milk protein it will not be possible to achieve plasma amino acid profiles identical to those found in breast-fed infants (Fig. 10) (20). If the purpose of formula design is to produce human milk substitutes that will produce plasma amino acid profiles similar to those found in infants fed human milk, the protein composition must be modified by increasing the bovine α-lactalbumin fraction (see the chapter by Heine) or human milk proteins must be produced by the transgenic technique in large quantities to be used for formula production.

PROTEIN CONTENT OF PRETERM BREAST MILK

Atkinson *et al.* 1978 (21) were the first to observe a higher total nitrogen content in milk from mothers having given birth to preterm infants when compared to that

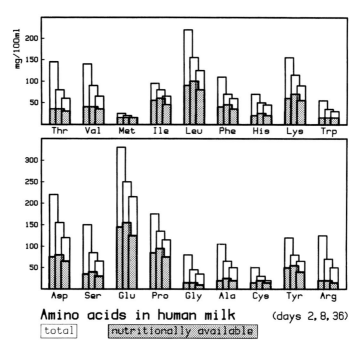

FIG. 9. Comparison of the nutritionally available amino acid concentrations with that of total protein of human milk at days 2, 8, and 36 of lactation. [From Harzer G, & Bindels JG *New aspects of nutrition in pregnancy, infancy and prematurity.* Amsterdam: Elsevier Scientific Publishers, 1987; 83–94.]

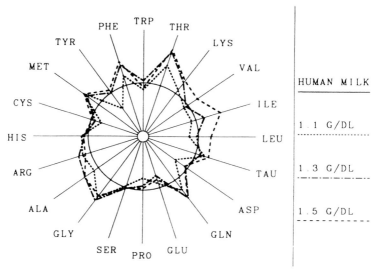

FIG. 10. Plasma amino acid profiles of infants fed either human milk (circle) or formulas with various concentrations of protein. [From Picone TA, et al. (20).]

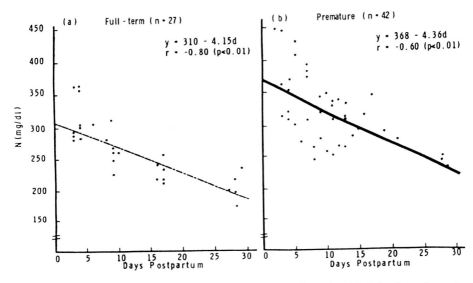

FIG. 11. Relationship of nitrogen concentration in human milk to day of lactation in mothers of full-term or premature infants. [From Atkinson SA, et al. *J Pediatr* 1978; 93: 67–69.]

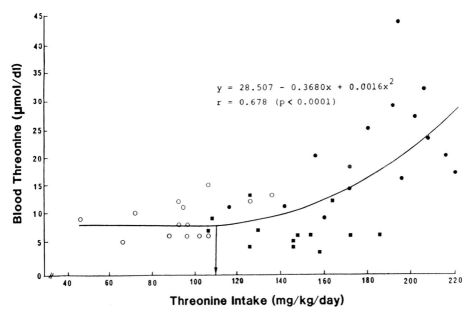

FIG. 12. Plasma threonine response in low-birthweight infants to feeding pooled breast milk (○), preterm breast milk (■), or standard infant formula (●). [From Atkinson SA, & Hanning RM Protein and non-protein nitrogen in human milk. Boca Raton, FL: CRC Press, 1989; 187–209.]

of mothers of term infants (Fig. 11). Subsequent studies (22–24) revealed that it was the protein nitrogen that was increased, whereas the NPN fraction was similar to that of term milk. One study (25), however, using a different method to determine the milk protein concentration, found no difference in the protein content between preterm and term breast milk.

Most of the increased concentration of protein in preterm milk may be due to higher concentrations of the protective proteins of breast milk. Considerably higher concentrations of lactoferrin, secretory IgA, and lysozyme have been documented in preterm than in term milk (26,27). Lemons *et al.* (24) have suggested that the higher protein content of preterm milk may represent a prolonged colostral phase in premature mothers who are establishing lactation by artificial means during periods of stress.

It may thus be that some of the protein which makes up the higher concentration of preterm milk is not nutritionally available to the infant. This is supported by the fact that plasma threonine concentration is not higher in infants receiving preterm milk containing more total milk proteins, especially more of the threonine-rich secretory IgA, than pooled banked human milk (Fig. 12) (28).

SUMMARY

1. Although total protein concentration is high in colostrum and transitional human milk, this is due primarily to an increased concentration of the protective proteins secretory IgA, lactoferrin, and lysozyme, which may not be fully absorbed and are thus partly unavailable to the infant as a protein or amino acid source.

2. The nutritionally available true protein content of mature human milk may be as low as 8.5 g/liter and the whey/casein ratio of those proteins is about 50:50 and thus not whey-predominant. The amino acid profile of the nutritionally available proteins of human milk does not change much during lactation. Because of the variability in digestion and absorption of human milk proteins, the amino acid composition differs in the hydrolyzed and absorbed protein. Plasma amino acid profiles in infants must thus be determined when improvements of formulations are tested. The discrepancy between threonine concentration of human milk and in the plasma in infants fed human milk is a typical example of this.

3. For a period of some weeks postnatally the breast milk produced by mothers of preterm infants has a higher total nitrogen content than the milk of mothers of term infants. This may reflect a prolonged colostral phase since the increased nitrogen content is due mainly to an increased concentration of secretory IgA and lactoferrin. The nutritional advantage of preterm milk needs to be studied more thoroughly.

REFERENCES

1. Linzell JL. Milk yield, energy loss in milk and mammary gland weight in different species. *Dairy Sci Abstr* 1972; 34: 351–60.

2. Bernhart FW. Correlation between growth rate of the suckling of various species and the percentage of total calories from protein in the milk. *Nature* 1961; 191: 358–60.
3. Palmiter RD. What regulates lactose content in milk? *Nature* 1969; 221: 912–4.
4. *Geigy scientific tables*. Basel: Ciba-Geigy, 1970.
5. Harzer G, Bindels JG. Main compositional criteria of human milk and their implications in early infancy. In: Xanthou M, ed. *New aspects of nutrition in pregnancy, infancy and prematurity*. Amsterdam: Elsevier Scientific Publishers, 1987; 83–94.
6. Kunz C, Lönnerdal B. Re-evaluation of the whey protein/casein ratio of human milk. *Acta Paediatr Scand* 1992; 81: 107–12.
7. Hambraeus L, Fransson G, Lönnerdal B. Nutritional availability of breast milk protein. *Lancet* 1984; ii: 167.
8. Räihä NCR. Nutritional proteins in milk and the protein requirement of normal infants. *Pediatrics* 1985; 75: 136–41.
9. Butte NF, Goldblum RM, Fehl LM, et al. Daily ingestion of immunologic components in human milk during the first four months of life. *Acta Paediatr Scand* 1984; 73: 296–9.
10. Lindh E. Increased resistance of immunoglobulin A dimers to proteolytic degradation after binding of the secretory component. *J Immunol* 1974; 114: 284–9.
11. Davidson LA, Donovan SM, Lönnerdal B, Atkinson ST. Excretion of human milk proteins by term and premature infants. In: Atkinson ST, Lönnerdal B, eds. *Protein and non-protein nitrogen in human milk*. Boca Raton, FL: CRC Press, 1989; 161–72.
12. Davidson LA, Lönnerdal B. Persistence of human milk proteins in the breast-fed infant. *Acta Paediatr Scand* 1987; 76: 733–40.
13. Prentice A, Ewing G, Roberts SB, et al. Nutritional role of breast-milk IgA and lactoferrin. *Acta Paediatr Scand* 1987; 76: 592–8.
14. Picone TA, Benson JD, Maclean WC, Sauls HS. Amino acid metabolism in human milk-fed and formula-fed term infants. In: Atkinson ST, Lönnerdal B, eds. *Protein and non-protein nitrogen in human milk*. Boca Raton, FL: CRC Press, 1989; 173–86.
15. Richards P. Nutritional potential of nitrogen recycling in man. *Am J Clin Nutr* 1972; 25: 615–9.
16. Fomon SJ, Bier OM, Matthews DE, et al. Bioavailability of dietary urea N in the breast-fed infant. *J Pediatr* 1988; 113: 515–20.
17. Heine W, Tiess M, Stolpe HJ, et al. Urea utilization by the intestinal flora of infants fed mother's milk and a formula diet, as measured with the ^{15}N-tracer technique. *J Pediatr Gastroenterol Nutr* 1984; 3: 709–12.
18. WHO. *Energy and protein requirements*. Technical Report Series No 724. Geneva: FAO/WHO/UNU, 1985; 64–112.
19. Janas LM, Picciano MF. Quantities of amino acids ingested by human milk-fed infants. *J Pediatr* 1986; 109: 802–7.
20. Picone TA, Benson JD, Moro G, et al. Growth, serum biochemistries, and amino acids of term infants fed formulas with amino acid and protein concentrations similar to human milk. *J Pediatr Gastroenterol Nutr* 1989; 9: 351–60.
21. Atkinson SA, Bryan MH, Anderson GH. Human milk: difference in nitrogen concentration in milk from mothers of term and premature infants. *J Pediatr* 1978; 93: 67–9.
22. Schanler RJ, Oh W. Composition of breast milk obtained from mothers of premature infants as compared to breast milk obtained from donors. *J Pediatr* 1980; 96: 679–81.
23. Gross SJ, David RJ, Bauman L, Tomarelli RM. Nutritional composition of milk produced by mothers delivering preterm. *J Pediatr* 1980; 96: 641–4.
24. Lemons JA, Moye L, Hall D, Simmons M. Differences in the composition of preterm and term milk during early lactation. *Pediatr Res* 1982; 16: 113–7.
25. Sann L, Bienvenu F, Lahet C, et al. Comparison of the composition of breast milk from mothers of term and preterm infants. *Acta Paediatr Scand* 1981; 70: 115–6.
26. Gross SJ, Buckley RH, Wakil SS, et al. Elevated IgA concentration in milk produced by mothers delivered of preterm infants. *J Pediatr* 1981; 99: 389–93.
27. Goldman AS, Garza C, Nichols B, et al. Effects of prematurity on the immunologic system in human milk. *J Pediatr* 1982; 101: 901–5.
28. Atkinson SA, Hanning RM. Amino acid metabolism and requirements of the premature infant: is human milk the "gold standard"? In: Atkinson SA, Lönnerdal B, eds. *Protein and non-protein nitrogen in human milk*. Boca Raton, FL: CRC Press, 1989; 187–209.

DISCUSSION FOLLOWING THE PRESENTATION OF DR. RÄIHÄ

Dr. Guesry: When you say that the entire 15 to 18 g per liter of protein that infant formula manufacturers put in the formula is available, it is a very nice compliment to the infant milk producers but we couldn't accept it as true. Bo Lönnerdal yesterday showed that on average only 75–80% was available, depending on the processing. We always take this margin of unavailable protein into account because it is probably less dangerous to incorporate a little bit more than a little bit less. Second, I should like to thank you for presenting Atkinson's data showing the protein content of the milk of mothers of preterm babies compared with mothers of term babies. Various people have advocated the use of preterm milk for feeding premature babies, but it should be borne in mind from Atkinson's data that by 10 days of age preterm milk has the same protein content as that of normal term milk. At this age a baby born weighing 1 kg or 1.2 kg is still very small and certainly needs a higher protein intake than that needed by a term baby.

Dr. Räihä: I agree that the preterm infant of less than 1500 g needs more protein than can be provided by his mother's milk.

Dr. Cooper: There are very limited data on the amino acid profiles in fetal life. You suggested that in the preterm baby fed human milk the amino acid profile should be the model for preterm formulas. What I am suggesting is that we should perhaps be looking more at the normal situation in the fetus, let's say at 26–28 weeks, rather than at the preterm baby fed human milk.

Dr. Räihä: This is a much debated question. We do not have and probably never will have a gold standard for the plasma amino acid profile in the preterm baby. However, some years ago we studied preterm babies fed on various amounts of exclusively human milk protein obtained by supplementing human milk with altered filtrated human milk protein. When very low birthweight infants achieve an intake of about 3.5–3.6 g/kg/d of this protein, their growth rate is maximal. We used the plasma amino acid levels obtained at this intake as a baseline and compared them with the profiles of formula-fed infants, and with fetal blood amino acid profiles, umbilical cord amino acid profiles, and amino acid profiles from normal breast-fed term infants. We found that the closest match was with the breast-fed term infant who was growing normally. So my view is that the gold standard for the preterm infant should be the amino acid profile of the breast-fed term infant. I don't think we should use the fetal values because the fetus *in utero* is physiologically quite different from the preterm baby.

Dr. Marini: The problem of reference amino acid values is the same for total parenteral nutrition. Some solutions for TPN use mature human milk as the reference composition. As you demonstrated, this is by no means the same as the plasma amino acid profile in the infant fed human milk. Cord blood or fetal blood obtained at about 32–24 weeks of gestation would probably be a better reference standard.

Dr. Rassin: Amino acid patterns in fetal blood or cord blood can't be reproduced by feeding formulas or breast milk because they reflect in part the more aerobic metabolism of the baby and several features of the profile are just not matchable. For example, cord blood has a much higher glutamate/glutamine ratio than you would ever see in a formula-fed or breast-fed baby. To feed a formula that would reproduce this ratio would be impossible because you would have to give far too much glutamate. Also, the lysine composition of fetal and cord blood is approximately 4 to 5 times higher than you would ever see in a formula-fed or breast-fed baby. We would have to make some very strange changes to formulas to reproduce those profiles, which reflect the transport mechanism of the placenta. For several amino acids the placenta produces maternal-to-fetal gradients of three- to fourfold, while for others there is almost no gradient, for example cystine. We are not going to produce these effects with any kind of parenteral feeding regimen without causing possible toxicity in the infant. So I think to suggest that cord blood or fetal blood is a profile to match is just not appropriate.

Dr. Marini: Battaglia carried out several studies in the normal human fetus and in the fetus with intrauterine growth retardation. He found that serine was quite high in the human fetus. Do you have an explanation?

Dr. Rassin: Each individual amino acid has a transport mechanism in the placenta and these transports probably determine the fetal amino acid concentrations.

Dr. Marini: But why should the fetus have more serine? Does it depend on muscle metabolism?

Dr. Rassin: I don't think we are in a position to answer why there are particular concentrations of individual amino acids. We just know that under certain circumstances and with particular transport mechanisms you develop certain concentrations of these compounds. We have tried very hard to determine the implications of this, but we are a long way from finding the answers.

Dr. Pettifor: The issue then is why do we use any amino acid profile as a gold standard for formula-fed or breast-fed infants. If you believe that we can't use fetal amino acid profiles as the gold standard, why should we use postnatal amino acid profiles either?

Dr. Räihä: I can only say that it is because we have done studies on them. I still think that even for the preterm infant who is orally fed the ideal protein is human milk protein because it is of the same species. When we feed very low birthweight babies with adequate amounts of human milk protein, in other words mother's own milk supplemented with ultrafiltrated human milk protein, such that we achieve a growth rate that is actually higher than the intrauterine one, we find that the blood amino acid profiles are close to what you find in breast-fed term babies. That is the only reason for thinking that it should be the norm.

Dr. Kashyap: In most studies in preterm infants, the aim has been to try to keep the amino acid concentrations at least in the safe range. In order to do this, people have come up with standards. Perhaps the aim should be to ensure that the concentrations are less than what the baby is exposed to *in utero*, on the assumption that they will then be in the safe range.

Dr. Pandit: There is evidence that a good deal of urea is secreted in breast milk. Can this be used for protein building?

Dr. Räihä: From studies done so far it seems that under normal conditions, when the baby is getting an adequate supply of nutritional protein, urea is not used extensively for protein synthesis. Under pathological conditions, if the baby is starved or when he has an infection, the situation may be different.

Dr. Heine: It is really interesting that human milk contains such a high amount of urea and the question is for what purpose this high urea concentration is excreted in the milk. Urea is normally an end product which presents a metabolic load for the baby. We carried out some investigations a few years ago in which we fed [^{15}N]urea together with human milk. We found variable utilization of the urea nitrogen for protein synthesis in the baby. The lowest utilization was 16% of the administered urea and the highest was 60% (1). When Fomon repeated these investigations some years later he found that only 13% of the urea nitrogen was used in healthy, well-growing babies. We believe that the variable utilization has something to do with catch-up growth and with a low supply of protein in the baby's diet. It is quite clear, however, that urea is utilized for protein synthesis, since the ^{15}N labeling was identifiable in the plasma proteins.

Dr. Pandit: Does this mean that we should consider this urea as a part of the utilizable protein in the theoretical figure that was presented?

Dr. Heine: This is in part true.

Dr. Fazzolari: Could you tell me if the diet of the mother influences the quality and quantity of human milk protein?

Dr. Räihä: As far as we know, breast milk protein composition is very resistant to change. The fat content is most likely to be affected by maternal diet, and after this the volume of the milk. The quality of the protein will probably not be much affected, at least from studies published so far.

Dr. Fazzolari: The reason for my question is that it has been shown that undernourished lactating mothers have very low taurine in their plasma amino acid profile.

Dr. Rassin: Please bear in mind that the taurine is not found in protein. It is much more responsive to diet than is protein. There are several studies showing that vegetarians will get almost no taurine in their diet and have low taurine in their milk. You can't really use that as a marker of protein response.

Dr. Guesry: Philippe Hennart did a study in eastern Zaire 3–4 years ago in severely malnourished mothers in which he analyzed their milk. As Niels Räihä has just said, it was mainly the volume of milk that was severely decreased (2). The malnourished mother could not produce more than about 250–400 ml of milk per day, but the protein quality was quite well preserved. It is also true that the fat was modified.

I should like to comment on Dr. Räihä's recommendation that infant food manufacturers use transgenic cows to produce so-called human milk. He rightly says that this would be quite expensive, not only because of the technology involved in creating such transgenic animals, but also because purification and separation of the human protein from the other proteins produced by the cow will cost a lot of money. However, it is not the price that is most likely to prevent us from doing it; it is more likely to be the law, since it is currently forbidden to give milk from a transgenic animal. It may also prove difficult to get the mother to accept the use of such milk. Public opinion is not yet ready to use transgenic milk for infant feeding. Progress in this will be made initially by the drug industry. Almost every diabetic now uses insulin produced by genetic engineering and it is accepted because it is life-saving and because it is a drug.

Dr. Räihä: This is what I said really. I personally think that it would be much wiser to use all that money to teach mothers to breast feed, but of course in some countries this is difficult. I can just say that in Sweden and Finland, almost 100% of mothers are breast feeding, at least when they go home from the maternity ward, and even at 6 months more than 40% are still breast feeding. I can also say that in Sweden and Finland there is not a single preterm baby who has not been fed exclusively on human milk. We would consider it unethical to give formula.

Dr. Marini: How do you manage if you have a very sick mother? Banked milk is not the same as milk coming from the baby's own mother. For example, if you don't use fresh human milk it is quite impossible to protect from infection.

Dr. Househam: One of the most important causes of low birthweight in South Africa, as in many developing countries, is fetal growth retardation. What information is available on the composition of the breast milk in such cases?

Dr. Lönnerdal: It seems unlikely that there would be any specific adaptation of the mother's milk that would benefit the infant in such cases, but what happened during gestation may of course affect lactation. Poor nutrition may have been one factor, of course, and another could be infection. We have done a study in Lima, Peru, looking at women who are sick during lactation, the reason being that infection is so common in developing countries that its effects on lactation need to be examined. We found there was a higher protein content in the milk of these women, probably due to the increase in the acute-phase reactants that we see both in plasma and in some cases in breast milk. Perhaps the reasons for fetal growth retardation may still be present during lactation and therefore affect the milk composition.

Dr. Pettifor: What are the factors that influence the changing concentrations of IgA, lysosyme, and lactoferrin in breast milk?

Dr. Marini: When you look at concentrations there are big differences between colostrum and mature milk, but when you look at quantities the difference is not so great. The amount of colostrum is probably around 300–350 ml/d. During mature lactation, the mother can produce almost a liter of milk a day, so the quantity of IgA is not very much reduced in terms of the amount consumed by the baby.

Dr. Lönnerdal: Although the volume consideration is important, there is more secretory IgA early on, even taking volume into account, than there is later. We need to consider milk protein gene expression. We know too little about this at present. We know in general that hormonal factors and nutritional factors will affect milk protein gene expression but we know little about how nutrition and stress and all the other factors that can affect the hormonal profile will affect milk protein gene expression.

Dr. Räihä: I still think that we have missed an important issue in this whole context and that is the question of the current protein content in standard formulas for normal babies. Nobody has discussed this. It is my personal feeling that most formulas in most countries have too much protein.

REFERENCES

1. Heine W, Tiess M, Wutzke KD. ^{15}N tracer investigations of the physiological availability of urea nitrogen in mother's milk. *Acta Paediatr Scand* 1986; 639–663.
2. Hennart PhF, Brasseur DJ, Delogne-Desnoeck JB, *et al*. Lysozyme, lactoferrin, and secretory immunoglobulin A content in breast milk: influence of duration of lactation, nutrition status, prolactin status, and parity of mother. *Am J Clin Nutr* 1991; 53: 32–9. Errata in Am J Clin Nutr 1991; 53: 988.

Nutritional Importance of Non-protein Nitrogen

Bo Lönnerdal

Department of Nutrition, University of California, Davis, California 95616-8669, USA

The class of compounds called *non-protein nitrogen* (NPN) is not a homogeneous group of similar chemicals but rather an operational term for the remainder of nitrogen in milk and formula once the protein fraction has been removed. NPN therefore consists of widely diverse types of compounds, ranging from urea to complex nitrogen-containing oligosaccharides. It is obvious that the capacity of an infant to utilize these compounds and the potential physiological effects that they can exert will also vary considerably.

CHARACTERIZATION OF NPN IN BREAST MILK

Human milk was early found to have a large proportion (about 20–25%) of total nitrogen in the form of NPN (1,2). It should be noted, however, that the *absolute* concentration of NPN in breast milk is similar to that of cow's milk; the reason for the high relative percentage is rather the low protein concentration of human milk, making total nitrogen considerably lower than in cow's milk (3).

The overall concentration of NPN in human milk does not vary significantly during the lactation period (4); the concentration is usually about 0.4–0.5 mg N/ml (Table 1). However, as a relative percentage, NPN comprises about 5–10% of the nitrogen in colostrum and in milk produced during early lactation, but 20–25% in mature milk.

TABLE 1. *Total nitrogen and NPN in human milk*

	Total N (mg N/liter)	NPN[a] (mg N/liter)	% NPN
Preterm milk	4000	600	15
Colostrum	2770	500	18
Mature milk	1660	400	24

[a] Mean values from several studies (reviewed in ref. 9).

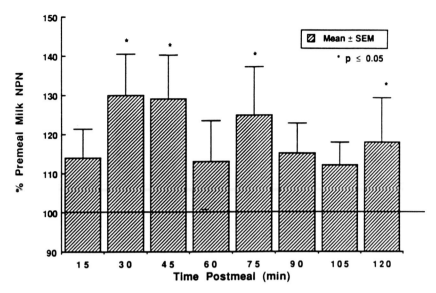

FIG. 1. Postprandial changes in milk NPN. Concentrations were normalized by the premeal values for each subject. Bars represent the mean (±SEM) for all subjects. [From Donovan SM, et al. (7).]

This is due to the high protein concentration of colostrum and early milk, not to a low concentration of NPN. Similarly, human milk from women delivering premature infants has a low percentage of NPN due to the higher protein concentration of the milk (so-called preterm milk) (5). Some effect of maternal protein intake on NPN has been observed: women having a high proportion of protein in their diet had a higher concentration of NPN in their milk, part of which was due to a higher level of urea (6). The time after ingestion of a meal by the mother can also affect milk NPN levels (7). Postprandial increases in milk NPN can be more than 30% above baseline levels (Fig. 1).

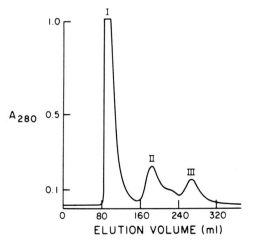

FIG. 2. Gel filtration chromatography of human whey on Sephadex G25. [From Donovan SM, & Lönnerdal B *Am J Clin Nutr* 1989; 50: 53–57.]

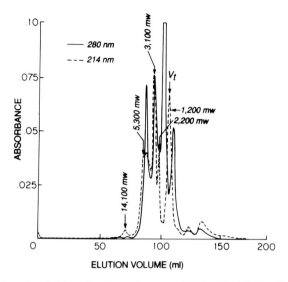

FIG. 3. Separation of peak 2 from Fig. 2 on a Superose 12B Fast Protein Liquid Chromatography (FPLC) column. [From Donovan SM, & Lönnerdal B *Am J Clin Nutr* 1989; 50: 53–57.]

Since NPN is operationally defined as non-protein, this milk fraction can be isolated and studied by removing all proteins from human milk. This can be achieved by gel filtration chromatography, ultrafiltration, or dialysis, but gel filtration has been shown to be the most efficient (8). It is important to note, however, that proteins are defined as polypeptides with molecular weights higher than about 6000 Da. This means that smaller polypeptides, dipeptides, and amino acids are part of the NPN fraction, even if the net result of digestion is that proteins *and* polypeptides in the NPN fraction all will yield dipeptides and amino acids available for absorption.

Practically, NPN is usually analyzed by precipitation of milk proteins by trichloroacetic acid (TCA) at 12% (wt/vol) and analysis of nitrogen in the supernatant (9). Results from this method are very similar to those obtained by gel filtration, while ultrafiltration and dialysis underestimate the NPN fraction (8). When separating human whey on Sephadex G-25, three peaks can be detected (Fig. 2). The first contains all the proteins and no NPN, while peaks 2 and 3 contain only NPN compounds. Of the NPN peaks, the first (peak 2 in the chromatogram) contains substances of higher molecular weight than the second and constitutes about 65% of all the NPN (Fig. 3). This fraction is likely to contain larger peptides (1000–6000 Da) and nitrogen-containing oligosaccharides. The last peak contains smaller compounds (<4000 Da), including free amino acids (among them taurine), small peptides, urea, and so on (Fig. 4). The identity of many of these NPN compounds remains to be established.

NPN IN INFANT FORMULA

Since cow's milk is generally considered to be low in NPN (2%), formulas based on skim milk powder and whey protein have often been assumed to contain only very

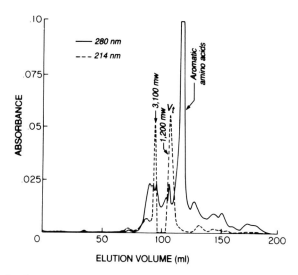

FIG. 4. Same as Fig. 3 but peak 3. [From Donovan SM, & Lönnerdal B (8).]

small amounts of NPN. While this is correct to some degree for casein-predominant formulas, which usually contain about 5–7% of total nitrogen as NPN, whey-adjusted milk formulas can have much more NPN, depending on the whey protein source used (10). Thus the NPN fraction of whey-predominant formula can vary from 5% up to 16%. Ultrafiltered whey (UF whey) contains the lowest proportion of NPN, 6–8%, followed by electrodialyzed whey (ED whey), 14–18% NPN, and finally, ion-exchanged whey with the highest NPN content, about 26% (Table 2). For the latter two whey protein sources, urea constituted 65–82% of the NPN content. Since the protein content of formula often is determined by the nitrogen content multiplied by a factor (6.25 or 6.38), the true protein content of a formula is overestimated considerably (Table 3) if a significant part of total nitrogen consists of NPN (11). As term infants cannot utilize urea nitrogen to any significant extent (see below), this overestimation of formula protein can have important implications with regard to fulfilling amino acid requirements of infants.

The proportion of peptide and free amino acid nitrogen in formulas also varies with the raw materials used. Some formulas using whey protein have relatively high

TABLE 2. *NPN in formula protein sources (%)*

Electrodialyzed (ED) whey	14–18%
Ultrafiltered (UF) whey	6–8%
Ion-exchanged whey	26%
Skim milk powder	6–7%

Adapted from Donovan S, & Lönnerdal B *Acta Paediatr Scand* 1989; 78 :497–504.

TABLE 3. *"True protein"* level of infant formulas (g/liter)

	Analyzed	Label claim
Enfamil	12.8	15
SMA	13.8	15
Similac	13.7	15
Milkotal 2	11.1	13
Babysemp 2	12.6	13

Adapted from Donovan S, & Lönnerdal B *Acta Paediatr Scand* 1989; 78: 497–504.

proportions of peptides, at around 100–160 μmol/liter, while casein-predominant formulas contain about 40–50 μmol/liter (10). As a comparison, human milk (and cow's milk) contains about 80–90 μmol of amino acid nitrogen as peptides per liter. The proportion of free amino acids also varies but is much less than of peptides, at about 7–12 μmol/liter. These values should be compared to the total concentration of amino acid nitrogen in formulas (1700–2500 μmol/liter) and in human milk (1300 μmol/liter).

UREA

The main constituent of NPN in human milk is urea; about 0.18 mg/ml urea nitrogen is present, which is about 50% of total NPN (10) (see above). During early lactation, urea levels can be somewhat higher (12). Cow's milk contains about 0.08 mg/ml urea nitrogen, while formulas have concentrations of 0.04–0.19 mg/ml, depending on which whey protein source is used.

Since the protein content of human milk is believed to be close to the estimated protein requirement of infants and urea nitrogen is a significant part (~12%) of total nitrogen in human milk, it has been speculated that urea could be a utilizable nitrogen source for the infant (13,14). Using stable isotope labeling (^{15}N,[^{15}N]urea) to trace the metabolic fate of dietary urea, it was found that healthy term infants retained about 13% of the labeled urea nitrogen (15,16). This retention was independent of whether the infant was given the labeled urea in breast milk (16) or in infant formula (15). It is possible that the proportion of urea nitrogen utilized by the infant is dependent on its nitrogen requirement. It has been shown that term infants recovering from infection (and therefore having higher nitrogen requirements than normal) can utilize a considerably higher fraction of urea nitrogen, around 39–43% (13,14). Low birthweight infants, who also have a high nitrogen requirement, were shown to retain about 28% of labeled urea nitrogen (17). Thus part of the urea nitrogen can be utilized by infants, although this fraction is relatively small in term infants (Table 4). The retained urea nitrogen can also be utilized by intestinal microorganisms (19), serving as a substrate for protein and nucleic acid synthesis by the gut microflora—up to 7% of labeled urea nitrogen was found incorporated into fecal bacteria in one study (13). However, part of urea is probably converted to ammonia by bacterial and mucosal enzymes and then metabolized by the infant. In summary, it is likely that the

TABLE 4. Comparison of [^{15}N]urea bioavailability from various studies

Reference	Infant	Diet	[^{15}N]urea dose (mg N/kg/d)[a]	Incorporation or bioavailability
Snyderman et al. (18)	Malnourished, term	Cow's milk	658	13% plasma protein N
Heine et al. (14)	Ill, term	Human milk	73	3% plasma protein N
	Ill, term	Human milk	8	43% ^{15}N retention
	Ill, term	Human milk	165	39% ^{15}N retention
Donovan et al. (17)	LBW	Formula	8	28% ^{15}N retention
Fomon et al. (16)	Term	Human milk	8	13% ^{15}N retention
Fomon et al. (15)	Term	Enfamil	8	13% ^{15}N retention

[a] Results presented as milligrams of nitrogen, assuming nitrogen constitutes 47% of the weight of urea.

degree of NPN utilization will be dependent on the growth rate of the infant, the amount and quality of dietary protein given, and the composition of the microflora.

NITROGEN-CONTAINING CARBOHYDRATES

The nitrogen-containing oligosaccharides and the smaller amino sugars in breast milk have received considerable attention, due primarily to their effect on intestinal bacterial growth, particularly on lactobacilli. Since the finding of a high degree of colonization of breast-fed infants with *Lactobacillus bifidus*, there has been an intense search for the so-called "bifidus factor" (20). Although the difference in intestinal bacterial flora between breast-fed and formula-fed infants may be less pronounced than was earlier believed, several amino sugars have been shown to stimulate the growth of *L. bifidus* (20). It should also be noted that there are many different oligosaccharides in human milk, some of which have very complex structures (21). Many of these have not yet been characterized and the potential biological activity has been evaluated for only a few.

All the milk oligosaccharides terminate with lactose or *N*-acetyl-lactosamine at the reducing end (20) and are classified as acidic or neutral, based on the presence or absence of *N*-acetylneuraminic acid (NANA, sialic acid). Many milk oligosaccharides contain the basic lacto-*N*-tetrose sequence (Gal-GluNAc-Gal-Glc, or a derivative), which has been shown to be a potent "bifidus factor" (20). As these oligosaccharides are not precipitated by acid, they are found in the acid-soluble NPN fraction of human milk.

The concentration of NANA in human milk changes considerably during lactation (22); NANA content in the NPN fraction is as high as 1100 mg/liter in early lactation

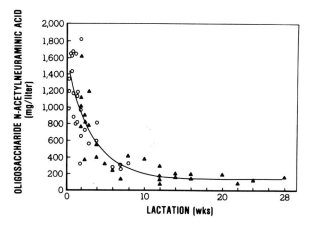

FIG. 5. Changes in oligosaccharide N-acetylneuraminic acid (NANA) in human milk during lactation. [From Carlson SE *Am J Clin Nutr* 1985; 41: 720–726.]

and then falls to about 100–150 mg/liter in mature milk (Fig. 5). N-Acetylneuraminic acid in human milk is found both in oligosaccharides and glycoproteins; oligosaccharide NANA appears to be the dominant form, particularly during early lactation (Table 5). Thus NANA constitutes about 10% of the NPN in colostrum but only 1.5% in mature milk. In addition to the possible role of NANA as a nitrogen source, it has been shown that brain ganglioside synthesis in young animals is modified by dietary NANA, suggesting a possible role for milk-derived NANA in early infancy (23). Liver enzymes needed for NANA biosynthesis have low activity in the newborn period (24), emphasizing a possible need for an exogenous supply of NANA. Since cow's milk and infant formulas contain much less NANA than human milk, breast-fed infants may benefit from this supply during early life.

Glucosamine nitrogen (GluN) contributes a significant part of NPN in human milk. Similar to NANA, the concentration of GluN is higher in early lactation than in mature milk and has been found to contribute as much as 30% of NPN (9); however, this value should be viewed with caution as no free glucosamine (or galactosamine)

TABLE 5. *Changes in N-acetylneuraminic acid in human milk during lactation*

Length of lactation (weeks)	Oligosaccharide NANA (mg/liter) ± SEM	Glycoprotein NANA (mg/liter) ± SEM
0–2	1138 ± 86	267 ± 24
2–4	706 ± 79	192 ± 33
4–6	348 ± 63	133 ± 27
6–8	258 ± 34	116 ± 20
10–28	135 ± 16	73 ± 8

Adapted from Carlson SE *Am J Clin Nutr* 1985; 41: 720–726.

is found when performing analysis of free amino acids in human milk. It is possible that the method of preparing the NPN fraction, in this case acid precipitation, liberates GluN from oligosaccharides and glycoproteins. Concentrations of GluN have been reported to be about 250 mg N/liter and that of galactosamine (GalN) to be about 15 mg N/liter (9,23). The considerably lower proportion of GalN is in agreement with its lower abundance in milk oligosaccharides and glycoproteins. Infant formulas contain very small quantities of GluN and GalN (8). This difference in GluN intake of breast-fed and formula-fed infants may also affect the growth of *L. bifidus*. The fate of GluN in the infant's gut remains uncertain. It is possible that it is absorbed and serves as a nitrogen source for synthesis of non-essential amino acids. Another possibility is that GluN is utilized within the enterocyte for intestinal glycoprotein synthesis. Furthermore, intestinal bacterial may utilize GluN for their growth and make this nitrogen source unavailable to the infant. Studies with labeled GluN are needed for a better evaluation of these different possibilities.

NUCLEOTIDES, NUCLEIC ACIDS, AND POLYAMINES

Nucleic acids have been reported to be present in human milk at concentrations of 100–5600 mg RNA/liter and 10–120 mg DNA/liter (25–28). Most of these nucleic acids are likely to originate from intact or lysed cells in the breast milk; in agreement with this, colostrum, which contains many more cells than mature milk, has a much higher nucleic acid concentration. Whether ribonuclease activity in human milk contributes to the degradation of RNA to nucleotides is not yet known. However, it is well known that human milk contains significant concentrations of preformed nucleotides (25–28). Most of these nucleotides appear to be monophosphates (AMP, CMP, IMP, GMP, UMP) and diphosphates (UDP, ADP, GDP), but cyclic nucleotides (cAMP, cGMP) are also present (25,26,29). Although the amount of nitrogen provided by nucleotides in human milk is low (~3 mg N/liter), it has been proposed that these nucleotides may play a physiological role under situations of nutritional stress. This aspect of nucleotides in human milk and their potential effect when added to infant formula is described in detail in the chapter by Uauy and Quan.

Polyamines such as putrescine, cadaverine, spermidine, and spermine have been described and quantitated in human milk (29). Although a physiological role of these compounds cannot be excluded, their concentrations are very low and the amount of nitrogen provided by them (0.05–0.2 mg N/liter) is minute. It is possible that their main role(s) is in milk synthesis within the mammary gland.

BIOLOGICALLY ACTIVE PEPTIDES

The peptide fraction of NPN constitutes about 60 mg of N/liter, which is about 4–5% of total amino acids in human milk (9,30). In this fraction, many peptides with known or suggested biological activity are found; epidermal growth factor (EGF), insulin, insulin-like growth factor (IGF), nerve growth factor, prolactin, calcitonin, and so on, have been reported in human milk at low but physiologically significant concentrations (31–34). Several of these have been shown to be absorbed by suckling rat pups (33); however, little is known about their fate in the gastrointestinal tract of human infants and the potential physiological effects in the infant. It should also be recognized that several of these peptides may exert their physiological effect within the mammary gland (34).

CHOLINE AND OTHER AMINO ALCOHOLS

Unesterified choline, phosphatidylcholine, and sphingomyelin have been found in human milk (35). About 3–9 mg N/liter of the NPN is derived from unesterified choline; the other forms are more likely to be associated with the lipid fraction. The concentration of choline (unesterified) is higher in early lactation (600–700 μmol/liter) than in mature milk (100–200 μmol/liter), but does not vary during a feed (35), showing that it is found in the water-soluble fraction. In contrast, choline-containing compounds are much higher in hind-milk, demonstrating their association with lipids (which also are higher in hind-milk). Concentrations of these compounds do not show the same pronounced changes during lactation as that of unesterified choline (Fig. 6). Choline from human milk provides an infant with about 6 mg/kg/d, which is similar to the adult diet. Choline has not yet been defined as an essential nutrient; however, it has been suggested that the premature infant may benefit from endogenous choline.

Other amino alcohols, such as phosphoethanolamine, phosphoglyceroethanolamine, and phosphoserine, have been detected in human milk (5,12). Accurate analysis

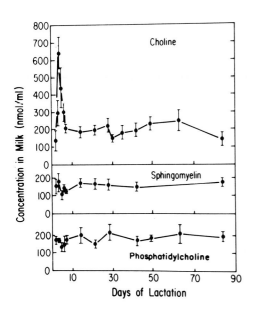

FIG. 6. Changes in concentrations of choline-containing compounds in human hind-milk during lactation. [From Zeisel SH, et al. J Nutr 1986; 116: 50–58.]

of these compounds is difficult and little is known about the physiological significance of the contribution of these compounds from the diet.

CREATININE, CREATINE, URIC ACID, AND AMMONIA

The presence of creatinine, creatine, uric acid, and ammonia in human milk has been well documented (9,30). The amount of nitrogen provided from all these compounds together is very low and their contribution to the total NPN fraction in human milk is minor. There have been no indications of any physiological significance of these compounds provided from the diet.

CARNITINE

Carnitine is found in the milk of all species and its concentration does not appear to be affected by maternal diet (36). It is a quarternary amine which is essential for the transport of long-chain fatty acids into mitochondria for β-oxidation (37). In human milk, most of the carnitine (80%) is present in free form. The amount of nitrogen provided by carnitine in human milk is about 1.5 mg N/liter. It has been suggested that breast milk or milk-based formula could supply adequate quantities of carnitine to infants with limited capacity to synthesize endogenous carnitine (37). Soy formulas are naturally low in carnitine, and infants fed such formulas were earlier found to have low plasma carnitine levels (38). This led to supplementation of soy formulas with carnitine.

CONCLUDING REMARKS

The NPN fraction of human milk contains many compounds of diverse chemical composition. Several of these, like some peptide hormones and nucleotides, may act as growth factors, while others, like oligosaccharides and amino sugars, may influence the intestinal microflora. Other compounds, such as urea, may be of little significance for the healthy term infant but could be of importance as a nitrogen source for the compromised infant. The biological activities of several NPN compounds remain to be explored.

REFERENCES

1. Denis W, Talbot FB, Minot AS. Non-protein nitrogenous constituents of human milk. *J Biol Chem* 1919; 39: 47–51.
2. Erickson BN, Gulick M, Hunscher HA, Macy IG. Human milk studies. XV. The non-protein nitrogen constituents. *J Biol Chem* 1934; 106: 145–59.
3. Lönnerdal B, Forsum E, Hambraeus L. The protein content of human milk. I. A transversal study of Swedish normal material. *Nutr Rep Int* 1976; 13: 125–34.
4. Lönnerdal B, Forsum E, Hambraeus L. A longitudinal study of the protein, nitrogen and lactose contents of human milk from Swedish well-nourished mothers. *Am J Clin Nutr* 1976; 29: 1127–33.

5. Atkinson SA, Anderson GH, Bryan MH. Human milk: comparison of the nitrogen composition in milk from mothers of premature and full-term infants. *Am J Clin Nutr* 1980; 33: 811–5.
6. Forsum E, Lönnerdal B. Effect of protein intake on protein and nitrogen composition of breast milk. *Am J Clin Nutr* 1980; 33: 1809–13.
7. Donovan SM, Ereman RR, Dewey KG, Lönnerdal B. Postprandial changes in the content and composition of nonprotein nitrogen in human milk. *Am J Clin Nutr* 1991; 54: 1017–23.
8. Donovan S, Lönnerdal B. Isolation of the nonprotein nitrogen fraction from human milk by gel-filtration chromatography and its separation by fast protein liquid chromatography. *Am J Clin Nutr* 1989; 50: 53–7.
9. Atkinson SA, Schnurr CM, Donovan SM, Lönnerdal B. The non-protein nitrogen in human milk: biochemistry and potential functional role. In: Atkinson SA, Lönnerdal B, eds. *Protein and nonprotein nitrogen in human milk*. Boca Raton, FL: CRC Press, 1989; 117–33.
10. Donovan S, Lönnerdal B. Non-protein nitrogen and true protein in infant formulas. *Acta Paediatr Scand* 1989; 78: 497–504.
11. Lönnerdal B, Zetterström R. Protein content of infant formula: how much and from what age? *Acta Paediatr Scand* 1988; 77: 321–5.
12. Harzer G, Franzke V, Bindels JC. Human milk nonprotein nitrogen components: changing patterns of free amino acids and urea in the course of early lactation. *Am J Clin Nutr* 1984; 40: 303–9.
13. Heine W, Tiess M, Stolpe HJ, Wutzke K. Urea utilization by the intestinal flora of infants fed mother's milk and a formula diet, as measured with the ^{15}N-tracer technique. *J Pediatr Gastroenterol Nutr* 1984; 3: 709–12.
14. Heine W, Tiess M, Wutzke KD. ^{15}N tracer investigations of the physiological availability of dietary urea nitrogen in mother's milk. *Acta Paediatr Scand* 1986; 75: 439–43.
15. Fomon SJ, Matthews DE, Bier DM, et al. Bioavailability of dietary urea nitrogen in infants. *J Pediatr* 1987; 111: 221–4.
16. Fomon SJ, Bier DM, Matthews DE, et al. Bioavailability of dietary urea nitrogen for the breast-fed infant. *J Pediatr* 1988; 113: 515–8.
17. Donovan SM, Lönnerdal B, Atkinson SA. Bioavailability of urea nitrogen for the low birthweight infant. *Acta Paediatr Scand* 1990; 79: 899–90.
18. Snyderman SE, Holt LE, Dancis J, Roitman E, Boyer A, Balis ME. "Unessential" nitrogen: a limiting factor in human growth. *J Nutr* 1962; 78: 57–72.
19. Rose WC, Dekker EE. Urea as a source of nitrogen for the biosynthesis of amino acids. *J Biol Chem* 1956; 223: 107–21.
20. Rose CS, Kuhn R, Zilliken R, György P. Bifidus factor. V. The activity of α- and β-methyl-*N*-acetyl-D-glucosaminides. *Arch Biochem Biophys* 1954; 49: 123–9.
21. Kobata A. The carbohydrates of glycoproteins. In: Ginsburg V, Robbins PW, eds. *Biology of carbohydrates*, Vol 2. New York: Wiley, 1984; 87–161.
22. Carlson SE. *N*-acetylneuraminic acid concentrations in human milk oligosaccharides and glycoproteins during lactation. *Am J Clin Nutr* 1985; 41: 720–6.
23. Carlson SE, House SG. Oral and intraperitoneal administration of *N*-acetylneuraminic acid: effect on rat cerebral and cerebellar *N*-acetylneuraminic acid. *J Nutr* 1986; 116: 881–6.
24. Kikuchi H, Tsuiki S. Activities of sialic acid-synthesizing enzymes in rat liver and rat and mouse tumors. *Biochim Biophys Acta* 1971; 252: 357–68.
25. Kobata A, Suzuki A, Kida M. The acid soluble nucleotides of milk. I. Quantitative and qualitative differences in nucleotide constituents in human and cow's milk. *J Biochem (Tokyo)* 1962; 51: 277–84.
26. Johke T. Acid-soluble nucleotides of colostrum, milk and mammary gland. *J Biochem (Tokyo)* 1963; 54: 388–97.
27. Skala JP, Koldovsky O, Hahn P. Cyclic nucleotides in breast milk. *Am J Clin Nutr* 1981; 34: 343–50.
28. Janas LM, Picciano MF. The nucleotide profile of human milk. *Pediatr Res* 1982; 16: 659–62.
29. Sanguansermsri J, György P, Zilliken F. Polyamines in human and cow's milk. *Am J Clin Nutr* 1974; 27: 859–65.
30. Hambraeus L, Lönnerdal B, Forsum E, Gebre-Medhin M. Nitrogen and protein components of human milk. *Acta Paediatr Scand* 1978; 67: 561–5.
31. Koldovsky O. Minireview: hormones in milk. *Life Sci* 1980; 26: 1833–6.
32. Jansson L, Karlsson FA, Westermark B. Mitogenic activity and epidermal growth factor content in human milk. *Acta Paediatr Scand* 1985; 74: 250–3.
33. Koldovsky O, Thornburg W. Peptide hormones and hormone-like substances in milk. In: Atkinson SA, Lönnerdal B, eds. *Protein and non-protein nitrogen in human milk*. Boca Raton, FL: CRC Press, 1989; 53–65.

34. Kidwell WR, Salomon DS. Growth factors in human milk: sources and potential physiological roles. In: Atkinson SA, Lönnerdal B, eds. *Protein and non-protein nitrogen in human milk*. Boca Raton, FL: CRC Press, 1989; 77–91.
35. Zeisel SH, Char D, Sheard NF. Choline, phosphatidylcholine and sphingomyelin in human and bovine milk and infant formulas. *J Nutr* 1986; 116: 50–8.
36. Sandor A, Pecsuvac K, Kerner J, Alkonyi T. On carnitine content of the human breast milk. *Pediatr Res* 1982; 16: 89–92.
37. Borum PR. Human milk carnitine. In: Hamosh M, Goldman AS, eds. *Human lactation*, Vol 2. New York: Plenum Press, 1986; 335–41.
38. Novak M, Monkus EF, Buch M, Lesmes H, Silverio J. The effect of a L-carnitine supplemented soybean formula on the plasma lipids of infants. *Acta Chir Scand (suppl)* 1983; 517: 149–55.

DISCUSSION FOLLOWING THE PRESENTATION OF DR. LÖNNERDAL

Dr. Garlick: I would like to comment on the ^{15}N experiments that have been reported where, from the fact that some of the labeled nitrogen appeared in amino acids and protein, it was concluded that urea nitrogen was utilized. This is equivalent to giving labeled bicarbonate or labeled CO_2—you will find the label in many body constituents but this doesn't mean that CO_2 is in any way being utilized as a nutrient. The fact that ^{15}N gets into the components of protein metabolism can be explained by exchanges during transamination; it does not mean that it has been used as a nutrient. Other experiments will be needed to prove that.

Dr. Lönnerdal: That is a valid point. While we are certainly no experts in the labeled isotope field, we used well-established methodology. We found only 20% utilization, and this is the maximum value, not the real value.

Dr. Räihä: I would like to point out that the very high non-protein nitrogen (NPN) you showed in the Swedish formula—I think it was 14% or 15%—was changed about a year ago when the manufacturers changed their source of whey protein. It is now under 10%. Your analysis was done with the old formula. This may be important. We have done studies with old formulas that had NPN values well over 10%, which means that the protein intake in the infant was probably much lower than we thought and they were still growing perfectly normally, with normal plasma amino acids and urea. When formulas are designed we need to look at the NPN fraction.

Dr. Lönnerdal: Many of these clinical studies just cite the protein level claimed on the label. It is impossible to judge the NPN from that.

Dr. Marini: You showed that urea was higher in whey protein formulas than in casein-based products. Could Dr. Guesry tell us how Nestlé prepare the whey?

Dr. Guesry: NPN (urea, creatinine, and so on) comes mainly with the whey and less with the casein. Since for term baby formulas, though not for premature formulas, we use 100% whey, this means that the quantity of NPN will be higher than in standard formulas containing a mixture of whey proteins and casein.

Dr. Marini: Another company has estimated that the urea content in a hydrolysate formula is about 5% of the NPN.

Dr. Uauy: When we consider NPN we should spend some time looking at the biological activity of some of these compounds as well as at their role as potential sources of nitrogen. What is your view as to the potential significance of some of these biologically active compounds? The fact that they are present doesn't mean that they play a nutritional or metabolic role. Which compounds do you think we should be focusing on? Should we consider the possibility of mass producing some of these compounds? Theoretically, they could be added to formulas.

Dr. Lönnerdal: I think there is strongest evidence for biological activity among the peptides (epidermal growth factor, EGF, for example) and the nitrogen-containing oligosaccharides. The latter are now being produced by genetic recombinant techniques and also by carbohydrate synthesis. We must await more clinical studies, but my bet would be that these compounds have biological importance. There is speculation that some non-protein compounds may be important for the premature newborn infant but I haven't seen any data that would convince me that they are important for the term infant.

Dr. Rassin: You raised the issue of glycosylation precursors. We need to think a little bit more about these in view of the suggestion that we may eventually achieve transgenic protein production. Even if we make a cow produce a human protein, a cow is still going to glycosylate like a cow. I think this illustrates some of the complications in trying to make human milk from another species. The glycosylation precursors and enzymes are going to be different in the cow than in the human. If you really want to make transgenic human milk, not only will you have to produce human caseins and human whey proteins but you will have to produce human glycosylation proteins and human glycosylation precursors. This is a good illustration of how complicated that kind of a product is going to be.

Dr. Lönnerdal: I agree. It is all becoming very complex, but it is quite certain that bovine recombinant glycosylated protein will never be like the human counterpart. We do, however, have proteins like α-lactalbumin which are neither glycosylated nor phosphorylated. Some of the caseins are phosphorylated and the bovine mammary gland can probably phosphorylate in a fashion similar to the human mammary gland. With regard to the glycosylated proteins, we have shown that the carbohydrate side chain is irrelevant for lactoferrin binding to the receptors, and in that sense it is not really important. On the other hand, the glycan of human lactoferrin may be part of the reason why human lactoferrin is less digested than bovine lactoferrin, by protecting it against proteases. It may not therefore be totally irrelevant.

Dr. Pettifor: What role do you believe the active peptides, particularly EGF, may play in maintaining or improving gut function, particularly in the premature infant, and do you have any comments on the role of some of the newer peptides, such as PTH-related peptide, which is present in high concentration in breast milk?

Dr. Lönnerdal: Starting with EGF, Leanna Read (unpublished) has done a nice set of experiments in premature newborn rhesus monkeys where she gave human recombinant EGF *in utero* and also postnatally and then looked at maturation. The gut mucosal protein growth (DNA and RNA) was significantly enhanced and lung maturation was improved. I think that the evidence is fairly strong that the EGF may enhance the maturation of certain organs.

The PTH-related peptide is very fascinating. Why is there so much in breast milk? I have not seen many studies on its potential function in the infant.

Dr. Uauy: One general point regarding the biologically active peptides is that, in order for them to be active, they need to be preserved from digestion and absorbed intact, otherwise you would have just a local effect. For EGF this makes sense because one can invisage it acting locally. What is your view about the potential for intact absorption of some of these peptides, for example insulin, insulin-like growth factor, thyroid hormones, and so on?

Dr. Lönnerdal: Many of these compounds do not occur in milk in the free form but are associated with binding proteins. I am just speculating here, but I look on these binding proteins as a kind of bait for proteases; they are attacked first, so to speak, and therefore may provide some protection for the active peptides down the gut. Most of the studies that have been done have been in suckling rat pups, and several peptide molecules have been able to survive gut transit in this species. I have seen very little yet to convince me that the same occurs in the human infant, but with duodenal intubation one could potentially look at the degree to which intact survival may occur in the upper gut, with the possibility of subsequent absorption.

Dr. Rassin: When we looked at some of our own amino acid constituents we found that a large proportion of the cysteine and glutathione is actually bound to the protein during TCA precipitation. Tryptophan could behave like this as well. Sometimes I wonder whether we underestimate the NPN fraction through losing some of it during TCA precipitation or in filtration types of separation. About 85–90% of glutathione in human milk would be precipitated by TCA and would not appear in the NPN fraction.

Dr. Bremer: The proteins of the fat globules contain some enriched nutrients such as selenium. No one has mentioned the proteins of the fat globules. Do they play a role in addition to selenium or are these constituents waiting to be investigated?

Dr. Lönnerdal: Fat globule proteins are specialized carbohydrate-rich high molecular weight structures with similarity to some of the glycocalix proteins. There is speculation that they could have an effect on both bacterial proliferation and bacterial attachment, but there is little evidence for this at present.

Dr. Pettifor: Would you like to speculate on what infant food manufacturers should be looking at as far as supplementation of cow's milk formula is concerned? Which of the non-protein nitrogen sources do you believe are the most important to be considered for inclusion in a cow's milk formula?

Dr. Lönnerdal: If I were a company I would step very carefully. They should wait until there is conclusive evidence for biological activity and only then consider supplementation. There is always likely to be the eternal dilemma that we had in the lactoferrin field: it is very difficult to show benefit in industrialized societies because the general level of health is high. You have to have huge cohorts in order to show biological activity. In developing countries, on the other hand, there is almost no limit to the number of confounders. Urea is not high on my list of possible additives, but there may be peptides and carbohydrates that will be proven to be of value.

Dr. Rassin: One NPN constituent that is widely supplemented is taurine. We could probably spend a couple of hours debating whether this is only window dressing. It is very difficult to prove that it has any useful function, yet it is a widely used supplement.

Dr. Marini: The same thing is happening with carnitine. We found that normal full-term neonates fed with soya milk, which is devoid of carnitine, can synthesize carnitine very well. The problem of carnitine deficiency is probably limited to preterm babies.

I should like to ask your opinion about the peptides and other substances. Some people say that polyamines may be important for closure of the intestinal barrier. It has been suggested that they could be of value in the prevention of allergy by promoting gut closure. I am also concerned about growth-promoting factors for infants born with intrauterine growth retardation. There is evidence that maternal blood levels of insulin-like growth factor (IGF) are low in pregnancies complicated by fetal growth retardation. It has been claimed that the possibility of catch-up growth in these babies can be predicted by IGF levels in the blood after 15 days or 1 month of age (1).

Dr. Lönnerdal: I believe, although I am not an expert on this, that gut closure is something that happens very quickly. Whether or not it can be accelerated by feeding formula with polyamines I do not know.

With regard to the other part of your question, there are more and more studies on IGF-I and IGF-II and their importance for growth. For example, synthesis of IGF-I is dependent on zinc supply and there may quite often be an inadequate zinc supply during pregnancy and lactation. Therefore, there may be confounding variables, but nevertheless, IGF appears to be a very important component.

Dr. Guesry: Cow's milk whey is full of carnitine and all infant formulas based on cow's milk contain a level of carnitine equal to or higher than breast milk. It is only in soy formulas that you need to add carnitine. To add it to cow's-milk-based formula is nonsense.

Dr. Heine: What are the practical consequences of your findings that the protein concentration in some formulas is below 12 g/liter, when most international committees on nutrition limit the lowest value of the protein concentration to around 12 g/L?

Dr. Lönnerdal: Most companies do not need to make any changes. They are at present fairly close to the level where they should be. Their protein values are lower than they believe them to be since they are used to thinking in terms of crude protein rather than true protein. Niels Räihä has evidence that even the lowest levels found in commercial formulas will be adequate. I think that based partly on Fomon's early studies, a protein concentration of 11 or 12 g/liter of true protein may well be the lower limit.

Dr. Rassin: With regard to soy formulas, how much does the non-protein nitrogen fraction reflect the methionine that has to be added to improve the protein quality?

Dr. Lönnerdal: It is part of it. I can't tell you how much.

Dr. Rassin: I think it is likely to result in a very different NPN fraction from that seen with other milks.

Dr. Heine: We have done some studies on the sialic acid concentration in different formulas. Soy formulas were found to have only 3% of the human milk sialic acid concentration. This may be crucial in the nutrition of low birthweight infants. You pointed out a connection between the development of the brain and the presence of sialic acid as a central constituent in the synthesis of gangliosides and glycoproteins.

Dr. Pandit: Since we have digressed from animal proteins to soya, I wonder what would be the status of other proteins, for example chicken- or meat-based protein. Would these be a good alternative? We have tried chicken-based formula in low birthweight babies and the clinical tolerance and weight gain have been excellent. The osmolarity comes to about 217 mOsm/liter and the protein content is 2.5 g per 100 ml. Protein utilization is almost 90%, and amino acid analysis is satisfactory. In India there is an enormous amount of poultry meat available and it is fairly cheap.

Dr. Marini: In Italy we have used lamb-based formulas over the past three decades, even in small infants as little as 2 weeks old, when there is severe food intolerance. We find it very effective.

Dr. Guesry: We are always on the lookout for good-quality chicken meat powder, but it is very difficult to find because it has either been heat treated (and the heat treatment has usually impaired the protein quality) or it has not been sufficiently heat treated and is full of salmonella.

Dr. Pandit: I have been working on this for the last 4 years. I can assure you it is possible to produce good-quality dried chicken meat protein with current technology without salmonella. However, I still don't know what its role is in human nutrition.

Dr. Househam: Developing countries cannot afford to buy commercial hydrolysates; they are simply too expensive to use, particularly for children who have chronic diarrhea who need rehabilitation. We have also made extensive use of chicken-based formula and the children do extremely well on it. I think this is an issue that needs to be addressed because chicken meat powder is cheap and can be used as the basis for a cheap formula which has a real application in the developing world.

Dr. Rassin: There are one or two things that you might want to look at in relation to chicken meat. One is that chicken muscle has more post-translational changes in its amino acids, for example histidine, than other kinds of proteins, and I have no idea what this does. Another thing is that chicken muscle has an enormous carnitine pool and the non-protein nitrogen pool originating from chicken sources would be very different from cow's milk.

REFERENCE

1. Thieriot-Provost G, Boccara JF, Francoval C, Badoval J, Job JC. Serum insulin-like growth factor 1 and serum growth-promoting activity during the postnatal year in infants with intrauterine growth retardation. *Pediatr Res* 1988; 24: 380–3.

Qualitative Aspects of Protein in Human Milk and Formula: Amino Acid Pattern

Willie E. Heine

Kinderklinik Medizinische Fakültät Universität Rostock, Rembrandt Strasse 16/17, 1855 Rostock, Germany

HISTORY OF COW'S MILK PROTEIN ADAPTATION TO HUMAN MILK

Since analytical data on the different composition of human milk and cow's milk became available at the beginning of this century, human milk has always been the gold standard for creation of infant formulas. When it became evident that cow's milk contains three times the protein concentration of human milk, the first steps of protein adaptation consisted in dilution of cow's milk at a 1:1 ratio with water and supplementation with sucrose and starch to compensate for dilution-related energy losses.

As methods for the determination of amino acid concentrations became more reliable, the considerable differences in almost all amino acid patterns in human milk and cow's milk proteins were recognized. Because of the comparatively low concentration in cow's milk of cystine, which was thought to be an essential amino acid in infancy, it was then recommended that cow's milk should be diluted with water in a ratio of 2:1 for the preparation of infant formulas in order to meet the infant's cystine requirements. These formulas, which contained 2.3% protein, were generally in use up to the 1950s. They contained the majority of all amino acids in abundance and represented a large excretory load in the form of urea and ammonia. At that time, attention was focused on the amino acid composition of the various fractions of human milk and cow's milk protein. The whey protein fraction of cow's milk was found to be rich in the essential amino acids deficient in whole cow's milk protein and the idea of whey protein supplementation of cow's milk came into being. The admixture of whey protein to cow's milk at ratios of 60:40 or 50:50 rendered it possible to adapt the amino acid pattern of cow's milk more closely to the amino acid composition of human milk protein and to avoid an excessive supply of aromatic amino acids and methionine, as well as avoiding cystine and tryptophan deficiency. The surplus of threonine, due to the whey protein admixture, was considered to be unimportant.

Whey protein-enriched formulas are one of the most important innovations in infant nutrition of modern times. Formulas adapted in this manner enabled a reduction of

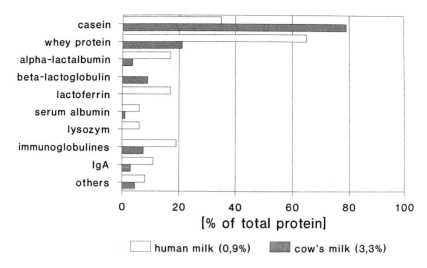

FIG. 1. Comparison of cow's milk protein fractions with human milk protein fractions.

the protein concentration to 1.8% and even to 1.5%. This seemed to be quite similar to the protein quantity of human milk when calculated by nitrogen content. However, human milk contains a substantial amount of non-protein nitrogen and its true protein concentration is apparently not higher than 0.9% (1). This means that whey protein adapted formulas still provide infants with an excess of most amino acids of up to 200%. Moreover, the various fractions of cow's milk protein differ substantially in quantity and quality compared with human milk (Fig. 1). In this context it is remarkable that β-lactoglobulin, which represents the main fraction of cow's milk whey protein, is completely absent in human milk. This may explain the relatively high prevalance of β-lactoglobulin-related allergies in formula-fed infants.

POTENTIAL HAZARDS RELATED TO HIGH OR LOW PROTEIN INTAKES IN EARLY LIFE

The potential consequences of feeding formulas rich in protein are not known in detail. High protein intakes may result in high protein accretion, leading to a completely different body composition from that found with human milk feeding. In this connection the question arises as to whether this is harmless in the long term or hazardous for certain at-risk groups, such as very low birthweight infants—and if so, whether we should continue the search for formulas more closely adapted to human milk than our adapted formulas are at present. Pediatric nutritionists would certainly approve the latter aim. The food industry, however, will always want a demonstration of the shortcomings of the current generation of infant formulas and of the advantages of intended improvements.

Lately, increasing evidence has emerged that the mode of infant, especially preterm infant, nutrition may have consequences for the neuropsychological outcome of infants in later periods of life (2). We are not certain whether this has anything to do with disturbances caused by excess or deficiency of amino acids or whether it may be related to a deficiency of long-chain polyunsaturated fatty acids, sialic acid, or other essential components of human milk. The neurological consequences of some inborn errors of amino acid metabolism such as phenylketonuria are well established. However, disturbances of plasma amino acid concentrations that place a newborn infant at risk are ill defined and probably vary from infant to infant.

Deviations in the amino acid pattern of infant formulas from the amino acid composition of human milk are reflected in the plasma amino acid levels of infants. In 1968, Snyderman and co-workers (3) studied the plasma amino acid alterations caused by protein intakes of between 1.1 and 9.0 g/kg/d in 15 normal infants 1–6 months old. The feeds were prepared from cow's milk which was either diluted or enriched with demineralized casein and was made isocaloric with dextrimaltose and corn oil. In comparison with a control group receiving 3–3.5 g protein/kg/d, protein intakes ranging between 1.1 and 1.5 g/kg/d were accompanied by a depression of the plasma concentrations of the branched-chain amino acids leucine, isoleucine, and valine and of lysine and tyrosine, whereas glycine and serine concentrations were found to be increased. The plasma concentrations of threonine, phenylalanine, ornithine, cystine, and proline were less strikingly decreased. The extremely high intake of 9 g protein/kg/d, corresponding to 32% of the total energy intake, resulted in a rise in the levels of most of the amino acids, with excessive elevations of methionine, valine, leucine, isoleucine, and proline while the level of glycine was depressed.

Low birthweight infants have been shown to be especially susceptible to such amino acid imbalances (4,5). The conversion of methionine to cysteine is apparently reduced in very low birthweight infants (6).

Cow's-milk-based formulas may induce prolonged elevations of valine, phenylalanine, and methionine in neonates or preterm infants (7). Whey protein predominant formulas have been shown to cause hyperthreoninemia in preterm infants (8). Feeding formulas that contain whey protein/casein mixtures in a 60:40 ratio may also cause elevations in the plasma concentrations of valine, phenylalanine, methionine, lysine, leucine, and isoleucine (9–11). Furthermore, in comparison with human milk feeding, plasma urea concentration is also increased on such feeds, indicating that these amino acids are administered in excess and are partially catabolized (Table 1). If the protein concentration of whey protein-enriched formulas is lowered into the range of human milk values to correct the excessive amino acid concentrations, this results in low plasma tryptophan and taurine concentrations and does not necessarily correct the plasma amino acid imbalances (12). Tryptophan is the precursor of the neurotransmitter serotonin. Deficiencies of tryptophan could conceivably influence conscious behavior and sleep patterns. In healthy infants tryptophan has been shown to promote the onset of serotonin-induced sleep while unmodified formula feedings have the opposite effect (13). Low plasma taurine concentrations suggest cysteine deficiencies.

While the protein tolerance of term infants, as reflected in their plasma amino acid patterns, is considered to be relatively high, low birthweight infants are apparently

TABLE 1. *Changes in the plasma amino acid pattern of neonates and preterm infants fed on cow's-milk-based formulas compared to breast feeding*

Bovine milk formula	Whey protein-enriched formula	Low protein whey protein-enriched formula
Valine ↑	Valine ↑	Valine ↑
Phenylalanine ↑	Phenylalanine ↑	Phenylalanine ↑
Methionine ↑	Methionine ↑	Methionine ↑
Urea ↑	Threonine ↑	Threonine ↑
	Lysine ↑	Lysine ↑
	Leucine ↑	Leucine ↑
	Isoleucine ↑	Isoleucine ↑
	Urea ↑	Tryptophan ↓
		Taurine ↓

Data from refs. 9 to 12.

put at risk by amino acid imbalances. In a follow-up study on 62 prematurely born infants, a significantly lower performance IQ was found in children who had had high plasma tyrosine levels in the neonatal period (4). However, the differences were true only of low birthweight infants weighing 2000 g or more; no adverse effects of high tyrosine levels were observed in a group of smaller preterm infants. Goldman and co-workers (14) conducted psychometric evaluations in 304 children aged 3 and 5–7 years of age who were born prematurely and randomly assigned to either a 2% or a 4% cow's milk protein-based diet in the premature nursery. The incidence of low IQ scores was significantly increased in the group of infants of birthweight below 1300 g, who had had high protein intakes. Comprehensive psychometric tests in 15 children who as full-term neonates had experienced transient neonatal tyrosinemia due to high protein milk formula feeding (5.7 g/kg/d) resulted in specific learning disabilities compared to a control group that had not received a high protein milk formula in early infancy (15).

It can be concluded from these studies that high protein intakes in the neonatal period in preterm infants and newborns at risk may induce neuropsychological disturbances in later life. It is currently not known whether this is due only to metabolic perturbations of aromatic amino acids, which are present in excess in casein predominant formulas that are high in protein. Deleterious effects may also originate from imbalances of other amino acids accumulating in the extracellular and intracellular space as a result of high intakes, disturbances of assimilation, different needs, low degradation rates, and impaired renal excretion. The introduction of whey protein-enriched formulas containing lower protein concentrations may have contributed to a reduced prevalance of amino acid imbalances in the neonatal period. However, preterm infant formulas still contain protein concentrations of between 2.1 and 2.3%, although such high requirements for protein are restricted to the period between the 28th and 34th week of gestational age if the fetal accretion rate is taken as a standard (16).

PROCESSING OF INFANT FORMULAS WITH IMPROVED BIOLOGICAL VALUE: α-LACTALBUMIN AS A KEY PROTEIN

The problems arising from deviations between the amino acid pattern of cow's milk protein and human milk protein cannot be solved by choosing other ratios of whey protein/casein mixtures. Further adaptation of the protein pattern of infant formulas to make it more like human milk protein composition would either require fortification with tryptophan, cystine, and other essential amino acids or a protein which is extremely high in these essential amino acids. Among all the fractions of cow's milk protein, α-lactalbumin is the only one that could fulfill these requirements (Table 1) (17,18). Most of the protein mixtures in bovine milk-based formulas are low in tryptophan and cysteine (19). α-Lactalbumin has an exceptionally high concentration in tryptophan and cystine and a low methionine concentration (Table 2) and is thus an ideal supplement to compensate for the deviations in the amino acid pattern of infant formula protein mixtures currently in use.

Fortification of low protein formulas with free amino acids seems to be an effective method to correct imbalances of protein amino acid patterns (20), although the absorption kinetics and the utilization rates might be different from those of protein-bound amino acids (21). Potential side effects of free amino acids and the high costs related to the fortification of infant formulas are other aspects that make the realization of this procedure difficult in practice. Considering the disadvantages of free amino acids, α-lactalbumin-enriched fractions of cow's milk whey protein seem to be a more suitable source of material for upgrading infant formula proteins. Whey protein fractions rich in α-lactalbumin are now being produced on a large scale and are offered by the food industry at reasonable prices.

One of the main problems in the adjustment of protein amino acid patterns in infant formulas to human milk protein values is related to the fact that we do not have reliable data on the amino acid composition of human milk protein. Comparing the

TABLE 2. *Percentages of essential amino acids in whole bovine milk protein, whey protein, α-lactalbumin, and human milk protein*

Amino acid	Whole bovine milk protein	Whey protein	α-Lactalbumin	Human milk protein
Tryptophan	1.3	1.9	5.9	1.9 (2.3)[a]
Threonine	4.6	7.3	5.0	4.6
Phenylalanine	4.7	3.5	4.0	4.4
Leucine	9.5	10.1	10.5	10.1
Isoleucine	5.8	6.2	6.1	5.8
Methionine	2.5	2.2	0.9	1.8
Lysine	7.6	9.0	10.3	6.2
Valine	6.2	6.2	4.3	6.0
Cystine	0.8	2.2	5.3	1.7

[a] The tryptophan value in human milk protein is apparently higher (2.3%) than assumed originally.

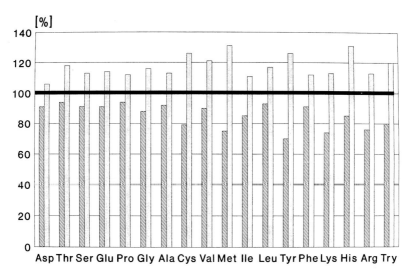

FIG. 2. Comparison of the values reported for amino acid composition of human milk protein.

values in published reports (22–25) shows substantial differences for the individual amino acids (Fig. 2). These differences are presumably due to the methods used for the determination of the amino acids and may also reflect a relatively broad range of amino acid concentrations in human milk protein. It is a matter of fact that there are considerable variations of single fractions, such as immunoglobulins, in human milk samples, related to individual particularities and to the lactation period. It is of special importance that the reference value for tryptophan in human milk is apparently considerably higher than originally assumed (Table 2). For practical purposes, it is sufficient to equalize the differences in the amino acid values of human milk protein, as published by different authors, by using mean values, which can serve as reference values for the creation of protein mixtures mimicking human milk protein. Such formulas make a reduction of the protein concentration in infant formulas to human milk protein values possible. Whether imbalances in the plasma amino acid patterns found in infants fed conventional low protein infant formulas can be avoided by such α-lactalbumin-enriched formulas has yet to be proven.

The experiences with tryptophan-supplemented low protein infant formulas have shown that these formulas produce plasma amino acid patterns in infants which do not differ from those found with human milk feeding (20). Since the consequences of amino acid imbalances in preterm infants and neonates at risk are currently not well known, we should continue the effort to formulate cow's milk protein mixtures with amino acid patterns closely resembling human milk. "Humanized" milk produced by transgenic cows is at present only a dream. Whether this dream will become true in the near future will depend on overcoming the numerous ethical, technical, and commercial problems connected with this new development.

CONCLUSIONS

1. Moderate deviations in the plasma amino acid levels from those found with human milk feeding are common in healthy infants fed on adapted formulas.
2. Preterm infants and neonates with limited functional capacity of the liver, the kidneys, and other organs are at risk of clinically relevant perturbations of the amino acid homeostasis.
3. The clinical significance of transient neonatal amino acid disorders is not well defined. Further studies are required to clarify potential side effects of neonatal amino acid imbalances on brain development and intellectual behavior.
4. Further adaptation of the amino acid pattern of infant formulas to that of human milk protein should become possible by admixtures of bovine α-lactalbumin to casein and whey protein.
5. Infant formulas fully adapted to human milk protein make a reduction of the total protein concentration below the currently recommended lowest limit possible. They could increase the protein efficiency rate, reduce the renal load, and thus provide more safety for infants at risk for transient neonatal disorders of amino acid metabolism.

REFERENCES

1. Hambraeus L, Lönnerdal B, Forsum E, Gebre-Medhin M. Nitrogen and protein components of human milk. *Acta Paediatr Scand* 1978; 67: 561–5.
2. Lucas A, Morley R, Cole TJ, Lister G, Leeson-Payne C. Breast milk and subsequent intelligence quotient in children born preterm. *Lancet* 1992; 339: 261–4.
3. Snyderman SE, Holt LE, Norton PM, Roitman E, Phansalkar SV. The plasma aminogram. I. Influence of the level of protein intake and a comparison of whole protein and amino acids diets. *Pediatr Res* 1968; 2: 131–44.
4. Menkes JH, Welcher DW, Levi HS, Dallas J, Gretaky NE. Relationship of elevated blood tyrosine to the ultimate intellectual performance of premature infants. *Pediatrics* 1972; 49: 218–24.
5. Rassin DK, Gaull GE, Räihä NCR, Heinonen K. Milk protein quantity and quality in low birth weight infants: effects on tyrosine and phenylalanine in plasma and urine. *J Pediatr* 1977; 90: 356–60.
6. Gaull GE, Rassin DK, Räihä NCR, Heinonen K. Milk protein quantity and quality in low birth weight infants. III. Effects on sulfur amino acids in plasma and urine. *J Pediatr* 1977; 90: 348–55.
7. Guesry PR, Secretin MC, Goyens Ph. Neue Aspekte der Ernährung von Neugeborenen mit niedrigem Geburtsgewicht. *Monatsschr Kinderheilkd* 1986; 134: 508–15.
8. Kashyap SK, Okamoto E, Kanaya S, et al. Protein quality in feeding low birth weight infants: a comparison of whey predominant versus casein predominant formulas. *Pediatrics* 1987; 79: 748–55.
9. Janas LM, Picciano MF, Hatch TF. Indices of protein metabolism in term infants fed human milk, whey predominant formula, or cow's milk formula. *Pediatrics* 1985; 75: 775–84.
10. Jarvenpää AL, Räihä NCR, Rassin DK, Gaull GE. Milk protein quantity and quality in the term infant. I. Metabolic responses and effects on growth. *Pediatrics* 1982; 70: 214–20.
11. Jarvenpää AL, Rassin DK, Räihä NCR, Gaull GE. Milk protein quantity and quality in the term infant. II. Effects on acidic and neutral amino acids. *Pediatrics* 1982; 70: 221–30.
12. Janas LM, Picciano MF, Hatch TF. Indices of protein metabolism in term infants fed either human milk or formulas with reduced protein concentration and various whey/casein ratio. *J Pediatr* 1987; 110: 838–48.
13. Yogman MW, Zeisel SH. Diet and sleep patterns in newborn infants. *N Engl J Med* 1983; 309: 1147–9.
14. Goldman HI, Goldman JS, Kaufman I, Liebman OB. Late effects of early dietary protein intake on low birth weight infants. *J Pediatr* 1974; 85: 764–9.

15. Mamunes P, Prince PE, Thornton NH, Hunt PE, Hitchcock ES. Intellectual deficits after transient tyrosinemia in the term neonate. *Pediatrics* 1976; 57: 675–80.
16. Pohlandt F, Kupferschmid C. The protein requirement of preterm infants. *Klin Padiatr* 1985; 197: 164–6.
17. Forsum E. Nutritional evaluation of whey protein concentrations and their fractions. *J Dairy Sci* 1974; 57: 665–70.
18. Heine WE, Klein PD, Reeds PJ. The importance of α-lactalbumin in infant nutrition. *J Nutr* 1991; 121: 277–83.
19. Sawar G, Botting HG, Peace RW. Amino acid rating method for evaluating protein adequacy of infant formulas. *J Assoc Off Anal Chem* 1989; 72: 622–7.
20. Fazzolari-Nesci A, Domianello D, Sotera V, Räihä NCR. Tryptophan fortification of adapted formula increases plasma tryptophan concentrations to levels not different from those found in breast-fed infants. *J Pediatr Gastroenterol Nutr* 1992; 14: 456–9.
21. Gropper SS, Gropper DM, Acosta PB. Plasma amino acid response to ingestion of amino acids and whole protein. *J Pediatr Gastroenterol Nutr* 1993; 16: 143–50.
22. Nayman R, Thomson ME, Scriver CR, Clow CL. Observations on the composition of milk substitute products for treatment of inborn errors of amino acid metabolism: comparisons with human milk. *Am J Clin Nutr* 1979; 32: 1279–89.
23. Harzer G, Bindels JG. Changes in human milk immunoglobulin A and lactoferrin during early lactation. In: Schaub J, ed. *Composition and physiological properties of human milk*. Amsterdam: Elsevier, 1985; 285–95.
24. Picone TA, Benson JD, Moro G, et al. Growth, serum, biochemistries, and amino acids of term infants fed formulas with amino acid and protein concentrations similar to human milk. *J Pediatr Gastroenterol Nutr* 1989; 9: 351–60.
25. Chavalittamrong B, Suanpan S, Boonvisut S, Chatranow W, Gershoff SN. Protein and amino acids of breast milk from Thai mothers. *Am J Clin Nutr* 1981; 34: 1126–30.

DISCUSSION FOLLOWING THE PRESENTATION OF DR. HEINE

Dr. Lönnerdal: I should like to return the question of the true protein level of infant formulas. Those studies that have found significantly lower tryptophan levels in the plasma of formula-fed infants all used low protein formulas with high non-protein nitrogen values. This means that the level of protein was lower than expected and I should call these very low protein formulas. When we did a study of a formula containing 15 g per liter of true protein, which I consider to be a low protein formula, we did not see any significant differences in tryptophan. Therefore, I am not especially concerned about the tryptophan level as long as we are talking about true protein levels.

Dr. Rassin: There is a paper looking at tryptophan supplementation (1). It is interesting that plasma tryptophan is about 5 μmol/dl without supplementation and at three different levels of supplementation it goes up by about 20%, although the rise does not correlate with the level of supplementation. However, some definite physiological and behavioral changes were found in association with the increase. This is one of the few papers I have seen that used the fluorometric technology to measure tryptophan. It is a very nice experimental design because the authors looked at the same formula with and without tryptophan supplementation.

Dr. Lönnerdal: In a study which we published (2,3), we suggested that bovine α-lactalbumin could be added to whey-predominant formula to change the plasma amino acid pattern. I think it is important to note that even if the α-lactalbumin in cow's milk is similar to that in human milk, they are not identical. There are significant differences in their amino acid composition, for example the antibodies do not cross-react with each other, and they are quite different with regard to composition, digestibility, and immunological reactivity. Our conclusions were that while tryptophan levels could be improved, the pattern of the other amino acids might well become less favorable.

Dr. Heine: The admixture of α-lactalbumin to casein/whey protein mixtures may compensate for low concentrations of cystine and aspartic acid and may reduce the surplus of threonine and methionine. The amino acid pattern obtained in this way are practically identical with the human milk pattern. The search for an improved amino acid composition in the protein of infant formulas is really only of importance for preterm infants and neonates at risk.

Dr. Räihä: I think this is a little bit too modest! I really believe that this could be an improvement for all newborn infants. Do we really need to wait for long-term effects to show up before we work to improve formulas? Taurine is a typical example of this. We say that exclusive breast feeding for the first 4, 5, maybe even 6 months should be the norm or the gold standard. If we agree on that, we should be trying to make formula as close as possible to breast milk, particularly as far as the infant's plasma amino acid pattern is concerned. There is enough evidence at present to make changes justified. We supplemented formulas with taurine long before we had any evidence that there are long-term effects of taurine deficiency. I don't think we should wait to find long-term pathology. Potential problems are very difficult to assess and as we know from the story of taurine, we often have to wait for highly sophisticated methods before we can identify pathological conditions for certain. They may well be there but we can't measure them. So I think this could be of benefit for all infants, not only the preterm and at-risk infants.

Dr. Heine: You may be right. We cannot be certain at the moment what the effects of individual deficiencies or abundance of particular amino acids may be. I appreciated Alan Lucas's paper very much (4) since it focused on the point that nutrition in early infancy may have consequences for the outcome of the infant. And if what Alan Lucas has published is indeed true, that the intelligence quotient is lower in preterm infants who are fed formulas as opposed to human milk, we must look seriously at the differences between the types of feeding. Niels Räihä has already mentioned that in Sweden all preterm infants are fed on human milk and we did the same in the eastern part of Germany. Pediatric nutritionists should think hard about improvements to formulas and have a close look not only at long-chain polyunsaturated fatty acids but also at the amino acid patterns, at the supply of sialic acid and other essential carbohydrates, and at the many factors that are present in human milk but absent in formula.

Dr. Räihä: I think your idea of modifying cow's milk protein even further than it is at present is an intriguing possibility. I really think this should be the next step before we start using transgenic proteins on a large scale. There may even be other advantages than just a better amino acid profile; for example, there is the possibility of decreasing cow's milk allergy by replacing β-lactoglobulins with α-lactalbumin. Do you have any comment on that?

Dr. Heine: I have some doubts on whether this is possible. The amount of β-lactoglobulin is not a very critical factor; allergies can be originated by very low concentrations of these antigenic proteins. Also the preparations now available are whey protein fractions, provided by the industry, which contain enriched concentrations of α-lactalbumin, not the pure protein. So I do not believe that we can solve all the problems connected with the allergy by these means.

Dr. Guesry: Not only is β-lactoglobulin allergenic in cow's milk but so is bovine serum albumin and also casein, which is as allergenic as β-lactoglobulin. So even if we replace part of the β-lactoglobulin by α-lactalbumin, we shall not change the allergenicity of the cow's milk.

Dr. Lönnerdal: Dr. Heine raised the question of the variation in amino acid content in human milk. Some of this may be methodological, but part could be due to the diet of the mother. I should also like to emphasize that there is substantial variation in plasma amino acid pattern standards among different laboratories, probably due to natural variation both in amino acid composition and in plasma amino acid patterns.

Dr. Heine: I was surprised to see how much difference there is in the amino acid concentrations reported by different authors. This makes it really difficult to formulate a gold standard for human milk.

Dr. Cooper: Are there any data, particularly from developing parts of the world where diets would be different, on individual amino acid compositions of human milk?

Dr. Lönnerdal: I should like to answer this question. Several people have looked at this, including us. It is a very complicated issue because the amino acids will reflect the maternal proteins. We did a study recently in the Gambia and looked at parity and its effect on breast milk composition. We found effects on the casein/whey ratio and therefore on the amino acid composition. So in developing countries, not only does the food supply affect the amino acid and protein composition to some extent, but so do other factors, such as parity.

Dr. Uauy: Should we not be looking at the tissue amino acid composition? Plasma will vary tremendously, but tissue will probably vary less.

Dr. Heine: That is a crucial question. Tissue amino acid values may differ considerably from the plasma values. However, plasma is readily available and I see some difficulties in determining tissue values in normal infants.

Dr. Uauy: In the field of adult nutrition, people are now suggesting that we should be looking at the amino acid composition of muscle as a representation of tissue protein. It would be interesting to see how the amino acid pattern of human milk reflects that of muscle. You mentioned that there were clinically relevant conditions in neonates relating to changes in dietary protein supply. To what were you referring specifically?

Dr. Heine: Menkes and co-workers (5) were the first to show lower IQ performances in school-age children who were prematurely born and who had high plasma tyrosine concentrations in their neonatal period. There have been two further publications in 1974 and 1976 by Goldman (6) and Mamunes (7) suggesting that there are differences in the neuropsychological development and behavior in children who were fed on high protein diets after birth. It is well known from phenylketonuria that such correlations exist, but the values quoted by these investigators are of course well below the critical values seen in phenylketonuria. I had the feeling, when I read these articles, that the conditions under which the studies were done were not sufficiently well defined. We need further randomized studies to clarify the situation and discover for certain whether there is a correlation between raised plasma amino acids, both essential and non-essential, in neonates and preterm infants and later outcome. The problem is that it may not only be amino acids that might be dangerous for the infant. We have seen that other components, sialic acid and so on, may also play a role, so it is very difficult to prove these correlations.

Dr. Bremer: I find it astonishing that nutritionists often neglect the results obtained by people working with metabolic diseases; for instance, in tyrosinosis very high levels of tyrosine occur over a long period of life and nothing occurs in the brain. There are results available for histidine and for other amino acids. I think it is important that these matters be discussed with people who have experience with high amino acid levels over long periods while treating children with metabolic diseases. I am not convinced that the concentration differences that are seen as a result of feeding different amounts and types of protein are likely to be of any relevance.

Dr. Rassin: One should nevertheless bear in mind some of the results of the maternal PKU collaborative studies, which suggest that the amounts of phenylalanine that can be toxic are much lower than we ever considered in true PKU. In fact, we keep lowering the cutoff point for what we consider toxic levels of phenylalanine. I appreciate that *in utero* exposure is a different model from *ex utero* exposure. It is important to remember that certain amino acids can become altered in the central nervous system without any particular indication of this from the plasma aminogram. Although the idea that muscle composition should form the amino acid standard is an interesting one, in terms of cognitive outcome the brain is where my money is.

I would like to support what Dr. Räihä said about the term infant. We have all been discussing Alan Lucas's study in the UK in relation to the outcome in preterm infants, but there were several earlier studies which showed that there are long-term cognitive benefits of breast feeding term infants. While not perfect studies, these have shown that there is a definite advantage to being breast-fed. When you look at all these studies together, there has never been one that has shown an advantage to being formula-fed. So, although the methodology is often weak, there has been a great deal of consistency about the findings. I think there is quite a lot to be done to improve full-term preparations as well as preterm preparations.

Dr. Räihä: You said that the brain is where your money is, but there are other organs we should look at in this context. There is certainly evidence that a high protein intake or amino acid intake may have effects on insulin secretion and I think there is evidence that breast-fed infants are less prone than formula-fed infants to developing type I diabetes in later life. There may also be long-term effects on renal function. Thus there may be many other organs that could be affected by either an excessive or a deficient supply of protein and amino acids.

Dr. Lönnerdal: I agree that there are long-term benefits of breast feeding, but I really don't think that the small deviations that we see in plasma amino acid patterns are responsible. Coming back to what I said earlier, I don't think that additional α-lactalbumin in formulas for the normal term infant would be likely to make any significant contribution. It is perfectly possible to reduce the protein level and produce a plasma amino acid pattern very close to the breast-fed one.

But the point I really want to pursue is the one of cost that was raised by our colleagues in South Africa and India. Even in the USA, and within the state of California, the price issue is a relevant one. We have a significant population of poor people in the state, and as soon as the price of the formula is raised they will go to evaporated cow's milk or whole cow's milk instead because it is much cheaper. This is something we don't want to see happening. There is no way that you can add a highly enriched α-lactalbumin fraction to regular term formula without increasing the price. For special groups, yes; preterm infants, yes; but not for the regular population.

Dr. Pettifor: One area we haven't discussed much is the effect of changing the protein constituents of cow's milk on mineral absorption. How important are metalloproteins, and do they have a role in the absorption of any of the minerals?

Dr. Heine: α-Lactalbumin is known to bind calcium and zinc and other metals, but its capacity to do so is pretty low, so it does not play a significant role in this field.

Dr. Lönnerdal: I don't have much to add. α-Lactalbumin is not really a significant contributor to calcium or zinc nutrition. Overall, when it comes to mineral absorption, the whey-predominant product is advantageous because it is more easily digestible. With casein you always have the problem of incomplete digestion and the possibility of forming aggregates which may make trace elements and to some extent calcium less available.

Dr. Guesry: This is not really a problem. Absorption of minerals is always quite good and when it is not good enough with a cow's milk formula, we can improve it either by increasing the quantity of the mineral or by adding other nutrients, for example vitamin C, which will improve iron absorption. With soya formulas there is more of a problem because there are some factors, such as phytate, which block the absorption of zinc and iron and perhaps calcium.

REFERENCES

1. Steinberg LA, O'Connell NC, Hatch TF, et al. Tryptophan intake influences infants' sleep latency. *J Nutr* 1992; 122: 1781–91.
2. Forsum E, Lönnerdal B. Protein evaluation in growing rats of breast milk and breast milk substitutes with special reference to the content of non-protein nitrogen. *J Nutr* 1979; 109: 185–92.
3. Forsum E, Lönnerdal B. Evaluation of breast milk and breast milk substitutes in growing rats. *Pediatrics* 1979; 64: 536–8.
4. Lucas A, Morley R, Cole TJ, et al. Breast milk and subsequent intelligence quotient in children born preterm. *Lancet* 1992; 339: 261.
5. Menkes JH, Welcher DW, Levi HS, et al. Relationship of elevated blood tyrosine to the ultimate intellectual performance of premature infants. *Pediatrics* 1972; 42: 218–24.
6. Goldman HI, Goldman JS, Kaufman I, et al. Late effects of early dietary protein intake on low birthweight infants. *J Pediatr* 1974; 85: 764–6.
7. Mamunes P, Prince PE, Thornton NH, et al. Intellectual deficits after transient tyrosinemia in the term neonate. *Pediatrics* 1976; 57: 675–80.

Protein Metabolism During Infancy, edited by
Niels C. R. Räihä. Nestlé Nutrition Workshop
Series, Vol. 33. Nestec Ltd., Vevey/
Raven Press, Ltd., New York © 1994.

Protein Requirements of Low Birthweight, Very Low Birthweight, and Small for Gestational Age Infants

Sudha Kashyap and *William C. Heird

*Department of Pediatrics, Columbia University College of Physicians and Surgeons and Babies Hospital (Presbyterian Hospital), New York, New York 10032, USA and *Children's Nutrition Research Center, Department of Pediatrics, Baylor College of Medicine, Houston, Texas 77030*

A definition of the protein requirement of any population must include consideration of the purposes for which the requirement is being defined. For the normal adult, the requirement addresses primarily the needs for maintenance (i.e., the exogenous protein needed to maintain nitrogen equilibrium and hence body protein stores). The protein requirement of any pediatric population, particularly that of low birthweight (LBW) infants, includes the needs for maintenance plus the additional needs to assure normal growth. In addition, if there has been a period of delayed or subnormal growth, either *in utero* or *ex utero,* and/or if actual weight loss has occurred, as is true for most LBW infants, there is an additional need to support catch-up growth. Another consideration in defining protein requirement is related to maintenance of plasma indices of protein adequacy and excess. In general, a protein intake that meets the maintenance needs of adults or supports normal growth of infants and children will also support normal rates of synthesis of all proteins and hence normal plasma concentrations of the proteins commonly considered indices of protein adequacy (e.g., albumin and transthyretin). Thus the concern is usually that the protein intake should not exceed metabolic capability, resulting in raised plasma indices of protein excess (e.g., urea and amino acids).

All of these considerations are relevant to defining the protein needs of LBW infants. Much of the discussion that follows is based primarily on the combined data of studies conducted by the authors and their colleagues over the past decade (1–4). In general, the LBW infants participating in these studies were assigned randomly to be fed a variety of protein and energy intakes. In all studies, the intake of each nutrient varied independently of the other. The outcome variables of all included growth and nutrient retention as well as plasma indices of protein adequacy and excess; all were monitored serially from the time the assigned intake was tolerated until the infant was discharged or weighed 2200 g. In total, 157 infants weighing

TABLE 1. *Clinical characteristics of LBW infants participating in studies of the effects of varying protein and energy intakes on growth, nutrient retention, and plasma indices of protein adequacy and excess*

Birthweight	750–1750 g
750–1250	29%
1251–1500	39%
1501–1750	32%
Gestational age	26–34 weeks
<30 weeks	16%
30–32 weeks	59%
>32 weeks	25%
Age at first feeding	1–10 days
Age at full intake	<28 days
Age birthweight regained	13–19 days
Respiratory distress	75%
Antibiotic therapy	74%
Parenteral nutrition	62%
Duration of total parenteral nutrition	7–14 days

Data from refs. 1 to 4.

between 750 and 1750 g at birth were enrolled into four studies. The characteristics of the total study population are summarized in Table 1. Protein intakes of the various groups ranged from 2.24 to 4.26 g/kg/d; energy intakes ranged from 99 to 150 kcal/kg/d.

Discussion of the protein requirement of LBW infants based on the combined results of these studies is preceded by a consideration of the protein requirement for maintenance which was not addressed by these studies. The discussion of the protein needs for growth, which relies heavily on the data obtained in the studies described above, is followed by discussion of the special needs, if any, of very low birthweight (VLBW) infants (i.e., infants who weigh less than 1250 g at birth) and infants with intrauterine growth retardation.

MAINTENANCE PROTEIN NEEDS OF LBW INFANTS

While it is important to have some idea of the maintenance protein needs of LBW infants, it is obvious that an infant who receives only enough protein for maintenance will not grow and thus will soon become "growth retarded" relative to a normal fetus of the same postconceptional age. For example, a 2-week-old infant whose weight was at the 50th percentile when born at 28 weeks gestation but whose weight is still the same as at birth is, by definition, "growth retarded" in comparison to the 30-week gestation fetus. Nonetheless, most would agree that maintaining such an infant's protein stores, regardless of body weight, is preferable to permitting losses of protein as well as water and fat.

Data from a number of studies (5–8) show that the LBW infant who receives no protein intake experiences urinary nitrogen losses of at least 150 mg/kg/d. Data from a recent study conducted by us in infants weighing less than 1250 g at birth are summarized in Tables 2 and 3. In this study, urinary nitrogen excretion was measured

TABLE 2. *Urinary nitrogen excretion of VLBW infants before and after addition of amino acids to parenteral nutrition regimens*

Infant	Birthweight (g)	Urinary nitrogen excretion (mg/kg/d ± SD)	
		Before[a]	After[b]
1	745	198	160 ± 6
2	1260	214	135
3	765	158	181 ± 75
4	1070	128	150 ± 36
5	970	162	325 ± 104
6	1100	271	199 ± 43
7	1070	142	188 ± 30
8	850	155	207 ± 28
9	1140	287	312 ± 86
10	985	117	153
Mean		183 ± 58	201 ± 66

[a] Duration ranged from 2 to 9 days (mean = 3 days); nitrogen intake = 0; mean energy intake = 30 ± 8.5 kcal/kg/d.
[b] Duration ranged from 2 to 9 days (mean = 4.8 days); mean nitrogen intake = 315 ± 58 mg/kg/d; mean energy intake = 50 ± 8 kcal/kg/d.

TABLE 3. *Blood urea nitrogen concentration (mg/dl) of VLBW infants before and after addition of amino acids to parenteral nutrition regimens*

Infant	Birthweight (g)	BUN concentration	
		Before[a]	After[b]
1	745	16	8.7 ± 1.5
2	1260	21	13.8 ± 1.9
3	765	33	21.3 ± 4.3
4	1070	8	11.8 ± 3.0
5	970	22	24.5 ± 6.4
6	1100	21	25.1 ± 2.6
7	1070	19	22 ± 1.8
8	850	23	20 ± 2.6
9	1140	21	28 ± 4.6
10	985	16	18 ± 0
Mean		20 ± 6	19.3 ± 6.2

[a] Duration ranged from 2 to 9 days (mean = 3 days); nitrogen intake = 0; mean energy intake = 30 ± 8.5 kcal/kg/d.
[b] Duration ranged from 2 to 9 days (mean = 4.8 days); mean nitrogen intake = 315 ± 58 mg/kg/d; mean energy intake = 50 ± 8 kcal/kg/d.

daily (or nearly so) over the first 10 days of life or until 10% of total intake was by the enteral route. During the initial few days, the infants received no amino acid intake; parenteral amino acid intake during the subsequent period was about 2 g/kg/d. The mean nitrogen excretion of these small infants during the first few days of life while they were receiving no exogenous nitrogen intake was about 180 mg/kg/d. This equates to a loss of approximately 1% of body protein stores daily. Interestingly, urinary nitrogen excretion did not increase dramatically when amino acids were added to the parenteral nutrition regimen. Blood urea nitrogen concentration also did not increase (Table 3). Moreover, plasma amino acid concentrations, which were extremely low prior to receiving nitrogen intake, were within an acceptable range after addition of amino acids (Table 4).

In total, the available data (5–8; Tables 2,3) suggest that the maintenance protein requirement of LBW infants is in the range 1–1.25 g/kg/d. However, parenteral amino acid intakes of twice this minimum amount appear to be retained and are not associated with disturbing metabolic consequences. Thus it seems reasonable to recommend that LBW infants receive an amino acid intake of at least 1–1.25 g/kg/d and as much as 2–2.5 g/kg/d as soon after birth as possible. In fact, as discussed below,

TABLE 4. *Plasma amino acid concentrations (μmol/dl) of VLBW infants before and after addition of amino acids to parenteral nutrition regimens*

Amino acid	Plasma amino acid concentration	
	Before	After
Threonine	11.5 ± 4.5[a]	21.4 ± 8.2
Valine	9.1 ± 3.3[a]	16.9 ± 3.7[a]
Leucine	5.8 ± 2.0[a]	11.4 ± 2.6
Isoleucine	2.0 ± 1.0[a]	5.4 ± 1.4
Lysine	11.6 ± 3.0[a]	15.1 ± 4.9[a]
Methionine	1.6 ± 1.1	2.7 ± 0.9
Cystine	1.2 ± 0.8	1.4 ± 0.8
Histidine	7.7 ± 1.9	9.8 ± 2.0
Phenylalanine	7.2 ± 2.2	7.8 ± 2.0
Tyrosine	9.4 ± 3.1	6.7 ± 3.8
Tryptophan	2.2 ± 0.9	2.9 ± 1.1
Serine	12.8 ± 5.1[a]	17.8 ± 5.3
Glutamine + glutamate	51.1 ± 28.3	45.0 ± 12.9[a]
Proline	9.1 ± 4.6[a]	12.8 ± 3.7
Glycine	22.8 ± 8.6[a]	26.8 ± 9.4
Alanine	12.9 ± 6.7[a]	15.0 ± 3.9
Arginine	5.1 ± 2.0	9.2 ± 3.2
Taurine	5.3 ± 3.4[a]	6.3 ± 2.1
Aspartic acid	5.8 ± 2.2	2.3 ± 1.3

[a] More than 2 SD lower than mean cord blood concentration of infants born at a mean gestational age of 29 weeks (9). The plasma concentration of no amino acid exceeded the mean cord blood concentration by more than 2 SD.

this simple change in current management may decrease the subsequent protein requirements for catch-up growth.

PROTEIN REQUIREMENT FOR GROWTH AND ACCEPTABLE INDICES OF PROTEIN ADEQUACY/EXCESS

Figures 1 to 5 depict the relationship between the mean protein intake of groups of infants ($n = 9-15$) fed protein intakes from 2.24 to 4.26 g/kg/d and, respectively, the mean rate of weight gain (Fig. 1), the mean rate of nitrogen retention (Fig. 2), the mean plasma concentrations of transthyretin (Fig. 3) and albumin (Fig. 4), and the mean blood urea nitrogen concentration (Fig. 5) of each group. From these, one can predict the protein intake likely to result, on average, in acceptable rates of weight gain and nitrogen retention as well as acceptable plasma indices of protein adequacy without evidence of protein excess.

The data suggest that the lowest mean protein intake likely to result in rates of both weight gain and nitrogen retention approximating intrauterine rates (i.e., ~15 g/kg/d and 300 mg/kg/d) is about 2.75 g/kg/d. This suggestion, of course, is based on the assumption that concomitant energy intake is adequate, as was the case in the studies giving rise to the data. It is important to note that every baby who receives this intake will not necessarily gain weight and retain nitrogen at intrauterine rates; the rates of some will be lower and the rates of others will be higher. Moreover, this mean intake does not appear to be sufficient to assure acceptable mean plasma indices of protein adequacy. For this purpose, a mean protein intake in excess of 3 g/kg/d appears to be required (Figs. 3 and 4). Further, as judged by the relationship between protein intake and blood urea nitrogen concentration, considerably higher intakes do

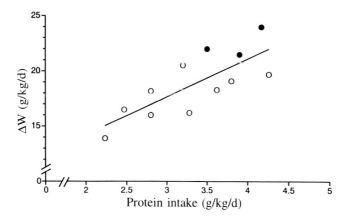

FIG. 1. Relationship between protein intake of LBW infants and rate of weight gain (Δw). Each point denotes the mean rate of weight gain of a group of infants ($n = 9-15$) fed the indicated protein intake (1–4). Concomitant energy intakes ranged from 100 to 150 kcal/kg/d; solid points denote concomitant energy intakes >120 kcal/kg/d. For the combined groups, $\Delta W = 3.44\ P_{in} + 7.34$; $r^2 = 0.6$.

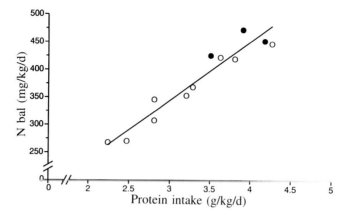

FIG. 2. Relationship between protein intake of LBW infants and nitrogen balance (Nbal). Each point denotes the mean nitrogen balance of a group of infants ($n = 9.15$) fed the indicated protein intake (1–4). Concomitant energy intakes ranged from 100 to 150 kcal/kg/d; solid points denote concomitant energy intakes >120 kcal/kg/d. For the combined groups, Nbal = 104 P_{in} + 33; $r^2 = 0.92$.

not appear to be excessive. For example, the mean blood urea nitrogen concentration of all groups that received protein intakes of less than 4 g/kg/d was below 10 mg/dl (Fig. 5). Similarly, as illustrated in Fig. 6 (for some amino acids), the mean plasma concentration of most amino acids in all groups whose protein intake was less than 4 g/kg/d was within the 95% confidence limits of the plasma concentration of infants fed their own mother's milk (3). An exception is the plasma concentration of threo-

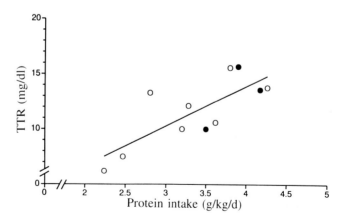

FIG. 3. Relationship between protein intake of LBW infants and plasma transthyretin concentration (TTR). Each point denotes the mean plasma transthyretin concentration of a group of infants ($n = 9-15$) fed the indicated protein intake (1–4). Concomitant energy intakes ranged from 100 to 150 kcal/kg/d; solid points denote concomitant energy intakes >120 kcal/kg/d. For the combined groups, TTR = 3.6P_{in} − 0.6; $r^2 = 0.6$.

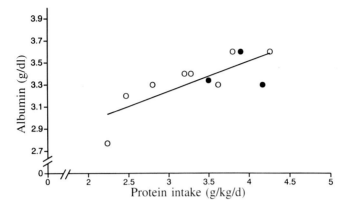

FIG. 4. Relationship between protein intake of LBW infants and plasma albumin concentration. Each point denotes the mean plasma albumin concentration of a group of infants (n = 9–15) fed the indicated protein intake. (1–4). Concomitant energy intakes ranged from 100 to 150 kcal/kg/d; solid points denote concomitant energy intakes >120 kcal/kg/d. For the combined groups, ALB = $0.18 P_{in} - 2.73$; $r^2 = 0.26$.

nine, which is known to be high in all infants who receive whey-predominant bovine milk protein formulas (10,11).

This set of data suggests that LBW infants, on average, require a minimum protein intake of 2.8–3 g/kg/d for maintenance plus growth. However, the data also suggest that considerably greater intakes are well tolerated. Moreover, these greater intakes result in greater rates of weight gain and nitrogen retention and also are more likely to assure that plasma indices of protein adequacy (e.g., plasma transthyretin and

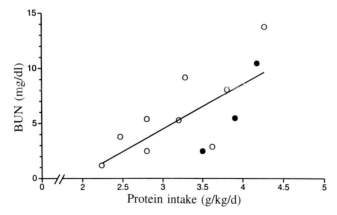

FIG. 5. Relationship between protein intake of LBW infants and blood urea nitrogen concentration (BUN). Each point denotes the mean blood urea nitrogen concentration of a group of infants (n = 9–15) fed the indicated protein intake (1–4). Concomitant energy intakes ranged from 100 to 150 kcal/kg/d; solid points denote concomitant energy intakes >120 kcal/kg/d. For the combined groups, BUN = $4.11 P_{in} - 7.8$; $r^2 = 0.5$.

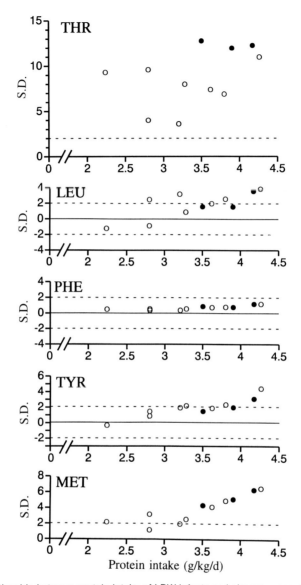

FIG. 6. Relationship between protein intake of LBW infants and plasma concentrations of threonine (THR), leucine (LEU), phenylalanine (PHE), tyrosine (TYR), and methionine (MET). The plasma amino acid data are plotted as the number of standard deviations by which the mean concentration of each group (n = 9–15) of infants fed the indicated protein intakes (1–4) differs from the mean concentration of infants fed predominantly their own mother's milk (3), that is, (mean plasma concentration of each group *minus* mean plasma concentration of infants fed their own mother's milk)/standard deviation of mean plasma concentration of control infants. Solid points denote concomitant energy intakes >120 kcal/kg/d. The solid and dashed lines, respectively, denote 0 ± 2 SD, or the 95% confidence limits of the plasma concentration of each amino acid observed in infants fed their own mother's milk.

albumin concentrations) remain within an acceptable range. As discussed in the following section, these greater rates of weight gain and nitrogen retention are likely to be particularly important for the infant with a need to experience considerable catch-up growth.

PROTEIN REQUIREMENTS FOR CATCH-UP GROWTH

The typical LBW infant with an uncomplicated course loses 10–15% of body weight over the first 4–5 days of life; thereafter, weight increases slowly and birthweight is usually regained between 2 and 3 weeks of age. The nature of this initial weight loss is not known with certainty but probably represents, at least in part, a disproportionate loss of extracellular fluid (12). If so, this component of the loss probably is of little consequence to the infant. Nonetheless, as illustrated in Fig. 7, the infant who gains weight at the intrauterine rate after birthweight is regained will remain 2–3 weeks behind a fetus of the same postconceptional age. Thus, in most cases, some catch-up growth is desirable. This is particularly true, as also illustrated in Fig. 8, for the infant who has experienced intrauterine growth retardation. In fact, most LBW infants—particularly VLBW infants, infants whose early neonatal course is complicated, and infants with intrauterine growth retardation—have protein needs for catch-up growth as well as for maintenance and normal growth.

There is no convincing evidence that the protein requirement for catch-up growth, *per se,* is any different from the requirement for normal growth. Forbes has illustrated

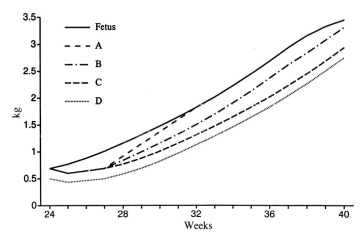

FIG. 7. Intrauterine growth curve (solid line) and theoretical growth curves of a variety of LBW infants. All infants lose 10% of body weight during the first week of life and regain this weight by 3 weeks of age. Infant A then gains weight at an accelerated rate relative to the fetus and by 8 weeks of age weighs the same as the fetus of the same postconceptional age. Infant B gains weight at a somewhat lower rate than infant A and does not "catch up" as quickly. Infant C gains weight at the intrauterine rate. Infant D is small for gestational age at birth and gains weight, after birthweight is regained, as described for infant C.

TABLE 5. *Protein and energy intakes predicted to result in different rates of catch-up growth*

Weight gain	Protein intake (g/kg/d)	Energy intake (kcal/kg/d)
690 g → 1830 g:		
In 56 days[a]	2.9	91
In 49 days	3.4	98
In 42 days	4.0	109
In 35 days	4.9	123
1830 g → 3450 g:		
In 56 days[a]	2.7	89
In 49 days	3.0	93
In 42 days	3.4	99
In 35 days	4.0	109

[a] Time required *in utero*.

that "catch up" occurs when the magnitude of growth above the normal growth curve equals the magnitude of growth below the growth curve as the growth deficit was incurred (13). In other words, the requirement for catch-up growth, which is additional to that for normal growth, is a function of the amount of catch-up to be achieved and the duration over which it is achieved. For example, if it is desired to exceed the normal growth rate by roughly 1 kg over 100 days, the requirement for catch-up is that required to produce an extra 10 g of weight gain daily. The daily requirement to achieve this catch-up growth in 50 days obviously will be greater; that for achieving it in 200 days will be less. Since some catch-up is likely to be desirable in almost every LBW infant, the greater protein intakes discussed above clearly seem desirable.

We have shown recently that it is possible to predict the protein and energy intakes required, on average, to produce both specific rates and specific compositions of weight gain and that the predicted intakes, on average, result in the expected outcomes (4,14). The exercise summarized in Table 5 illustrates application of the same principles in an infant who weighs 690 g at birth and regains birthweight at 3 weeks of age. For complete catch-up, that is, to weigh the same and have the same body composition as a fetus of the same postconceptional age by the time of discharge for example, when weight reaches about 1830 g, the infant must increase in weight from 690 to 1830 g in 35 rather than in 56 days. The predicted protein and energy intakes are 5 g/kg/d and 123 kcal/kg/d. A lower protein intake and a higher energy intake conceivably could produce the desired rate of weight gain, but the fat content of this weight would be vastly in excess of that *in utero*.

Based on the data summarized in Figs. 5 and 6, a protein intake of 5 g/kg/d clearly will not be tolerated. However, at least some of the 3-week period required to regain birthweight can be recouped before the baby is ready for discharge. Theoretically, a protein intake of 3.4 g/kg/d and an energy intake of approximately 100 kcal/kg/d will support rates of growth of sufficient magnitude to allow a week of this time to be recouped prior to discharge (1830 g). Continuing these protein and energy intakes

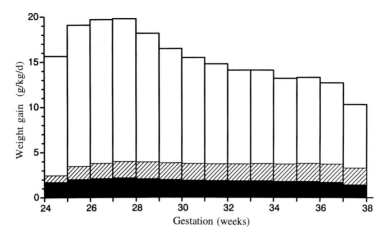

FIG. 8. Fetal rates of weight gain and accretion of protein and fat from 24 to 38 weeks' gestation. The height of each bar represents the increase in weight (g/kg/d) during each week of gestation. The solid and hatched portions of each represent concurrent increases, respectively, in protein and fat (g/kg/d). Calculations are based on data of Ziegler et al. *Growth* 1976; 40: 329–341.

after discharge should make it possible to recoup the remaining 2-week delay in regaining birthweight by the time the infant reaches a postmenstrual age of 40 weeks.

THE VERY LOW BIRTHWEIGHT INFANT

It is commonly believed that VLBW infants, particularly those weighing less than 1000 g at birth, have greater protein requirements than larger infants (15). In part, this belief is based on the fact that the rate of protein accretion of the fetus, expressed on a bodyweight basis, declines as gestation proceeds (16). These data, as well as the rates of accretion of body fat and total weight from 24 to 38 weeks' gestation, all expressed as g/kg/d, are shown in Fig. 8. Using these data and linear regression equations relating protein retention to protein intake and fat retention to energy intake (4,14,17), the protein and energy intakes required to support intrauterine rates of protein and fat accretion can be calculated. These calculations, summarized in Table 6, suggest that the protein requirement, expressed as g/kg/d, increases from 24 to 28 weeks and then declines gradually from about 2.94 g/kg/d at 28 weeks to 2.4 g/kg/d at 36 weeks. Interestingly, the energy requirement estimated in this way changes minimally over this period.

The predictions of the exercise described above are based on duplication of intrauterine rates of accretion. Since it is likely that the water content of weight gain *ex utero* is less than that *in utero,* accretion of protein and fat at the intrauterine rates will not necessarily assure that the rate of weight gain also will be the same as the intrauterine rate. Thus, if the infant has not regained birthweight immediately or has experienced other needs for catch-up growth, the actual requirements are likely to be greater than the minimal requirements depicted in Table 6.

TABLE 6. *Calculated protein and energy requirements for intrauterine rates of protein and fat accretion*

Gestation (weeks)	Protein (g/kg/d)	Energy (kcal/kg/d)
24–25	2.26	72.25
25–26	2.7	83.9
26–27	2.84	87.5
27–28	2.94	89.8
28–29	2.81	89.6
29–30	2.65	88.7
30–31	2.56	88.0
31–32	2.46	87.7
32–33	2.46	87.7
33–34	2.47	88.3
34–35	2.39	87.9
35–36	2.39	88.9
36–37	2.26	88.3
37–38	1.85	84.3

Unfortunately, there are virtually no data concerning the actual response of VLBW infants to specific protein and/or energy intakes. Thus the validity of the predictions summarized in Table 6 is uncertain. A major factor concerns the VLBW infant's digestive, absorptive, and/or metabolic capacities relative to those of larger, more mature infants. In an effort to understand the impact of potentially confounding variables such as birthweight, gestational age, size for gestational age, and so on, on growth and nutrient retention incident to specific nutrient intakes, we have performed a multiple regression analysis of the data summarized in Figs. 1 to 6 plus additional data (18). The results of this analysis suggest that these potentially confounding variables, including birthweight, are less important determinants of growth and nutrient retention than are the actual nutrient intakes. However, this possibility remains to be proven for all infants, particularly for VLBW infants, data from whom comprised a small percentage of the total data base subjected to regression analysis.

THE SMALL FOR GESTATIONAL AGE INFANT

In contrast to the situation with the VLBW infant, only a few of whom survived until very recently, considerable data are available concerning growth of SGA *vs* appropriate for gestational age (AGA) infants. Although both greater and lower rates of weight gain have been observed in SGA *vs* AGA infants, most studies report greater rates of weight gain as well as greater rates of increase in length and head circumference in SGA *vs* AGA infants (19). The greater growth rates of SGA infants appear to be related to greater intakes (i.e., infants who are fed *ad libitum* consume much more milk per unit of weight than do AGA infants). Unfortunately, there are few published studies in which intakes were controlled carefully (i.e., in which intakes

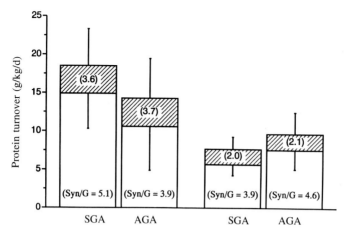

FIG. 9. Rates of endogenous protein synthesis and protein breakdown observed in SGA and AGA infants by different investigative groups. Data of Pencharz et al. (23) are shown in the left set of bars; data of Cauderay et al. (21) are shown in the right set. The height of each bar depicts endogenous protein synthesis and the height of the clear portion depicts endogenous breakdown. The hatched area (i.e., synthesis *minus* breakdown) depicts protein gain. Syn/G is the ratio of endogenous synthesis to protein gain, a measure of the efficiency of protein utilization (21).

of SGA and AGA infants were identical or very nearly so). Thus, whether the nature of growth in SGA *vs* AGA infants differs is not clear.

Interestingly, virtually all studies show that the energy expenditure of SGA infants is greater than that of AGA infants (20–22). This has been attributed to the fact that the brain of these infants, the energy expenditure of which is relatively high compared to that of other organs, comprises a larger percentage of body mass and hence contributes proportionally more to total energy expenditure (22). In contrast to the relative constancy of existing data concerning energy expenditure of SGA *vs* AGA infants, data concerning protein requirements or the efficiency of protein utilization are often conflicting. Two such conflicting sets of data are summarized in Fig. 9. One suggests that SGA infants utilize protein more efficiently than AGA infants (20); the other suggests that SGA infants utilize protein less efficiently (22).

We interpret the bulk of the data concerning the protein needs of SGA *vs* AGA infants as indicating that the two types of infants probably respond similarly to similar intakes. In other words, the rates of weight gain and nitrogen retention of AGA *vs* SGA infants fed the same intakes, on average, are likely to be similar. However, the SGA infant's nutrient needs for catch-up growth are much greater than those of the AGA infant. Whether these infants can metabolize the apparently greater needs more efficiently than AGA infants remains to be determined.

REFERENCES

1. Kashyap S, Forsyth M, Zucker C, Ramakrishnan R, Dell RB, Heird WC. Effects of varying protein and energy intakes on growth and metabolic response in low birth weight infants. *J Pediatr* 1986; 108: 955–63.

2. Kashyap S, Schulze KF, Forsyth M, et al. Growth, nutrient retention, and metabolic response in low birth weight infants fed varying intakes of protein and energy. *J Pediatr* 1988; 113: 713–21.
3. Kashyap S, Schulze KF, Forsyth M, Dell RB, Ramakrishnan R, Heird WC. Growth, nutrient retention, and metabolic response of low-birth-weight infants fed supplemented and unsupplemented preterm human milk. *Am J Clin Nutr* 1990; 52: 254–62.
4. Kashyap S, Schulze KF, Ramakrishnan R, Dell RB, Heird WC. Evaluation of protein and energy intakes predicted by a mathematical model to produce specific rates and composition of weight gain in low birth weight (LBW) infants. *Pediatr Res* (in press).
5. Rivera A, Gell EF, Stegink LD, Ziegler EE. Plasma amino acid profiles during the first three days of life in infants with respiratory distress syndrome: effect of parenteral amino acid supplementation. *J Pediatr* 1989; 115: 465–8.
6. Saini J, MacMahon P, Morgan JB, Kovar IZ. Early parenteral feeding of amino acids. *Arch Dis Child* 1989; 64: 1362–6.
7. Mitton SG, Garlick PJ. Changes in protein turnover after the introduction of parenteral nutrition in premature infants: comparison of breast milk and egg protein-based amino acid solutions. *Pediatr Res* 1992; 32: 447–54.
8. Anderson TL, Muttart C, Bieber MA, Nicholson JF, Heird WC. A controlled trial of glucose vs. glucose and amino acids in premature infants. *J Pediatr* 1979; 94: 947–51.
9. Pittard WB, Geddes KM, Picone TA. Cord blood amino acid concentrations from neonates of 23–41 weeks gestational age. *J Parenter Enteral Nutr* 1988; 12: 167–9.
10. Järvenpää A-L, Rassin DK, Räihä NCR, et al. Milk protein quantity and quality in the term infant. II. Effects on acidic and neutral amino acids. *Pediatrics* 1982; 70: 221–30.
11. Kashyap S, Okamoto E, Kanaya S, et al. Protein quality in feeding low birth weight infants: a comparison of whey-predominant versus casein-predominant formulas. *Pediatrics* 1987; 79: 748–55.
12. Van Der Wagen A, Okken A, Zweens J, Zulstra WG. Composition of postnatal weight loss and subsequent weight gain in small for dates infants. *Acta Paediatr Scand* 1985; 74: 57–61.
13. Forbes GB. A note on the mathematics of "catch-up" growth. *Pediatr Res* 1974; 8: 929–31.
14. Heird WC, Kashyap S. Nutrition, growth and body composition. In: Salle BL, Swyer PR, eds. *Nutrition of the low birth weight infant*, Vol 32. New York: Raven Press (in press), 1993; 169–82.
15. Hay WW. Nutritional requirements of the extremely-low-birth-weight infant. In: Hay WW, ed. *Neonatal nutrition and metabolism*. St. Louis: CV Mosby, 1991; 361–91.
16. Ziegler EE, O'Donnell AM, Nelson SE, Fomon SJ. Body composition of the reference fetus. *Growth* 1976; 40: 329–41.
17. Heird WC, Kashyap S, Schulze KF, Ramakrishnan R, Zucker C, Dell RB. Nutrient utilization and growth. In: Goldman A, Atkinson SA, Hanson LA, eds. *Human lactation 3: The effects of human milk on the recipient infant*. New York: Plenum Press, 1987; 9–21.
18. Heird WC, Ramakrishnan R, Kashyap S, Schulze KF, Dell RB. Effect of a variety of confounding variables on growth and metabolic response of low birth weight infants fed a variety of protein and energy intakes. Unpublished data.
19. Davies DP. Growth of "small-for dates" babies. *Early Hum Dev* 1981; 5: 95–105.
20. Chessex P, Reichman B, Verellen G, et al. Metabolic consequences of intrauterine growth retardation in very low birthweight infants. *Pediatr Res* 1984; 18: 709–13.
21. Cauderay M, Schutz Y, Mitchell J-I, Calame A, Jequier E. Energy–nitrogen balances and protein turnover in small and appropriate for gestational age low birthweight infants. *Eur J Clin Nutr* 1988; 42: 125–36.
22. Sinclair JC. Heat production and thermoregulation in the small-for-date infant. *Pediatr Clin N Am* 1970; 17: 147–158.
23. Pencharz PB, Masson M, Desgranges F, Papageorgiou A. Total-body protein turnover in human premature neonates: effects of birth weight, intra-uterine nutritional status and diet. *Clin Sci* 1981; 61: 207–15.

DISCUSSION FOLLOWING THE PRESENTATION OF DR. KASHYAP

Dr. Uauy: You have showed us the potential for predicting fat stores and protein accretion. I know that you have attempted to test this. Do you get the expected result or do you find, as in many studies, that sometimes babies acquire extra fat rather than the expected protein?

Dr. Kashyap: We have tested this. We predicted protein and energy intakes for different rates of weight gain with the composition of weight gain similar to that of *in utero* protein to fat store ratio. One of the predicted rates of weight gain was to achieve partial catch-up growth. The predicted intake was 3.4 g of protein and 100 kcal. The intake was well tolerated; the weight gain and the composition of weight gain were as predicted. For higher rates of catch-up growth we had predictions of 4.5 g of protein and 120 kcal. In this group the fat stored was also at the predicted level, as was the weight gain, but the protein stored was less than predicted. I think this was because these babies were not able to utilize the nitrogen intake at the given energy intake. Their energy intake somehow became limiting for them. Increasing energy intake would result in increased fat stores. We felt that to achieve better nitrogen retention either the quality of the protein or the quality of the energy had to be changed.

Dr. Marini: Did you evaluate urinary potassium excretion and net renal acid excretion? We did a study on preterm babies of average weight 1800 g on a very low intake of protein, no more than 2 g/kg/d, with two different levels of potassium. We found better growth on the higher potassium formula and also a reduced net renal acid excretion. To make a motoring analogy, these babies worked like diesel cars! They produced more work, say 20% more, for the same amount of energy.

Dr. Kashyap: We did measure potassium excretion. These babies were all in positive potassium balance and were receiving at least the amount of potassium required to maintain potassium accretion at the intrauterine rate. This is how the formula was designed. We have extensive data on net acid excretion, and this was measured in most of our studies. We also followed the acid-base balance status of these babies every week. Not a single baby became acidotic on the intakes we used.

Dr. Marini: We didn't find any acidotic babies, but we found marked changes in net renal acid excretion. Work is required to produce an increase in net renal acid excretion and this work will be at the expense of growth.

Dr. Kashyap: Our babies were gaining weight, and they were all in positive net base balance. I think a large contribution to net base came from the diet itself. The undetermined anion fraction of most preterm formulas is quite high because of the large amounts of citrate that are added with the calcium. This gets converted to bicarbonate and balances out the increased acid production resulting from the high protein and high calcium intake (calcium releases hydrogen ions when it is deposited in bone).

Dr. Rothberg: How do you recommend that one estimate body composition?

Dr. Kashyap: There is no easy way to estimate body composition in LBW infants. It can be estimated by total body K_{40}. Our composition of weight gain measurements are based on determinations of nitrogen balance and energy balance (determined from energy intake minus energy expenditure and fecal and urine energy losses), and from these we calculate protein and fat stored. We also performed dynamic skinfold measurements and observed that with higher energy intake the rate of increase of skinfold thickness was greater.

Dr. Räihä: You studied infants receiving both breast milk and formulas with various protein concentrations. In the curves you showed relating protein intake to growth parameters, you did not differentiate between the babies who got human milk and those who received protein-fortified human milk or formula. We did a similar study with infants fed exclusively on human milk protein, and compared with your data we found that they achieved the same weight gain with less protein. Have you looked at this?

Dr. Kashyap: Infants who received only human milk had a protein intake of approximately 2.5 g/kg/d and energy intake of 130 kcal/kg/d. They gained weight at the intrauterine rate, but their protein nutritional status as determined by plasma transthyretrin was lower than the accepted range and their nitrogen retention was below the intrauterine rate. These babies had a significantly lower ratio of protein stored to fat stored compared to other diets that we studied.

Dr. Räihä: If you gave your babies exclusively human milk protein at rates of 3 g or 3.2 g/kg/d and compared them with formula with the same protein content, do you think they would grow better?

Dr. Kashyap: At the same protein and energy intakes the growth rates were similar.

Dr. Räihä: So you don't think there is any advantage in giving human milk?

Dr. Kashyap: Nutritionally, from the point of view of nitrogen retention and weight gain, our babies did not behave differently.

Dr. Räihä: I would like to stress that I am convinced, as I am sure are many others, that there are definite advantages of feeding very low birthweight and extremely low birthweight infants with human milk, supplemented of course with protein and other nutrients in order to give them an adequate intakes. I think human milk is very important, especially for the very low birthweight infant. We have recently done a survey of all infants born in Sweden below 1000 g, all of whom are fed human milk, and the frequency of necrotizing enterocolitis was under 1%.

Dr. Kashyap: By saying that there is no advantage nutritionally for growth, I am not saying that human milk is not good. Most neonatologists encourage mothers to provide breast milk for their infants because we know that it has so many advantages. Human milk is a living active fluid with all sorts of immunological and developmental advantages and I don't think we shall ever be able to replace it. The question really is what should breast milk be supplemented with, and by how much.

Dr. Marini: In our hospital, as in many others, we have numerous high risk pregnancies with mothers taking substantial amounts of drugs. Also, we are unable to screen the mothers for HIV. So how can we have fresh human milk? When we store human milk we must sterilize it, which damages it. It is a mystery to me how can manage to collect such large amounts of human milk in Scandinavia.

Dr. Pettifor: When you talk about growth rates, in fact what you are looking at is weight gain. In our studies a large number of the babies have been small for gestational age. By increasing protein and energy intake we certainly got an increase in weight gain, but when we looked at linear measurements we did not show any difference in growth. I think one has to be very careful about what one means when one talks about growth rates. Weight may be put on, but is there proportional growth in length?

Dr. Kashyap: The early study by Babson in LBW infants (1) showed that there was a relationship between protein intake and growth in length. Our first study showed a definite effect of protein intake on increase in length. It is quite true that investigators and caretakers must remember that linear measurements are an important part of anthropometry that shouldn't be forgotten. Protein intake is very important for linear growth. If LBW infants are fed high energy diets with inadequate protein intake, the result will be short fat babies.

Dr. Rey: But I would like to return to fat because I think this is the main question in your work. I am not sure that the protein/fat ratio remains constant during growth in your babies. I am not sure that nitrogen and energy balances provide reliable enough data on longitudinal changes in body fat content. It is well known since Widdowson's work that the protein increment is a linear function of time in fetuses but fat increases exponentially. It is not satisfactory to estimate fat increase by dividing the weight by the weight of the baby. If the fat increase is different from the protein increase, if it is greater, or if the ratio is different, you cannot be sure about your data, especially in small for gestational age babies, because these babies usually have a lower fat content. You should express your data not according to body weight but according to fat-free mass. If you divide by the fat-free mass, you will probably find no difference between small for gestational age infants and appropriate for gestational age infants.

Dr. Kashyap: We calculated the protein/fat stored ratio from 30 weeks' gestation to 38 weeks' gestation, and each week that ratio will differ because the fat accretion increases and the protein accretion declines with gestational age. We took what the accretion would be on average and our predictions were based on that.

Dr. Rey: You conclude that the minimum protein requirement is 2.8 g/kg/d. What do you mean by "minimum protein requirement"? Is this the lowest value of the group? The highest value? The median value?

Dr. Kashyap: The smallest value that will achieve intrauterine rates of weight gain and nitrogen retention and maintain the plasma indices of protein nutritional status was 2.8 g/kg/d.

Dr. Räihä: I agree with Dr. Kashyap on this. From our study where we used exclusively human milk protein our conclusion was exactly the same. To achieve intrauterine growth rate we needed a minimum of 2.8 g protein/kg/d.

Dr. Uauy: When you are saying 2.8 g/kg, you are not really saying that this is the minimum but that this is the mean for the group. Dr. Rey is correct to query this.

Dr. Räihä: When you draw a curve relating weight gain to protein intake, and you increase protein intake from 2 g up to 4 g/kg/d, you find that weight gain increases as you increase protein intake up to a level of 2.8 g/kg. Above this you reach a plateau. We therefore take this to be the minimum protein intake.

Dr. Rothberg: Thus all infants fed less than 2.8 g/kg fail to achieve intrauterine growth rates?

Dr. Rey: The point where the slope changes is a statistic point, so it represents the mean of the group; 2.8 is not the minimum value of the group nor the highest value of the group. We usually consider the minimum requirement to be the highest value of a group, not the mean. Then what does 2.8 g/kg mean for a baby of 1 kg or less, and what does it mean for a baby of 1.7 kg? For me these are completely different situations. You are mixing different types of problems when you conclude that the minimum intake is 2.8 g/kg. It is my conviction that you never divide the protein intake by the weight of the baby.

Dr. Axelsson: Did you heat the human milk before you gave it to the babies?

Dr. Kashyap: No, neither mother's own milk nor mother's milk with supplement was heat treated.

Dr. Cooper: You mentioned that you did balance studies. What was the variation in stool losses between babies?

Dr. Kashyap: The stool losses were in the range 10–12 kcal/d with a standard deviation in the range of about 10–20%.

Dr. Marini: What was the percentage of MCT in your formula?

Dr. Kashyap: All the formulas that we studied had 40–50% of fat as MCT.

Dr. Pandit: I have followed over 500 newborns of less than 2 kg birthweight, and the study is still going on. Most of them were small for gestational age. I have no metabolic studies such as you have presented, but I have basic anthropometry and protein intake measurements on these children. I would like to comment on the figure of 2.8 g protein/kg/d. Most of our babies were either on human milk or on a formula, depending on the mother's lactation at the start. Ninety percent of the babies have not caught up (the oldest are now 9 years of age) in spite of receiving the best intakes and the best support we could give. The protein intake ranged from 2.8 g to as high as 4.5 g/kg/d. The energy intake was as high as 200 kcal/kg. I still don't know what to recommend for these small for gestation infants. Such infants constitute 2.5% of all the infants born in our country. What should their protein requirement be?

Dr. Kashyap: I don't think that protein requirements for the small for gestational age infant have been worked out completely. We haven't been able to come up with any recommendations because the requirements for catch-up are definitely much greater than the requirements for matching intrauterine growth rate.

Dr. Cooper: We have also been involved in a number of follow-up studies. In a study using a special formula with high nutrient density we found a rapid equalization of weight gain once the babies had graduated form the formula, such that by about 2–3 months thereafter one could not distinguish which babies had been fed on the special formula and which on standard formula. I would agree that whatever you do with small-for-dates babies in hospital they seem to go back to some sort of preprogrammed size and end up very small later.

Dr. Kashyap: Do you think that if you continued with the special formula the outcome would have been different?

Dr. Cooper: I think one might prolong that rate of weight gain perhaps for another few weeks or even a month or two, but I think that as soon as you stop it, the infants will soon return to the lower growth trajectory. I have seen no long-term advantages in terms of growth in the babies I have studied.

Dr. Rothberg: We are talking about rather short-term benefits in looking at weight gain, but I am quite impressed by the UK data showing that the incidence of diabetes, for example, is high in small for gestational age babies (2). There is concern about what may be the effects on organs such as the pancreas of attempting to obtain good weight gain. We may achieve external growth at the expense of optimal organ growth and development.

Dr. Uauy: As we look at the published work we find that there are studies of groups of subjects and we come up with ranges of values, with individual babies at different gestational ages and progressing at different rates. However, we tend to use formulas or feeds of fixed composition. I think we need to rescue the concept of adjusting the intake individually to the baby's needs. Do you have any suggestions on how can we do that? Some babies will have greater tolerance and will have lower requirements; other babies will have lower tolerance and higher requirements. How do we deal with that on a day-to-day basis?

Dr. Kashyap: I don't know how you would deal with it on a day-to-day basis unless you have analysis facilities, especially if you are dealing with human milk. So far as formulas are concerned, are you suggesting that there should be a separate formula for infants in different birthweight groups as their requirements may be different? One for infants <1000 g, another for between 1000 and 1250 g, then another for 1250 to 1500 g, another for 1500 to 1800 g, and another for those above 2000 g? I am not sure how you would deal with this in practice. I think it could be sensible to have two separate formulas, one for very low birthweight infants and another for low birthweight infants.

Dr. Räihä: I think Dr. Uauy's comment is very important. We have to individualize protein intake. In our nurseries we measure the protein, energy, and fat content in all mother's milk using readily available techniques. Then we monitor plasma and urinary urea on a day-to-day basis, sometimes even several times a day. In this way we individualize the protein intake. When urea levels are high and the excretion is high, we decrease the protein intake; when they are low we increase the protein intake. This has been published in *Acta Pediatrica Scandinavia* (3).

Dr. Uauy: Somehow we need to adjust our regimes to the needs of individual babies.

Obviously, not all centers will be able to do urinary urea twice a day, but periodic monitoring of growth rate, albumin, and urea nitrogen in serum and urine are simple indicators that can be performed on a daily basis or less frequently.

REFERENCES

1. Babson SG, Bramhall JI. Diet and growth in the premature infant. *J Pediatr* 1969; 74: 890–900.
2. Hales CN, Barker DJP. Non-insulin dependent (type II) diabetes mellitus: thrifty phenotype hypothesis in fetal and infant origins of adult disease. Barker DJP, ed. London: British Medical Journal Publications, 1992; 258–72.
3. Polberger SKT, Axelsson IE, Räihä NCR. Urinary and serum urea as indicators of protein metabolism in very low birthweight infants fed varying human milk protein intakes. *Acta Paediatr Scand* 1990; 79: 737–42.

Protein Requirement of Healthy Term Infants during the First Four Months of Life

Niels C. R. Räihä

Department of Pediatrics, University of Lund, Malmö General Hospital 21401 Malmö, Sweden

APPROACHES USED TO DETERMINE PROTEIN REQUIREMENT

Protein Intake in Growing Breast-Fed Infants

The approach most commonly used has been to estimate the intake of protein by exclusively breast-fed infants who are maintaining satisfactory growth. It is assumed on an evolutionary basis that this intake is likely to be near the requirement. The intakes of protein by male breast-fed male infants during the first 4 months of life as estimated by the Joint FAO/WHO/UNU Expert Consultation (1) are 2.46, 1.93, 1.74, and 1.49 g protein/kg/d during the age intervals from birth to 1, 1–2, 2–3, and 3–4 months. The calculations are based on a nutritionally available protein concentration in human milk of 11.5 g/liter. Fomon (2) has recently published estimations based on human milk intakes reported from two separate studies from birth to 4 months (3,4), and on a decreasing bioavailable protein content of human milk from 14 g/liter during the first month to 8.6 g/liter during the interval from 3 to 4 months (Table 1). It is observed that these intake figures are less than those presented by the FAO/WHO/UNU Expert Committee (1), due mainly to a previous overestimation of the nutritionally available protein content of human milk.

Factorial Method for Estimation of Protein Requirement

The factorial method is another approach often used to estimate the protein requirement in infants. Calculations based on the factorial approach for estimating protein requirement during various ages are derived by adding the quantity of protein (nitrogen × 6.25) used for growth to that needed to replace losses through urine, feces, and the skin. A male reference infant fed on cow's milk formula is used for the calculations (5). In the calculations by Fomon *et al.* (5) the conversion of dietary protein to body protein is assumed to be 90% efficient. The younger the infant, the higher the growth component of the requirement. The growth component is about

TABLE 1. *Estimated protein intake of breast-fed infants*

Age (months)	Milk consumption (ml/d)	Protein intake (g/kg/d)
0–1	630	2.09
1–2	773	1.59
2–3	787	1.18
3–4	810	1.06

From Fomon S. *Pediatr Res* 1991; 30: 391–395.

two-thirds of the total requirement in the 1- to 2-month-old infant and decreases to one-third in a 12-month-old infant. The requirement for growth is estimated from the daily rate of weight gain and the nitrogen concentration in the body. The daily increments in body protein by the male reference infant has been provided by Fomon *et al.* (2) and are presented in Table 2. The protein accretion for a female reference infant is on average 9% less than that estimated for the male infant during the first 2 months of life.

The Joint FAO/WHO/UNU Expert Committee (1) has also estimated the protein requirement using the factorial method (Table 3). These estimations are somewhat

TABLE 2. *Protein requirement estimated by the factorial approach*[a]

	Protein (g/kg/d)		
Age (months)	Growth	Losses	Requirement
0–1	1.03	0.95	1.98
1–2	0.78	0.93	1.71
2–3	0.56	0.90	1.46
3–4	0.38	0.89	1.27

From Fomon S. *Pediatr Res* 1991; 30: 391–395.
[a] Assuming 90% efficiency in converting dietary protein.

TABLE 3. *Average protein requirement and safe level of intake calculated by the factorial method by FAO/WHO/UNU (1)*

	Protein (g/kg/d)	
Age (months)	Requirement[a]	Safe level[b]
1–2	2.25	
2–3	1.82	
3–4	1.47	1.86

[a] Efficiency of protein utilization assumed as 70%; 50% added to theoretical body protein increment.
[b] 35% added to requirement for variability in body protein increment.

higher than those reported by Fomon *et al.* (5). This is explained by two differences in the calculations of the two authorities. The FAO/WHO/UNU Committee used an efficiency factor of 70% instead of 90% for the conversion of dietary protein to body protein, and assumed a 50% increase in the body increment for protein used for growth.

Beaton & Chery (6) have also estimated protein requirement in infants of 3–4 months using the factorial method, and from the analysis presented in their paper they conclude that the FAO/WHO/UNU (1) estimation for this age interval of 1.47 g/kg/d is an overestimation and that the actual mean protein requirement of breast-fed infants aged 3–4 months is on the order of 1.1 g/kg/d or even slightly lower. They also suggest that a formula providing 17 g of protein per 100 kcal is adequate for this age group.

Other Approaches to Measuring Protein Requirements

The plasma amino acid pattern reflects dietary protein intake as well as concurrent protein synthesis, protein breakdown, and endogenous amino acid catabolism, synthesis, and excretion (7). Thus plasma amino acid profiles give some indication of the balance between intake and utilization and can be used to estimate excessive or inadequate intakes and hence the requirements. As a complement to plasma amino acid concentrations, serum urea concentration can be used (8). Another approach to the estimation of protein requirement is the use of stable isotope techniques (9).

RDI of Protein or "Safe Level of Protein Intake"

When suggesting recommended dietary intakes (RDI) for protein in healthy infants the requirement for growth and losses is increased by a factor of between 35% and 10%, due to variability in the body protein increment. FAO/WHO/UNU (1), using 35%, has suggested a "safe level of protein intake" of 1.86 g/kg/d for infants between 3 and 5 months (Table 3). Fomon (2) has published somewhat lower RDI values, as seen in Table 4, using a 26% supplement to allow for variability in infants during the first 2 months and 10% for infants older than 2 months.

TABLE 4. *Recommended dietary intake (RDI) for protein*

Age (months)	RDI	
	g/kg/d	g/100 kcal
0–1	2.6	2.2
1–2	2.2	2.0
2–3	1.8	1.8
3–4	1.5	1.6

From Fomon S. *Pediatr Res* 1991; 30: 391–395.

RECOMMENDATIONS FOR PROTEIN CONTENT IN INFANT FORMULAS

The European Society for Pediatric Gastroenterology and Nutrition (ESPGAN) (10) proposes a protein content in standard infant formulas between 1.8 and 2.8 g per 100 kcal, whereas the Committee on Nutrition of the American Academy of Pediatrics (11) gives a wider range from 1.8 to 4.5 g per 100 kcal and the US Food and Drug Administration (FDA) gives the same limits as the American Academy (12). The upper limit of 4.5 g/per 100 kcal has been criticized by Young & Pelletier (13) and Ziegler & Fomon (14) as being too high. They suggest that it should be decreased to 3.2–3.5 g per 100 kcal. Fomon (2) has recommended a minimum concentration of protein in infant formula of 2.2 g per 100 kcal for infants less than 3 months of age and a content of 1.6 g per 100 kcal for infants over 3 months. This is very similar to the recommendation of Beaton & Chery (6) of 1.7 g per 100 kcal for infants aged 3–4 months. The average intake of protein at the age of 3–4 months from such formula would be about 1.75 g/kg/d.

CLINICAL STUDIES WITH DIFFERENT PROTEIN LEVELS IN FORMULA

Studies with Reduced Protein Formulas

Most standard infant formulas marketed in Europe and in the USA are based on bovine milk and have a total protein content of 2.2 g per 100 kcal or 1.5 g/dl in a 67 kcal/dl formula. This is much higher than in mature human milk which has a nutritionally available protein content of only about 8.5 g/dl (see Chapter 9), and thus the intake of protein in infants on a standard infant formula is much higher than that of breast-fed infants. If breast-fed healthy term infants are considered to be the gold standard during the first 4 months of life, then the intake of protein is too high in infants fed a standard 2.2 g per 100 kcal formula or one containing more protein (15).

Räihä and co-workers (16,17) published a study in 1986 in which normal term infants were fed human milk, a standard formula, or a formula in which the total protein was reduced to 1.8 g per 100 kcal (1.2 g/dl). Growth rates were normal in all groups from birth to 3 months of age. There was a statistically significant increase in the growth rate from 2 to 12 weeks in the infants receiving the standard formula compared to the breast-fed infants: weight gain 250 g/week *vs* 203 g/week, and crown-rump length 0.67 cm/week *vs* 0.57 cm/week, respectively. This increased growth rate may be mediated by protein/amino acid-induced insulin secretion (see the chapter by Axelsson). Protein intake was significantly higher in the standard formula group than in the breast-fed and reduced protein formula groups (Fig. 1). Blood urea nitrogen and urine total nitrogen were also significantly higher in the standard formula group than in the breast-fed and reduced protein formula groups (Fig. 2).

Plasma amino acid concentrations also indicated that there was a high protein intake in the standard formula group compared to the breast-fed group. Most plasma amino acid concentrations were closer to the breast-fed group in the reduced-protein

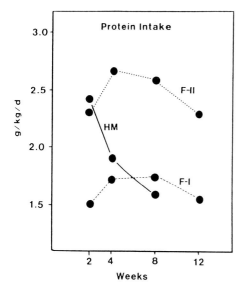

FIG. 1. Protein intake in infants fed human milk (HM), formula containing 1.8 g of protein per 100 kcal (F-1) or standard formula containing 2.2 g of protein per 100 kcal (F-II). [From Räihä N, et al. Acta Paediatr Scand 1986; 75: 881–886.]

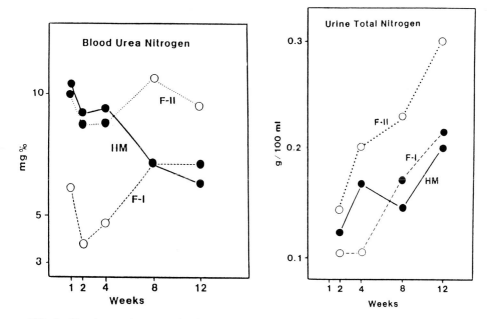

FIG. 2. Blood urea nitrogen and urine total nitrogen in infants fed human milk (HM), formula containing 1.8 g of protein per 100 kcal (F-I) or standard formula containing 2.2 g of protein per 100 kcal (F-II). (From Räihä N, et al. Acta Paediatr Scand 1986; 75: 881–886.]

formula group, however. The results suggested that a protein content of 1.8 g per 100 kcal is adequate for feeding healthy term infants and that growth and indices of nitrogen metabolism when such a formula is fed are more similar to those found in infants fed human milk. However, the non-protein nitrogen content of such "low" protein formulas must be considered as well.

Subsequently, these findings were confirmed by Janas *et al.* (18), Picone *et al.* (19), and Lönnerdal & Chen (20). All studies indicated that standard formulas containing 2.2 g of protein per 100 kcal produced indices of nitrogen metabolism suggestive of excessive protein intake and that formulas with 1.8 g per 100 kcal, or even as low as 1.6 per 100 kcal (19), resulted in many indices of protein metabolism and growth more characteristic of human milk feeding.

The Swedish Experience

Since 1986 only two kinds of standard infant formulas have been marketed in Sweden: one formula for infants from birth to 2 months with a total protein content of 2.2 per 100 kcal (1.6 g/dl) and another from 2 months on, with a total protein content of 1.8 per 100 kcal (1.2 g/dl). Thus, although the frequency of breast feeding is high in Sweden, several thousands of infants have been fed exclusively on formula with a protein content at the minimum level of international recommendations from 2 to 4 months of age without any problems.

CONCLUSIONS

1. Exclusive breast feeding by a healthy mother should be the feeding standard for the first 4 months of life in normal term infants. This provides protein intakes from 2.09 g/kg/d during the first month to 1.06 g/kg/d during the fourth month.
2. During the first 2 months of life a standard formula should have a minimum milk protein content of 2.2 g per 100 kcal or 1.5 g/dl for a 67 kcal/dl formula.
3. After the second month of life a protein content of 1.8 g per 100 kcal or 1.2 g/dl in a 67 kcal/dl standard formula is adequate. These protein concentrations in standard formulas will provide approximately 25% more nutritionally available protein to the infant than breast feeding at the respective age intervals.
4. Clinical studies in term infants show that formulas with more protein will not give any advantages but will stress the metabolic and excretory functions of the infant. Excessive protein intakes during critical periods of development may have harmful long-term effects on the infant.

REFERENCES

1. Joint FAO/WHO/UNU Expert Consultation. *Energy and protein requirements.* WHO Technical Report Series No 724. Geneva: WHO, 1985.

2. Fomon S. Requirements and recommended dietary intakes of protein in infancy. *Pediatr Res* 1991; 30: 391–5.
3. Neville MC, Keller R, Seacat J, et al. Studies in human lactation: milk volumes in lactating women during the onset of lactation and full lactation. *Am J Clin Nutr* 1988; 48: 1375–86.
4. Whitehead RG, Paul AA. Infant growth and human milk requirements: a fresh approach. *Lancet* 1981; ii: 161–3.
5. Fomon SJ, Haschke F, Ziegler EE, Nelson SE. Body composition of reference children from birth to age ten years. *Am J Clin Nutr* 1982; 35: 1169–75.
6. Beaton GH, Chery A. Protein requirements of infants: a reexamination of concepts and approaches. *Am J Clin Nutr* 1988; 48: 1403–12.
7. Heird WC. Interpretation of the plasma amino acid pattern in low birth weight infants. *Nutrition* 1989; 5: 145–6.
8. Taylor LS, Scrimshaw NS, Young VR. The relationship between serum urea levels and dietary nitrogen utilization in young men. *Br J Nutr* 1974; 32: 407–11.
9. Young VR. Tracer studies of amino acid kinetics: a basis for improving nutritional therapy. In: Tanaka T, Okada A, eds. *Nutritional support in organ failure*. Amsterdam: Excerpta Medica, 1990; 3–34.
10. ESPGAN Committee on Nutrition. Guidelines on infant nutrition I. Recommendations for the composition of an adapted formula. *Acta Paediatr Scand* (suppl) 1977; 262: 3–20.
11. American Academy of Pediatrics Committee on Nutrition. Commentary on breast-feeding and infant formulas, including proposed standards for formulas. *Pediatrics* 1976; 57: 278–85.
12. Food and Drug Administration Rules and Regulations. Nutrient requirements for infant formulas (21 CFR, Part 107). *Fed Reg* 1985; 50: 45106–8.
13. Young VR, Pelletier VA. Adaptation to high protein intakes, with particular reference to formula feeding and the healthy, term infant. *J Nutr* 1989; 119: 1799–809.
14. Ziegler EE, Fomon SJ. Potential renal solute load of infant formulas. *J Nutr* 1989; 119: 1785–8.
15. Räihä NCR. Milk protein quantity and quality and protein requirements during development. *Adv Pediatr* 1989; 36: 347–68.
16. Räihä N, Minoli I, Moro G. Milk protein intake in term infants. I. Metabolic responses and effects on growth. *Acta Paediatr Scand* 1986; 75: 881–6.
17. Räihä N, Minoli I, Moro G, Bremer HJ. Milk protein intake in term infants. II. Effects on plasma amino acids concentrations. *Acta Paediatr Scand* 1986; 75: 887–92.
18. Janas LM, Picciano MF, Hatch TF. Indices of protein metabolism in term infants fed either human milk or formulas with reduced protein concentration and various whey/casein ratios. *J Pediatr* 1987; 110: 838–48.
19. Picone TA, Benson JD, Moro G, et al. Growth, serum biochemistries, and amino acids in term infants fed formulas with amino acid and protein concentrations similar to human milk. *J Pediatr Gastroenterol Nutr* 1989; 9: 351–60.
20. Lönnerdal B, Chen CL. Effects of formula protein level and ratio on infant growth, plasma amino acids and serum trace elements. *Acta Paediatr Scand* 1990; 79: 257–65.

DISCUSSION FOLLOWING THE PRESENTATION OF DR. RÄIHÄ

Dr. Lönnerdal: You clearly showed that several metabolic indices in the infants fed on 10.9 g of protein per liter were below those of breast-fed infants; in that they can certainly be judged as inadequate. In the second study you fed either 13 or 15 g/liter, or a combination of these, and you showed growth data and metabolic indices. What metabolic indices in the infants fed 13 g/liter protein were lower than in the breast-fed infants? Why did you conclude that a diet of 15 g/liter protein was preferable? In the data you showed I did not see anything to suggest that 1.3 g/liter was inadequate. Are you aiming to include a safety margin?

Dr. Räihä: Yes. I am sure that it would be quite adequate to feed a baby on a 1.3 g protein formula from birth, and some of your studies have shown that this is adequate. But there are two reasons why we suggest a higher intake during the first 2 months. One is the theoretical estimations, which clearly show that there is a higher requirement during the first months. The second reason is that in the first study we did with the very low protein formula, which was low because of the high non-protein nitrogen fraction, there were some nutritional indices that were lower than in breast-fed infants. This was why this recommendation was made—we wanted to be quite safe.

Dr. Lönnerdal: You have used two different approaches. One is an arithmetic one—the factorial approach—and the other is the practical one, where you feed a group of infants and look at the results, as you have been doing. In practise, whatever the theoretical considerations, your results show that the protein intake was adequate, though in both types of approach one can always raise the issue of safety margins. However, I have doubts about the factorial approach. Often what we consider to be scientific truth is what we have heard repeated so many times that we tend to believe it. I have a feeling that this may be true of some of Fomon's data. The numbers of infants were relatively small. How sure can we be about their accuracy?

Dr. Uauy: The Fomon data were obtained longitudinally from a cohort of infants who were measured weekly over the whole first year of life. These are probably the only longitudinal data with weekly measurements. There were between 300 and 500 children, so the sample was not that small, and for some of those infants nitrogen balance data were obtained. Thus it is a very coherent data set. It may not be the largest growth data set but it has the great advantage of being longitudinal.

Dr. Lönnerdal: I have no problems over the growth data. I just wonder about the protein and body composition data. They certainly weren't directly determined, I am sure.

Dr. Axelsson: Fomon didn't measure total nitrogen concentration in the body; he measured total body potassium and total body water and he used a regression relationship between total body potassium and total nitrogen. These are the only data that the body composition values are based on.

Dr. Räihä: I think we need new growth curves which differentiate between breast-fed and formula-fed infants, and perhaps then we shall need to make new theoretical calculations.

Dr. Rey: I would like to explain my ideas about the factorial method because we differ from Fomon and the WHO/FAO on the requirements for growth and maintenance. For inevitable losses Fomon has used a value of 0.9 g/kg body weight, but he overestimated the losses. They should have been measured without any protein intake, and this was not the case. Others have used a value of 0.7 g/kg, and this is probably the best value of the maintenance requirement.

For growth, Fomon and WHO have used a value of 12% protein in the weight increment, according to their work on body composition of reference children from birth to age 10 years (1). My feeling is that it is wrong; 12% is the value at birth. In his paper Fomon calculated that the protein content of the baby between birth and 1 year remains constant at 12%. But in another paper published 25 years ago (2) the value changed from 12% to 15%, and expressed in relation to fat-free mass the protein content was 19% at 1 year of age. You cannot explain the biochemical maturation of the baby and the increase in protein in the baby's body if the protein concentration in the weight increment is 12%. It should be 16% or 18%. If it is less than this, the protein content of the body should not increase. Fomon used a coefficient of efficiency of 90% with 12% protein in new tissue, so he has underestimated the growth requirement. FAO/WHO start from the same value but they multiply it by 1.5 to take into account the variability from day to day (although I don't know precisely what they mean by this) and they use a coefficient of efficiency of 70%. My feeling is that they have overestimated the growth requirement substantially, while in the Fomon calculation there is probably an underestimation. I fully agree with Niels Räihä that we cannot use the FAO value as a reliable estimate of requirements. I have one question. Kathryn Dewey *et al.* have published some very interesting data last year (3) showing that the growth in length of breast-fed babies is not different from that of formula-fed infants. The same conclusion were reached by Leena Salmenperä et al. some years ago, at least for infants in the first 6 months of life (4). I am therefore puzzled by your statement that growth is different during the first 4 months of life between breast-fed and formula-fed babies. Are you convinced of this?

Dr. Räihä: I did not say it was statistically different. I said that in our study the weight and length measurements became more different after 3 months of age.

Dr. Uauy: I don't think anybody should consider the values published by WHO/FAO to be the definitive answer. It was the best answer that could be arrived at in 1981 by the group of people that met. The 70% value for utilization came from data in infants closer to 12 months of age. There are no available month-by-month data on infants over the first year, but there was a lot of data from several studies throughout the world at 12 months. The issue of efficiency is complicated by two factors. As you come closer to being deficient in protein you improve efficiency, and as you become replete you decrease efficiency, so the 90% value of Fomon is correct if you feed a low protein diet, and the 70% value of WHO/FAO is correct if you feed a more generous diet. Efficiency can go as low as 50%. So the answer will be somewhat related to the model you have used to test your hypothesis; obviously, you can even improve over 90%.

The concept of the variability of growth is, I think, still valid because a baby will not grow in the same way every day. You can derive a mathematical average of weight gain per day; however, on one day the baby may grow at the average rate, while on the next day or so he may not grow at all, and then he may grow at twice the average rate for a period. The protein required for this growth cannot be stored during periods when growth is below average. That is the concept. It may or may not be right. In fact, I think that the most recent data would suggest that there are indeed growth spurts. There is an article in *Science* suggesting that if you measure babies on a daily basis, as the authors did, with the precision of 0.5 mm, you find that babies are growing 10–15% of the time during the first 24 months (5). So when you come up with recommended values you have to take into account that growth in length and weight gain are not continuous in a way that allows you to use a daily average value.

Dr. Rey: There are so many papers on the growth of babies that it is virtually impossible to determine the truth. You can find anything you want to. When you see differences in weight from day to day, is it not likely to be water? Are you sure it is protein? And I do not believe you can measure the length of a baby from one day to the next and say you have found variation.

Dr. Uauy: The precise measurement of linear growth in babies is a science and requires very special anthropometry. The investigators measured 19 female and 12 male infants during their first 21 months of life and were able to determine daily changes in length with a precision of 0.5 mm and intra- and interobserver variability of measurements of close to 1 mm. As far as weight is concerned, I agree, nobody knows if weight gained is water or whatever, but if you take the average for the month and divide it by 30, you are not going to find that babies gain $\frac{1}{30}$ of the monthly total on a daily basis, or $\frac{1}{365}$ of the yearly total. Thus, allowing that babies grow at variable rates, when they are growing I want to give them as much protein as they will need to gain not $\frac{1}{365}$ of the yearly total but $\frac{1}{365}$ plus 35% or whatever.

Dr. Rassin: The type of feeding certainly affects the growth patterns. How do you determine what the pattern of growth should be? Growth is really not something that happens in the same way day by day. We have growth spurts throughout the growth period. It is a very interesting problem to try to understand these differences.

Dr. Räihä: The fact that growth occurs in spurts may point to another advantage of breast feeding. At times when the baby needs to grow a little more, he just asks for more milk. This would be a most natural thing to do. With formula feeding every 4 or 5 hours you may miss the critical time.

Dr. Rey: I agree that the breast-fed baby takes what he wants and regulates his intake as a function of his requirements, but consider the formula-fed infant. We increase his requirement because on some days he grows and on some days he does not, but the indices of adequacy will be different from day to day. If he has a growth spurt, his indices will be good, and if he has no growth, they will be bad because he received too much protein. So which value will you choose in a formula-fed baby if this value can change from day to day because of variations in growth spurting?

Dr. Garlick: I would have thought that it would make no difference to your assessment of protein requirement whether or not there was spurting growth, simply because spurts will be occurring anyway during your balances, so you will already have included them in the requirements.

Dr. Uauy: The usual nitrogen balance studies are too short term. During nitrogen balance studies of the usual 1 or 2 weeks' duration, you don't expect to find that growth parallels nitrogen accretion.

Dr. Garlick: But some of the children will be growing and some won't be at the time you make the measurement.

Dr. Uauy: Yes, but the measurement is usually based on the last 3 or 4 days of the balance period. In general, we accept that at that time there is growth and that the nitrogen retention value, if there is growth, will be sufficient for growth, but the growth component is not used when you derive your nitrogen balance data.

Dr. Garlick: I missed the point that the measurements were carried out over such a short period of time.

Dr. Marini: Do you have any behavioral data—sleeping habits, heart rate, respiration rate during sleep, and so on—which might show differences between different protein intakes?

Dr. Räihä: No we don't, because these were all healthy babies who were fed at home, so it would have been very difficult to get such data. The reason for having alternative formulas was to accommodate babies who need more protein, after an infection for example. The pediatrician in charge of the infant can then have a choice of giving a higher protein formula to an infant who may need more for catch-up growth. We had no behavioral motives for making a higher protein formula available.

Dr. Marini: Garza studied heart rate in the first month of life in breast-fed and formula-fed babies (I don't know the protein content) (6). He found that in the first month of life breast-fed babies have almost similar heart rates but later their heart rates are lower. Heart rate was correlated significantly with energy intake but not with sleeping metabolic rate or total daily energy expenditure. Heart rate is a simple indirect way to estimate metabolism, although possibly a double product such as heart rate × blood pressure would be better. Also to measure growth in weight and length alone may not be sufficient. We should estimate arm circumference or skinfold in these babies. Another question: Do you have any data on creatinine excretion?

Dr. Räihä: No, we only looked at plasma amino acids, urine urea excretion, nitrogen excretion and the blood urea nitrogen, plus the anthropometric measurements. There is a limit to what you can do in babies who are healthy and at home. All these studies have to be passed by an ethics committee and there are quite strict limits to what you can do.

Dr. Heine: We did some experiments in the 1960s where we fed babies on a low protein diet (1.3% protein) and compared the data with babies fed on 2.1% protein. We found when running long-term nitrogen balances in these babies that the more nitrogen the baby received, the more it retained. We explained this at the time on the basis of chemical maturation of the cells, but when we continued measuring nitrogen balance over months we found that positive balance persisted. Now it is suspected that babies fed on protein-rich formulas have a higher lean body mass, with greater synthesis of muscle and other proteins. Can you speculate on the potential hazards of these differences in body composition?

Dr. Räihä: No, I would not speculate on that. I can only say that high protein intakes will affect hormonal release, insulin secretion, and growth. But I would not like to speculate more.

Dr. Wharton: Although I think the move toward reducing the protein content of formulas is probably correct overall, I am not convinced that there is at present enough evidence to tell formula manufacturers to carry this into practice, although you are obviously convinced in Sweden that there is. Breast milk protein, based on a very large nationally collected sample of milks in the UK, was found to be 13 g/liter. Very careful attention was paid to the methods of collection, to full expression of a breast, to the time of the day when collection was made, and so on, when this study was made.

When assessing such data it is very important to look carefully into the methods used for collection. In so many studies of breast milk composition, the investigators dispose of the method of collection in two lines before giving the biochemical methods in extreme detail. Often one knows very well how the milk was collected—somebody went along with a little tube and asked the mothers to squirt a few drops into it! This is a major source of variability. Then we have the uncertainty about the quantity of non-protein nitrogen. Just on basis of the breast milk model I begin to wonder about the wisdom of taking the protein down so low.

It cannot be claimed that all the biochemical variables remain unchanged on very low protein formula; they do not. When Dr. Räihä uses these low protein formulas, he finds low plasma tryptophan levels, so the infants are not biochemically the same as those fed on higher protein intakes. Whether this matters is of course unknown, but I suggest it is something to be cautious about.

Dr. Rassin: The question of variability is an important one. We spend a lot of time trying to come up with single values for protein content and so on—which everyone likes because then they have a single goal—but in fact there is no such thing as a single value. I would guess that we would be lucky if the variability in protein content of breast milk wasn't at least as much 10–20%, which means there is still a lot of scope for the neonatologist and anybody else to make some real decisions within the normal range of protein intakes. I wonder if we are not doing a disservice by relying so much on single numbers rather than looking at the ranges of protein that are present in normal milk.

Dr. Marini: It is not only the variability of mother's milk but also the variability of the baby that must be taken into account. The potential for growth is not the same on any particular intake, so we cannot necessarily give the same level on intake.

Dr. Rey: In relation to non-protein nitrogen, I agree with your calculations but when we consider "protein" for babies we should remember that the baby doesn't need protein. He needs nitrogen and essential amino acids. It is dangerous to subtract non-protein nitrogen when we discuss the nitrogen requirements of the baby.

Dr. Guesry: I think everyone will be in agreement after this interesting discussion that there is a lot of variation from day to day, from baby to baby, from breast milk to breast milk, so it is impossible to give a single value; we have to take a range. In fact, all the recommendations coming from the EC, from ESPGAN, and from Codex Alimentarius always give a range, not a single figure. Infant food manufacturers, who have to cater for groups of babies of possibly millions, are not permitted to introduce variations and so have to provide for the greater number. This is why I think it would be very dangerous to follow the recommendations of the people who want a decrease in protein to extremely low values. This is all very well when you are doing a study and monitoring the growth of the baby; if anything happens you can react. This is certainly not the case for a manufacturer who has to provide for the general population.

Dr. Cooper: I was going to make a similar point from the perspective of developing countries where one has a much higher incidence of infection, particularly gastrointestinal, and where there is also the danger of overdiluting the feed. To lower the current level of protein in standard formula could be very dangerous in our situation.

Dr. Räihä: I think there has been some misunderstanding here. I don't think anyone wants an extreme decrease in protein. Nor have we in these studies referred at all to developing countries. After the age of 4 months or 6 months, when the baby starts to get supplementary foods, we have to look at total intake of protein and not only what the baby is getting from breast milk, formula, or follow-on formula. This of course is important when we are dealing with developing countries because there the supplementary food may be deficient in protein, and in those places the baby needs a higher protein formula or follow-on formula. When discussing the premature infant we talked about individualization. I think we have to individualize from country to country and even inside one country from one population to another.

REFERENCES

1. Fomon SJ, Haschke F, Ziegler E, Nelson SE. Body composition of reference children from birth to age 10 years. *Am J Clin Nutr* 1982; 35: 1169–75.
2. Fomon SJ. Body composition of the male reference infant during the first year of life. *Pediatrics* 1967; 40: 863–70.
3. Kathryn G, Heinig MJ, Nommsen LA, et al. Growth of breast-fed and formula-fed infants from 0 to 18 months: the Darling Study. *Pediatrics* 1992; 89: 1035–41.
4. Salmenperä L, Perheentupa J, Siimes MA. Exclusively breast-fed healthy infants grow slower than reference infants. *Pediatr Res* 1985; 19: 307–12.
5. Lampl M, Veldhuis JD, Johnson ML. Saltation and stasis: a model for human growth. *Science* 1992; 258: 801–3.
6. Butte NF, O'Brian Smith E, Garza C. Heart rates of breast-fed and formula-fed infants. *J Pediatr Gastroenterol Nutr* 1991; 13: 391–6.

Protein Needs During Weaning

Irene E. M. Axelsson

Department of Pediatrics, University of Lund, Malmö General Hospital, S-21401 Malmö, Sweden

Weaning, the period in which non-milk foods are introduced into the diet, is one of the most critical nutritional events in the life of mammals. It marks the transition from pure milk feeding to more adult nutritional habits. In the rat, metabolic adaptations such as increased enzymatic activity in the liver and kidney follow the increase in protein intake at this time (1,2). The quality and quantity of various dietary proteins available affect neuronal protein synthesis in the brain (3). It has been fairly well established through animal experiments that early diet has a lasting effect on metabolism and development.

Studies in humans are still surrounded by a number of uncertainties. Few studies have addressed the nutritional needs of infants at the time of weaning. Most estimations are extrapolated from data obtained in younger or older children. In this chapter the term "weaning" is defined as the time from the introduction of non-milk foods, regardless of whether the infant is breast-fed or formula-fed, until the time when breast feeding or formula feeding is discontinued.

In this presentation the methods by which the international recommendations for protein intake in infants have been determined will be discussed. I shall also review the existing data on nutrient intakes during the weaning period and the effects of different nutrient intakes on growth in breast-fed and formula-fed infants. Finally, some possible consequences of a high-protein intake during this period of life will be discussed.

PROTEIN REQUIREMENTS AND RECOMMENDATIONS

Different approaches have been taken to estimate protein requirements during infancy and to give recommendations for protein intake. There are certain limitations implicit to these approaches, and I would like to emphasize that there is still a great deal of uncertainty as to what the protein requirement during this period actually is. The most authoritative international recommendations available are contained in the report by the Food and Agriculture Organization/World Health Organization/United Nations University (FAO/WHO/UNU) Committee on Energy and Protein Requirements released in 1985 (4). In this the protein requirement during the first 4 months of life was determined by measuring the protein intake of healthy breast-fed infants maintaining satisfactory growth. Possible reasons why the theoretically calculated protein requirements of formula-fed infants aged 2–5 months are in excess of the estimated protein intake of breast-fed infants are discussed elsewhere (5; and the chapter by Räihä).

The protein requirement after 4 months of age was determined by combining two components, the requirement for maintenance and the requirement for growth, according to the factorial method (4). The requirement for maintenance is based on short-term nitrogen (N) balance studies where energy intake is kept constant at a level assumed to be adequate, and varying amounts of protein are given (6). In this calculation obligatory nitrogen losses from skin, feces, and urine are taken into consideration. During the first year of life 120 mg nitrogen/kg/d is required for maintenance (Table 1) [i.e., 0.75 g protein/kg/d (including an interindividual variability of 12.5%)].

The mean rate of nitrogen accretion during growth can be estimated from the expected daily rate of weight gain and the nitrogen concentration in the body. Nitrogen accretion during growth has been determined by Fomon et al. (7). It is assumed that the growth component is an additional 50% of the theoretical amount of nitrogen laid down per day. Growth does not always proceed at exactly the same rate from day to day and the body has a very limited capacity for storing amino acids in order to utilize them for growth later. Furthermore, the assumption is made that only 70% of the dietary protein is utilized. According to the factorial method the total protein requirement (maintenance and growth) is 1.25 g/kg/d between 6 and 9 months of age and 1.15 g/kg/d between 9 and 12 months (Table 1). A safe intake is defined as the mean protein requirement +2 standard deviations. The estimated safe levels of protein intake for the period of 3–12 months of age as determined by this method are

Table 1. *Average protein requirements of infants calculated by the factorial method*

Age (months)	N increment[a]	N increment × 1.5[b]	Corrected for efficiency[c] (mg/kg/d N)	Maintenance[d]	Total as protein	Fomon milk protein[e] (g/kg/d protein)	Fomon protein from beikost[f]
4–5	44	66	94	120	1.34	1.18	1.26
5–6	41	62	89	120	1.30	1.18	1.26
6–9	37	56	80	120	1.25	1.17	1.24
9–12	30	45	64	120	1.15	1.14	1.20

From WHO. *Technical Report Series No 724*. Geneva: 1985; 11: 64–112.
[a] From ref. 7.
[b] See the text.
[c] Efficiency of utilization assumed to be 70%.
[d] See the text.
[e] Efficiency of utilization assumed to be 90% (5).
[f] Efficiency of utilization assumed to be 70% (5).

Table 2. Safe level of protein intake (milk or egg protein) of infants

Age (months)	Safe level protein (g/kg/d)[a]	RDI Fomon protein (g/kg/d)[b]
3–5	1.86	1.5
5–9	1.65	1.4
9–12	1.48	1.3

[a] From ref. 4.
[b] From ref. 5.

given in Table 2. Between 5 and 9 months of age it is 1.65 g/kg/d and between 9 and 12 months, 1.48 g/kg/d. Recommendations by other authorities differ slightly from those given by the FAO/WHO/UNU. The Food and Nutrition Board recommends a protein intake of 1.6 g/kg/d for infants 6–12 months (8) and the Canadian Committee an intake of 1.9 g/kg/d from 6 to 11 months (9).

The calculations made by Fomon (5) are not completely consistent with those of the FAO/WHO/UNU group. The figures for nitrogen increment in the body (Table 1) are the same in both cases. Fomon, however, does not add 50% of the growth component. It is difficult to assess fluctuations in growth rate and the variation in protein requirement due to these fluctuations. Even during periods as short as 12 h without food, the nitrogen balance becomes negative. The Committee therefore found it unlikely that amino acids provided on a day when there is no growth can be utilized for growth later. Furthermore, balance studies in children often show that they retain more nitrogen than the theoretically determined amounts (10). Since Fomon reckoned with higher nitrogen losses (141 mg/kg/d) than the expert group (120 mg/kg/d), the two calculations resulted in the same total requirement (Table 1). The expert group and Fomon assumed a 70% efficiency in the conversion of protein from beikost to body protein. According to Fomon, for infants on milk-based formulas this is too low and a 90% conversion efficiency was used instead. To arrive at a safe level of protein intake (Table 2) for virtually all healthy children, FAO/WHO/UNU applied a 35% coefficient for variation in growth. As seen from Table 2, the recommended daily intake (RDI) according to Fomon is lower, only 10% above the estimated protein requirement. Fomon justifies the use of the lower figure by pointing out that since the protein requirement of younger infants was overestimated, it may well be too high in the case of older babies as well. Furthermore, he believes that the variability in length gain is of more significance during the first month of life than after 6 months of age.

As a complement to the methods discussed above, concentrations of free amino acids in plasma and the concentration of urea in serum, which reflect dietary protein intake and composition, have been used to estimate the protein requirement (11–13). The relationship between serum urea concentrations and the efficiency of protein utilization in human subjects was evaluated by Taylor et al. (13). The protein intake

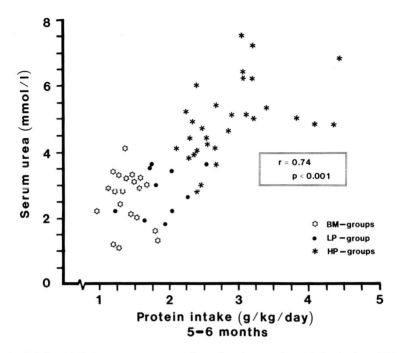

FIG. 1. Relationship between serum urea at 6 months of age and protein intake ($r = 0.74$; $p < 0.001$). [From Räihä NCR, & Axelsson IEM. *Carnation Nutrition Education Series,* Vol. 2. New York: Raven Press, 1991; 175–197.]

was directly correlated to serum urea and the net protein utilization was inversely correlated to serum urea. In our studies we have found that serum urea was related to protein intake (Fig. 1). The mean concentration of urea in serum was significantly higher in infants fed high protein formulas than in infants fed human milk or low protein formula ($p < 0.001$) (14,15). These findings suggest that net protein utilization of infants fed high protein formulas is lower than that of the other groups (13,16). We also found high concentrations of amino acids in the plasma of formula-fed infants 4–6 months of age on high protein intakes (3.6 g/kg/d) compared with the levels in infants on low protein intakes (1.9 g/kg/d) and the levels in infants fed human milk (Fig. 2) (11,17).

The high concentrations of amino acids together with other indices suggest that the high intake provides an excessive protein load. Amino acid kinetic data have also been used to estimate protein requirements and specific amino acid requirements (18).

RECOMMENDATION FOR INFANT FORMULAS AND FOLLOW-ON FORMULAS

The most favorable feeding is human milk with appropriate supplementation throughout the first year of life (19,20). An alternative to human milk is an iron-fortified commercially prepared standard infant formula. This should resemble human

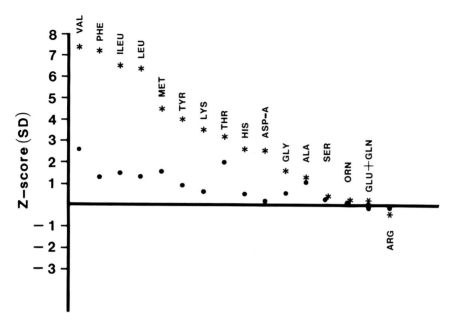

FIG. 2. Z-scores of plasma amino acid concentrations of infants fed high protein formula marked with * (protein content 2.7 g per 100 ml) and low-protein formulas marked with filled circles (protein content 1.3 g per 100 ml) in relation to plasma amino acid concentrations of infants fed human milk. [From Räihä NCR, & Axelsson IEM. *Carnation Nutrition Education Series,* Vol. 2. New York: Raven Press, 1991; 175–197.]

milk as much as possible and is satisfactory during the first 12 months of life according to international recommendations from the American Academy of Pediatrics (AAP), the U.K. Department of Health and Social Security (DHSS), and the European Society for Pediatric Gastroenterology and Nutrition (ESPGAN) (19–21). ESPGAN proposes a protein content in standard infant formulas of 1.8–2.8 g per 100 kcal (22), the AAP 1.8–4.5 g per 100 kcal (23), and the DHSS 1.2–2 g per 100 ml (20). The Scientific Committee for Food (SCF) suggests 1.8–3 g per 100 kcal (24). Clinical experience and experimental data confirm that a formula with 2.2 g of protein per 100 kcal together with cereals and supplementary food is sufficient until 12 months (25,26).

We have demonstrated that a formula with a protein content of 1.3 g per 100 ml (1.8 g per 100 kcal) together with diversified supplementary foods is safe between 4 and 6 months of age (14) according to the standards set by FAO/WHO/UNU. A formula with this protein content provides a mean protein intake of 1.72 g/kg/d at 4–5 months and 1.43 g/kg/d at 5–6 months of age. Together with the supplementary food the protein intakes are 1.88 and 1.89 g/kg/d, respectively. Unpublished preliminary results indicate that such a formula can be used throughout infancy. The upper limit of 4.5 g per 100 kcal is far too high. It should be decreased because it results in a high renal solute load (27) and greatly exceeds the amount of protein necessary for growth.

The so-called "follow-on" formulas were introduced on the market with the intention of replacing the use of cow's milk. These formulas should be less expensive than standard infant formula and are iron-fortified. The incidence of iron deficiency anemia at 12 months is higher in infants fed cow's milk than in infants fed infant formulas (28). The iron deficiency may be due to gastrointestinal blood loss (29) or to inhibition of the absorption of iron by the relatively high calcium and phosphorus content of cow's milk (30). Other consequences of feeding infants cow's milk is a low intake of linoleic acid (31), as well as a high intake of sodium (31–33) and of protein (31–34).

ESPGAN has proposed the use of a follow-on formula for infants between 5 and 12 months (21) containing 3.0–4.5 g of protein per 100 kcal. SFC suggests a follow-on formula with a protein content of 2.25–4.5 g per 100 kcal (24). In respect to protein, it has not been shown that there is a need for follow-on formulas during the second part of infancy in countries with high-quality supplementary foods. The reports available (15,25) suggest that these formulas together with supplementary foods give a mean protein intake of between 2.6 and 4.0 g/kg/d, which is more than the recommended intake and almost as much as consumed by infants drinking cow's milk (31–34).

SUPPLEMENTARY FOODS

Recommendations for formulas to be used during the second half of infancy cannot be made without discussing the quantity and quality of supplementary foods. There are international guidelines on the composition of commercially prepared weaning foods (35,36). Many countries have adopted Codex Alimentarius guidelines. The organizations involved give advice about purity requirements, food additives such as thickening agents, contaminants, hygiene, and labeling. Vitamins and minerals may be added in accordance with the legislation of the country in which the food is sold. The maximum value for sodium content in canned baby foods is 200 mg sodium per 100 g and in cereals 100 mg per 100 g of food (35). ESPGAN's minimum values for protein are 6.5 g per 100 kcal in fish and meat preparations and 4.2 g per 100 kcal in mixtures of meat and fish with vegetables, rice, potatoes, and so on (complete dishes) (36). These recommendations for the protein content of baby foods are given under the assumption that infants are receiving breast milk or standard infant formula but not cow's milk or follow-on formula.

The protein intake from supplementary foods is 0.07 g/kg/d at 4–5 months and 0.5 g/kg/d at 5–6 months of age in breast-fed infants (14). Formula-fed infants receive 0.5–0.6 g/kg/d from supplementary foods at 5–6 months (14), irrespective of the protein content of the formula used. Hence even if most infants are fed beikost before 6 months (20), the proportion of nutrients provided by the beikost remains small and an infant less than 6 months of age rarely derives a major proportion of nutrients from it.

In the second half of the first year of life the proportion of energy and protein

TABLE 3. *Protein content (g/100 kcal) in baby foods*

From 5 months		From 8 months	
Veal, potatoes, corn	4.4	Beef, rice, vegetables	4.7
Chicken, potatoes, carrots	4.4	Beef, potatoes, vegetables	4.7
Beef, potatoes	4.4	Chicken, potatoes, peas	5.0
Rump steak, potatoes, carrots	4.4	Meatballs, vegetables	4.4
Beef Stroganoff, vegetables	3.7	Beef Stroganoff, potatoes	4.2
Beef, potatoes, carrots	4.4	Minced liver, potatoes	3.9
Ham, potatoes, carrots	3.8	Spiced beef, potatoes	4.7
Cod, dill sauce, potatoes	4.7	Minced meat, spaghetti	5.0
Porridge (rice)	2.6	Boiled mutton veal	5.0
From 6 months		Pot of vegetables, beef	4.0
Porridge, (gluten)	2.5	Pot of beef	4.4
		Haddock, potatoes	5.6
		Cod, vegetables	5.6
		Porridge (wholemeal)	5.3

provided by supplementary food increases. In a summary from four national surveys Ernst *et al.* demonstrated that at 7–8 months about 42% of the protein sources are table foods, infant cereals, and baby foods (37). The corresponding figures for 9–10 and 11–12 months are 48% and 56%, respectively. The remaining protein is derived mainly from cow's milk and formula.

The protein content in commercial baby food used in Sweden from 3 to 12 months of age is shown in Table 3. The baby foods used after 5 months are rich in protein except for those containing fruit and berries. If these commercial baby foods are used together with porridge and the follow-on formulas available on the market in Sweden and recommended by the Child Health Centers from 6 to 12 months, the total protein intake will be about 2.8 g/kg/d at 6–8 months and about 3.6 g/kg/d at 8–12 months.

Ziegler has reported (38) that the protein content of beikost in the USA is 2.4 g per 100 kcal for infants of 6 months of age and 3.4 g per 100 kcal at 10 months. These values are lower than those presented in Table 3 (Swedish baby foods). Together with a standard infant formula the total protein intake is calculated to be 2.3 and 2.8 g per 100 kcal at 6 and 10 months, respectively. From this calculation it can be concluded that formulas for young infants in combination with beikost meet the nutritional needs with respect to protein during the second half of the first year of life.

Heinig *et al.* have recently studied nutrient intake during the first year of life in breast-fed and formula-fed infants in relation to the timing of the introduction of supplementary foods (39). In the formula-fed group (standard infant formula) the time at which solid food was introduced was unrelated to protein intake. At 6 months, male breast-fed infants had significantly higher protein intake (1.22 g/kg/d) if they had already been given solids than did exclusively breast-fed infants (0.85 g/kg/d). At 9 months protein intake was significantly higher in breast-fed infants given solids before 6 months of age (females 1.77 g/kg/d and males 2.12 g/kg/d) than in infants first given solids after 6 months of age (females 1.08 g/kg/d, males 1.14 g/kg/d). At 12 months there was no significant difference in protein intake between breast-fed and formula-fed infants and between breast-fed infants given solids early or late.

TYPE OF MILK FEEDING: EFFECTS ON PROTEIN INTAKE

At 6 months 20% to 30% of the infants in Europe and the USA are breast-fed (20,40). The frequency decreases during the second half of infancy, being 12% in the UK at 9 months (20) and 10% in the USA at 12 months (40). Formula feeding and cow's milk feeding are consequently the predominant forms of milk feeding during the second 6 months of life, despite the fact that in some countries (e.g., Australia) a much higher frequency of breast feeding is reported (41). The Social Board in Sweden reports a frequency of breast feeding of 47% at 6 months for infants born in 1988.

The mean total protein intake from different studies in exclusively breast-fed (EB) and partially breast-fed (PB) infants and in infants fed formula (FF) or cow's milk (CF) are given in Fig. 3 (4-8 months) and in Fig. 4 (9-12 months). The total protein intake varies from 0.9 to 1.2 g/kg/d in exclusively breast-fed infants from 4-12 months of age (5,42-44). When supplementary foods are introduced the mean protein intake increases to 1.27 g/kg/d (14,32,39) for the first period (Fig. 3) and to 2.0 g/kg/d (39) for the second period (Fig. 4). When human milk is replaced by formula or cow's milk, protein intake increases markedly to 2.74 g/kg/d in the FF group and 4.75 in the CF group for the first period. After 9 months the corresponding figures are 3.1 and 4.35, respectively. Formulas with different protein content were used in these studies. Infants given formulas with a low protein content of 1.3 to 1.5 g per 100 ml have total protein intakes from 1.9 to 2.4 g/kg/d for the period between 4 and 7 months (14,25), close to the figures for partially breast-fed infants of 9-12 months of age. At 12 months Heinig (39) reports a mean protein intake of 2.67 g/kg/d in boys and girls fed standard infant formula. The highest reported values for formula-fed infants are 3.8 g/kg/d in Australia (41), and 3.6 g/kg/d from a formula with a protein content of 2.7 g per 100 ml used in Sweden (15). These formulas with high protein content are no longer used in Sweden for babies under 6 months but are still used during the second half of infancy.

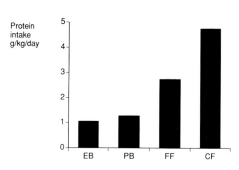

FIG. 3. Mean total protein intake (g/kg/d) in exclusively breast-fed, EB (mean from refs. 5, 42–44); in partly breast-fed, PB (mean from refs. 14, 32, 39); in formula-fed, FF (mean from refs. 14, 15, 25, 31–34, 44); and in cow's milk-fed infants, CF (mean from refs. 31–34) at 4–8 months.

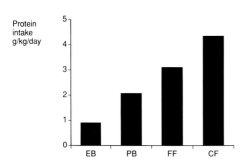

FIG. 4. Mean total protein intake (g/kg/d) in exclusively breast-fed, EB (42); and in partly breast-fed, PB (39); in formula-fed, FF (mean from refs. 31–34, 41, 44); and in cow's milk-fed infants, CF (mean from refs. 31–34) at 9–12 months.

Cow's milk feeding results in high figures for protein intake: 4.75 g/kg/d at 4–8 months and 4.35 g/kg/d at 9–12 months (31–34). One recent study reported even higher values. In a nutritional survey of 1-year-old infants conducted in Milan, the protein intake was remarkably high: 5.09 g/kg/d (45). More than 50% of the protein came from milk, yogurt, pasta, rice, cereals, and veal.

The American Academy of Pediatrics recently recommended that cow's milk should not be used during the second half of infancy, due to the resulting high renal solute load and a protein intake by far in excess of that needed (19). Optimal nutrition during infancy is provided by breast milk together with appropriate solid foods, according to this authority. The only acceptable alternative to human milk is an iron-fortified standard infant formula (19).

CONSEQUENCES OF DIFFERENT MILK DIETS

Several studies have shown that breast-fed infants grow more slowly after 3 months of age than do formula-fed infants (15,41,42,46). Different standards of growth are available. The weight/length curves provided by the National Center for Health Statistics (47) have become the most widely accepted, but they are based mainly on data from artificially fed infants given high solute formulas and receiving solids earlier than is usual today. Thus it can be argued that these standards reflect the growth of infants who nowadays would be considered overfed. The weight gain reported from healthy breast-fed infants during the first year of life has remained similar over a period of more than 30 years (41–43), in distinct contrast to the changes in growth pattern seen in artificially fed infants. Even in recently published studies (46) where infants were fed modern formulas, breast-fed infants were leaner. One possible reason for the difference in weight gain could be the lower energy intakes in breast-fed infants compared to formula-fed infants (17,46). Heining *et al.* examined how the timing of introduction of solid foods is related to growth and intake (39). Her study revealed that total energy intake was not significantly different whether breast-fed infants were given solids before or after 6 months of age and that there was no difference in weight gain between these groups (apart from a 3-month period from 6 to 9 months of age with lower weight gain in breast-fed male infants given solids late).

The increased weight gain in formula-fed infants may be the result of higher protein intake. We have studied growth in breast-fed and artificially fed infants with different protein content in the formulas given (14,15). There was no statistically significant difference in weight gain between infants fed low protein formula (protein content 1.3 g per 100 ml) and those fed human milk. However, those fed high protein formulas (protein content 1.8, 1.9, or 2.7 g per 100 ml) had significantly higher weight gains (Fig. 5). The energy intake was similar in the different formula groups. High protein intake results in increased concentrations of branched-chain amino acids (11), which in turn stimulate insulin secretion (17,48). We found a significantly positive correlation between plasma concentrations of the branched-chain amino acids valine, leucine, and isoleucine, and the concentration of C-peptide in plasma (Fig. 6). Furthermore, C-peptide excretion was positively correlated to weekly weight gains ($r = 0.51$, $p < 0.01$). Multiple linear regression analysis suggests that the protein-induced weight gain could be mediated by an increase in insulin secretion (17). Insulin pro-

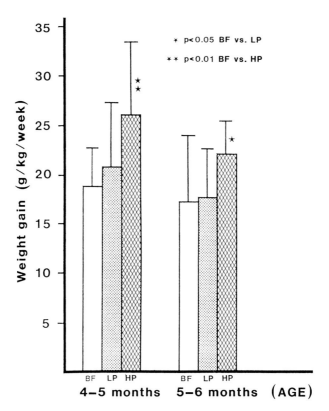

FIG. 5. Mean weight gain expressed as g/kg/week in breast-fed infant, BF, in infants fed low-protein formula, LP (protein content 1.3 g per 100 ml); and in infants fed high protein formula, HP (protein content 1.8 g per 100 ml). There were no differences between BF and LP groups, but there were significant differences between the BF and HP groups from 4 to 5 months ($p < 0.01$) and from 5 to 6 months ($p < 0.05$). Values are mean ± SD. [From Axelsson IE, et al. Pediatr Res 1988; 24: 297–301.]

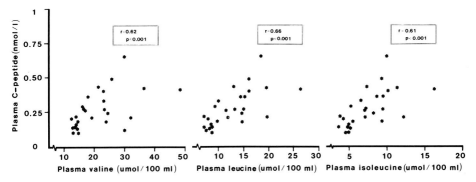

FIG. 6. Relationship between plasma C-peptide concentration and plasma concentrations of the branched-chain plasma amino acids (valine, $r = 0.62$, $p < 0.001$; leucine, $r = 0.66$, $p < 0.001$, isoleucine, $r = 0.61$, $p < 0.001$). [From Räihä NCR, & Axelsson IEM (15).]

motes growth by facilitating the transport of amino acids over the cell membrane, which results in increased protein synthesis and increased lipogenesis (49).

Another consequence of feeding cow's milk and formulas with high protein content is the increased risk of hypernatremic dehydration due to a high potential renal solute load (PRSL) (27). The PRSL is the sum of dietary nitrogen, sodium, potassium, chloride, and phosphorus. Protein is the major contributor to PRSL. The PRSL of whole cow's milk is 308 mOsm/liter (27), that of the follow-on formula used in one of our studies 276 mOsm/liter, and that of breast milk 99 mOsm/liter. Other follow-on formulas and formulas used for gastroenteritis available on the Swedish market contain between 208 and 277 mOsm/liter. These high values increase the risk for hypernatremic dehydration if the amount of available water decreases due to fever, diarrhea, or reduced intake of water. Ziegler and Fomon recommended that PRSL for formulas should not exceed 221 mOsm/liter (27), a value lower than the PRSL which most follow-on formulas provide.

CONCLUSIONS

Feeding with cow's milk or follow-on formula with high protein content provides protein intakes far above the requirement and the recommended intake. Adequate volumes of human milk together with appropriate supplementation may be the preferable type of feeding during the second half of infancy and should not imply protein deficiency (42). If the milk diet is based on formula, a protein content of 1.8 g per 100 kcal is sufficient for the 4- to 6-month-old infant, providing a total protein intake of 1.9 g/kg/d including supplementary food (14). No secure recommendation for the lower limit of protein content in a formula can be given since there is insufficient information. Standard infant formulas containing 2.2 g per 100 kcal with supplementary food of high protein quality promote normal growth and provide a protein intake within the safe range (4,26,39). This diet can be recommended during the second part of infancy under conditions where adequate supplementary food is available and where catch-up growth is not required.

REFERENCES

1. Schimke RT. Adaptive characteristics of urea cycle enzymes in the rat. *J Biol Chem* 1962; 237: 459–63.
2. Rane S, Aperia A. Ontogenity of Na,K-ATPase activity in the thick ascending limb of Henle and of the urinary concentrating capacity in rats. *Am J Physiol* 1985; 249: F723–8.
3. Yokogoshi H, Hayase K, Yoshida A. The quality and quantity of dietary protein affect brain protein synthesis in rats. *J Nutr* 1992; 11: 2210–7.
4. *Energy and protein requirements.* WHO Technical Report Series No 724. Geneva: FAO/WHO/UNU Expert Consultation. 1985; 64–112.
5. Fomon S. Requirements and recommended dietary intakes of protein during infancy. *Pediatr Res* 1991; 30: 391–5.
6. Huang PC, Lin CP, Hsu JY. Protein requirements of normal infants at the age of about 1 year: maintenance nitrogen requirements and obligatory nitrogen losses. *J Nutr* 1980; 110: 1727–35.
7. Fomon SJ, Haschke F, Ziegler EE, et al. Body composition of reference children from birth to age 10 years. *Am J Clin Nutr* 1982; 35: 1169–75.
8. Food and Nutrition Board, National Research Council. *Recommended dietary allowances,* 10th ed. Washington, DC: National Academy Press, 1989.
9. Ad Hoc Consultative Group on the Dietary Standard for Canada. *Recommended nutrient intakes for Canadians.* Ottawa: Health and Welfare Canada, 1983; 7–47.
10. Torun B, Cabrera-Santiago MI, Viteri FE. Protein requirements of preschool children: milk and soybean protein isolate. In: Torun B, Young VR, Rand WM, eds. *Protein–energy requirements of developing countries: evaluation of new data.* Tokyo: United Nations University. *Food Nutr Bull* 1981; Suppl 5: 210–22.
11. Axelsson I, Borulf S, Abildskov K, et al. Protein and energy intake during weaning. III. Effects on plasma amino acids. *Acta Paediatr Scand* 1988; 77: 42–8.
12. Axelsson I, Borulf S, Räihä N. Protein intake during weaning. II. Metabolic responses. *Acta Paediatr Scand* 1987; 76: 457–62.
13. Taylor LS, Scrimshaw NS, Young VR. The relationship between serum urea levels and dietary nitrogen utilization in young men. *Br J Nutr* 1974; 32: 407–11.
14. Axelsson IE, Jakobsson I, Räihä NC. Formula with reduced protein content: effects on growth and protein metabolism during weaning. *Pediatr Res* 1988; 24: 297–301.
15. Räihä NCR, Axelsson IEM. Protein requirements during weaning: an evaluation of a standardized weaning program with varying protein intakes. In: Heird WC, ed. Nutritional needs of the six to twelve months old infant. *Carnation Nutrition Education Series,* Vol 2. New York: Raven Press, 1991; 175–97.
16. Bodwell CE. Biochemical indices in humans. In: Bodwell CE, ed. *Evaluation of proteins for humans.* Westport, CT: AVI Publishing, 1977; 119–48.
17. Axelsson IEM, Ivarsson SA, Räihä NCR. Protein intake in early infancy: effects on plasma amino acid concentrations, insulin metabolism, and growth. *Pediatr Res* 1989; 26: 614–7.
18. Young VR, Joaquin C. Protein and amino acid requirements of healthy 6 to 12 month-old infants. In: Heird WC, ed. *Nutritional needs of the six to twelve month old infant.* Carnation Nutrition Education Series. New York: Raven Press, 1991; 149–74.
19. American Academy of Pediatrics Committee on Nutrition. The use of whole cow's milk in infancy. *Pediatrics* 1992; 89: 1105–9.
20. Department of Health and Social Security. *Present day practice in infant feeding.* Report on Health and Social Subjects No 32. London: DHSS, 1988; 1–66.
21. ESPGAN Committee on Nutrition. Comment on the composition of cow's milk based follow-up formulas. *Acta Paediatr Scand* 1990; 79: 250–4.
22. ESPGAN Committee on Nutrition. Guidelines on infant nutrition. I. Recommendations for the composition of an adapted formula. *Acta Paediatr Scand* 1977; Suppl 262: 3–20.
23. American Academy of Pediatrics Committee on Nutrition. Commentary on breast-feeding and infant formulas, including proposed standards for formulas. *Pediatrics* 1976; 57: 278–85.
24. Commission of the European Communities. Modified proposal for a council directive on the approximation of the laws of the member states relating to infant formula and follow-up milks. *COM* 1986; 564: 1–54.
25. Lönnerdal B, Chen C-L. Effects of formula protein level and ratio on infant growth, plasma amino acids and serum trace elements. II. Follow-up formula. *Acta Pediatr Scand* 1990; 79: 266–73.

26. Saarinen UM, Siimes A. Role of prolonged breast feeding in infant growth. *Acta Paediatr Scand* 1979; 68: 245–50.
27. Ziegler EE, Fomon SJ. Potential renal solute load of infant formulas. *J Nutr* 1989; Suppl. 119: 1785–8.
28. Tunnessen WW, Oski FA. Consequences of starting whole cow milk at 6 months of age. *J Pediatr* 1987; 111: 813–6.
29. Ziegler EE, Fomon SJ, Nelson SE, *et al*. Cow milk feeding in infancy: further observations on blood loss from the gastrointestinal tract. *J Pediatr* 1990; 116: 11–8.
30. Monson ER, Cook JD. The effects of calcium and phosphate salts on the absorption of nonheme iron. *Am J Clin Nutr* 1976; 29: 1142–8.
31. Martinez GA, Ryan AS, Malec DJ. Nutrient intakes of American infants and children fed cow's milk or infant formula. *Am J Dis Child* 1985; 139: 1010–8.
32. Horst CH, Obermann-de Boer GL, Kromhout D. Type of milk feeding and nutrient intake during infancy. *Acta Paediatr Scand* 1987; 76: 865–71.
33. Montalto MB, Benson JD, Martinez GA. Nutrient intakes of formula-fed infants and infants fed cow's milk. *Pediatrics* 1985; 75: 343–51.
34. Martinez GA, Ryan AS. Nutrient intake in the United States during the first 12 months of life. *J Am Diet Assoc* 1985; 85: 826–30.
35. Codex Alimentarius Commission: Codex standards for foods for infants and children. *Codex Alimentarius*, Vol IX. Rome: FAO/WHO, 1982; 24–36.
36. ESPGAN Committee on Nutrition. Guidelines on infant nutrition. II. Recommendations for the composition of follow-up formula and beikost. *Acta Paediatr Scand* 1981; Suppl 287: 4–25.
37. Ernst JA, Brady MS, Rickard KA. Food and nutrient intake of 6- to 12-month-old infants fed formula or cow milk: a summary of four national surveys. *J Pediatr* 1990; 117: 86–99.
38. Ziegler EE. Milks and formulas for older infants. *J Pediatr* 1990; 117: 76–9.
39. Heinig JM, Nommsen LA, Peerson JM, Lönnerdal B, Dewey KG. Intake and growth of formula-fed infants in relation to the timing of introduction of complementary foods: the Darling Study. *Acta Paediatr Scand* 1993; 82: 999–1006.
40. Fomon S. Reflections on infant feeding in the 1970s and 1980s. *Am J Clin Nutr* 1987; 46: 171–82.
41. Hitchcock NE, Gracey M, Gilmour AI, Owles EN. Nutrition and growth in infancy and early childhood. A longitudinal study from birth to five years. In: Falkner F, Kretchmer N, Rossi E, eds. *Monographs in Paediatrics*, Vol 19. New York: S Karger, 1986; 1–54.
42. Salmenperä L, Perheentupa J, Siimes MA. Exclusively breast-fed healthy infants grow slower than reference infants. *Pediatr Res* 1985; 19: 307–12.
43. Butte NF, Wong WW, Ferlic L, *et al*. Energy expenditure and deposition of breast-fed and formula-fed infants during early infancy. *Pediatr Res* 1990; 28: 631–40.
44. Kylberg E, Hofvander Y, Sjölin S. Diets of healthy Swedish Children 4–24 months old. III. Nutrient intake. *Acta Paediatr Scand* 1986; 75: 937–46.
45. Bellu R, Ortisi MT, Incerti P, *et al*. Nutritional survey on a sample of one-year-old infants in Milan: intake of macronutrients. *Nutr Res* 1991; 11: 1221–9.
46. Dewey KG, Heinig MJ, Nommsen LA, Peerson JM, Lönnerdal B. Growth of breast-fed infants from 0–18 months: the Darling Study. *Pediatrics* 1992; 89: 1035–41.
47. Hamill PVV, Drizd TA, Johnson CL, *et al*. Physical growth: national center for health statistics percentiles. *Am J Clin Nutr* 1979; 32: 607–29.
48. Fajans SS, Floyd JC. Stimulation of islet cell secretion by nutrients and by gastrointestinal hormones released during digestion. In: Steiner DF, Freinkel N, eds. *Handbook of physiology*. Washington DC: American Physiological Society, 1972; 473–93.
49. Hill DJ, Milner RDG. Insulin as a growth factor. *Pediatr Res* 1985; 19: 879–86.

DISCUSSION FOLLOWING THE PRESENTATION OF DR. AXELSSON

Dr. Rey: I should like to give additional explanation about the differences between the recommendations on the concentration of protein in different formulas provided by ESPGAN, the Scientific Committee for Food, and the American Academy of Pediatrics. The 1977 ESPGAN recommendation for infant formula is a recommendation for a starting formula, to be given from birth to 4–6 months of age. The recommendation of the American Academy of Pediatrics is a recommendation for a formula to be taken by the child between birth and 1 year of age. This explains the difference between the value of 4.5 g per 100 kcal given by the American Academy of Pediatrics and the value of 2.8 g per 100 kcal given by ESPGAN. The Americans did not recognize the need for follow-on formulas and they pushed for only one type of infant formula for the entire first year. I remember a great deal of conflict in the Committee of the Codex Alimentarius in Germany between the Canadians and the Americans and some of the European countries about the possibility of using a less specialized, nutritionally incomplete formula for infants over 4–6 months of age.

When you recommend that an infant formula containing 2.2 g of protein per 100 kcal is suitable for infants from 6 months to 1 year of age, I fully agree. But we have tried in the Codex, in ESPGAN, and in the Scientific Committee for Food to promote a formula for children in the second part of the first year which is less sophisticated and does not contain all the vitamins and all the minerals. We have tried in this way to promote a cheaper formula. But I recognize that industry has not been able to put such formulas on the market at a cheaper price.

Dr. Räihä: If we look back at the inception of follow-on formulas, they were really developed because they were going to be used instead of cow's milk. It was therefore an aim of such a formula that it should have a lower protein content than cow's milk. It is true that one also wanted to make it cheaper than standard infant formula, which is a complex product. But today, at least in most of the industrialized countries and especially in Scandinavia, we no longer use cow's milk before 1 year of age and we now use a high quality protein supplementary food starting from 6 months. I see no reason today to continue with a high concentration of protein in follow-on formulas because you can make them even cheaper if you decrease the protein to a level of 1.2 or 1.3 g per 100 ml, which is certainly enough. And in the USA, of course, the idea of follow-on formulas never caught on; regular standard formula is used up to 1 year of age. It makes no sense to give the baby more protein as he gets older when his rate of protein synthesis per kilogram of body weight is diminishing.

Dr. Marini: A basic question: When should we introduce solid foods? When you say 4–6 months, this is a wide range for me. If we follow nature, and the breast-fed infant is the most correct example, I should say that breast milk composition remains almost constant for up to 6–7 months after delivery. So this means for me that the milk is probably nutritionally satisfactory for the baby. In the past the pediatricians used to say that we should start giving solid food when the baby could stand up with the head held high, and when teeth first appear.

I think that the bad habit of introducing solid foods early came from the USA in the 1950s when were considered out of date if we were not giving homogenized meat by 6 weeks of age. It is difficult to change this bad habit. If you tell the mother that she can give meat at 3 months, fish at 4 months, and eggs at 6 months, then when the baby is 8 months old the mother says when can I give cow's milk? There are many data in the literature showing that early introduction of solid food is a most important cause of induced allergy. I am of course talking about developed countries; I don't know about underdeveloped countries.

Dr. Axelsson: I cannot give an answer about when it is proper to introduce solid foods. An infant can starve at the breast at 3 months and an infant can grow well at 9 months entirely on breast milk. Siimes showed in his paper (1) that there is a risk of iron deficiency at 9 months in exclusively breast-fed infants. I am not sure that there are any convincing data regarding the appropriate time to introduce solid foods to prevent allergy.

Dr. Marini: I would not propose that infants should go without solid food up to 9 months of age, but I would not introduce solid food before 6 months of age.

Dr. Axelsson: We have the same recommendation in Sweden if the parents are allergic.

Dr. Marini: But I would recommend this for the normal population, not for allergic subjects only.

Dr. Guesry: Dr Axelsson, you stated that due to the high potential renal solute load of 250 mOsm/liter there is a risk of hypernatremic dehydration. You quote Ziegler, who recommends a value of 220 mOsm/liter, but this recommendation relates to starter infant formula, not to follow-on formula; the figure of 250 mOsm you showed was for follow-on formula. The concentrating power of the kidney increases with age. By 4 months the average concentrating power of the kidney is around 700 mOsm/liter, which is quite enough to deal with a potential renal solute load of 250 mOsm/liter. Thus I doubt whether you would really be likely to see hypernatremic dehydration in Fomon's older babies on this account alone. In addition, as you know, the kidney is a very adaptable organ, and the more protein you give, the higher the concentration of urea in the extracellular fluid and the more the concentrating ability increases. I agree that during diarrhea there is always a risk of hypernatremic dehydration, but I am not sure that you will see more of this in babies fed with the follow-on formula than with babies fed with a lower-protein formula. Maybe the situation would be reversed because the kidney has been able to develop increased concentrating ability.

Dr. Axelsson: If the infant has fever and a reduced intake by about 25%, and also has diarrhea, he will be unable to concentrate sufficiently. If he takes in a formula with a PRSL of 277 mOsm/liter, he would need to concentrate to about 1700 mOsm/kg, which many babies cannot do. In my daily work I have not infrequently seen babies with hypernatremic dehydration, and there have been reports in a paper from *Journal of the Swedish Medical Association* about it (2). When I have calculated the intakes of such babies, they have in some cases been receiving about 4.5 g/kg/d of protein on the day of admission from formulas designed for gastroenteritis. I suspect that this is far too high. Some of these formulas are designed for adults.

Dr. Guesry: Vis and co-workers described the high levels of sodium in oral rehydration fluids given to infants with hypernatremic dehydration seen in Belgium, which I think is representative of Europe (3,4). It seemed that when the baby developed diarrhea the mother stops giving milk and gives oral rehydration salt instead. So it was not the milk that caused the problem but the 90 mmol/liter of sodium of the oral rehydration salt.

Dr. Räihä: Although this is an interesting discussion, it is not really to the point, which is that whichever way you look at it, infants are still getting a protein load that is way above their requirement. We know that the kidney is developing up to the age of 2–3 years and that enzymatic adaptation and development are occurring, and we also know from animal experiments that high protein loads in the kidney will have permanent effects on renal function. So I see no reason to overload these babies with protein while they are in a critical period of their development. There is also, of course, no reason to give them too little, but I think that in the industrialized countries we have for several years now been giving them too much, both in the follow-on formulas and in supplementary foods.

Another area that we are going into now is the question of infants over 1 year of age. There are virtually no reports or information on intakes in the age range 1–3 years. It is important that we study this age group because of the type of development that is still occurring at this age.

Dr. Househam: The majority of children in the world come from developing countries but we keep hearing discussion relating to the effects of marginally high protein levels in infant formulas. I am not aware that there is a major problem in the developed countries of children with important learning deficits, brain damage, and other complications resulting from receiving a little too much protein. On the other hand, in developing countries you are dealing with children who have protein deficiency, and you are dealing with children who are hyponatremic, so discussion of hypernatremic dehydration is irrelevant to us. I wonder whether we could concentrate on recommendations for developing and less affluent areas. If we are looking at this issue, we need to look at it on a national basis, looking at recommendations for infant formulas for different countries. This would presumably be entirely feasible for infant formula manufacturers.

I would like this issue to be addressed. The discussion is being dominated by the Eurocentric–North American axis and those of us who are from developing countries feel excluded and on the periphery.

Dr. Axelsson: I agree with you that the problem is not the same in the developing countries. I have not done research in these parts of the world. In less privileged populations where supplementary foods contain no or only low protein, a formula with more protein should be used. The protein content of the formula should be adjusted so that total protein intake is sufficient.

Dr. Räihä: I think Dr. Axelsson made it clear that her study was done in Sweden, where we have no malnutrition. She was reporting a controlled study on babies at an age where there are at present few if any published studies. She also said at the end that you have to look at the total protein intake.

As for the developing countries, I would say that if the protein intake or quality from supplementary foods is poor or inadequate, then certainly you need to give a formula that is high in good-quality protein. You must always look at the total intake of the infant at any age, regardless of whether the protein comes from formula, breast milk, follow-on formula, or supplementary food. This is certainly clear. I think that in the future we shall need to specialize more, and produce particular products for particular countries or for certain populations in those countries. It is really up to the pediatricians in any area to look at the need for such products, and they have to know enough about nutrition to be able to understand what babies need more of and what babies need less of.

Dr. Uauy: This field has traditionally been dominated by people who have done studies in developing countries. The first protein recommendation came from the studies of Cicely Williams and the first international recommendations were the result of the need to treat kwashiorkor. Work in Jamaica, South Africa, Chile, North Africa, and Lebanon has fully documented that if the population is malnourished, there is a proportionally greater need for protein than for energy, and I think this was translated into worldwide recommendations during the 1950s and 1960s to deal with what was called the protein gap.

Eventually, metabolic studies were conducted and a more experimental approach was developed, though without forgetting the continued need for field studies. We now have good documentation of the increased protein needs during diarrheal disease, infection, fever, and so on. I don't think the needs of the developing world have been neglected. However, I think the next step is precisely as Dr. Househam has outlined: we now need national bodies to take local realities into account and make the appropriate recommendations.

Dr. Cooper: We know that too little protein does harm, but what evidence is there that marginally high levels of protein intake do harm, other than producing biochemical differences and perhaps predisposing to hypernatremic dehydration?

Dr. Axelsson: We have no evidence for any harmful effect in the infants we have studied. We see different metabolic profiles but their significance is unknown. We have only hypotheses at present, for example relating to diabetes.

Dr. Räihä: You are right, there is very little evidence that it does harm. There is also no evidence that it does good. What I think is important is that we start to realize that we need long-term studies like those of Alan Lucas and David Rassin, which are among the first to look at long-term effects. Such studies will probably have to be done in countries where it is easy to follow up babies, where the population does not move very much, and where there are people who are willing and able to do such long-term studies, which are neither easy nor nice to do.

Regardless of whether or not we know of harmful effect, there are *potential* harmful effects that are already known, which include such things as high insulin levels, effects on kidney function and effects on blood pressure, coronary artery disease, and so on.

Dr. *Uauy:* There is a potential effect of high protein intake on renal disease or kidney maturation and development. There is evidence from animal experiments that the protein load in early life influences the way the kidney ages. Rats fed high protein diets may develop lethal glomerulonephritis. Reducing the protein intake in human adults with renal disease results in a decrease in the rate of progression of the disease. These data are relevant to discussion of protein intakes in children.

We have undertaken a study in children, precisely on this basis, in collaboration with Dr. Malcolm Holiday and the South West Pediatric Nephrology group, which is a group of 15 centers in the southwestern USA. Subjects were babies diagnosed at birth or soon after with obstructive neuropathy, multicystic disease, or renal cortical necrosis. All patients had plasma creatinine values of 0.6 mg/dl or more in the first 6 months of life, and a creatinine clearance of less than 10%. From 6 to 8 months of age the patients received a formula providing 2.0 g of protein per 100 kcal. At 8 months they were randomized to a diet containing 1.5 g/kg body weight according to FAO/WHO, or to the customary management of 2.5 g protein/kg. The study was blinded. The children were followed with anthropometry monthly and biochemical measurements every 2 months. Creatinine clearance was repeated at 18 months to assess whether there was progression of the renal failure. Solid foods were all of very low protein content so that they would not alter the intake by more than 5%. Energy intake for both groups was around 100 kcal/kg.

Selected aspects of the results were as follows. After randomization to a low or higher protein diet, we found that creatinine in the higher protein diet group fell in comparison with the low protein group, perhaps because more renal reserve was recruited with the higher protein intake. Follow-up anthropometry showed a decline in linear growth velocity in the low protein diet group which became significant at 12–18 months. These data suggest the need to reexamine the effects of varying the protein intake. Obviously, we need long-term data. However, in terms of renal function these children seemed to be more at risk from the lower protein intake than from the higher.

I am not saying that we should not think about lowering the protein intake, but we should be following linear growth very carefully, as Dr. Axelsson proposed. I think that we still need this kind of research to find out in the long term what may be the potential benefits and potential adverse effects of such a policy.

Dr. *Marini:* But these were, of course, not normal babies.

Dr. *Uauy:* Nevertheless, they were not very sick. The highest creatinine was about 2.5 mg/dl and they have been followed very closely for morbidity.

Dr. *Wharton:* I do not think there is as much difference between us as we think, as long as we take geography into account. If we take Sweden, there seems very good evidence from this study that using a low protein milk has no disadvantages and may well have advantages; long-term studies would be attractive but the argument is at present very convincing. In the UK I expect the situation to be the same, but I would like to take the argument a bit further. In the UK solid food provides 43–64% of energy in the 6- to 12-month age group and this solid food already contains 3.3 g of protein per 100 kcal (this is from a recent nationwide survey). So you could argue that no milk is necessary at all in the weaning diet in the UK to achieve a satisfactory protein and energy intake, although of course the function of milk in the diet is not just to provide protein—it is a matter of providing energy, iron, possibly zinc, and so on.

The other alternative is to try to reduce the protein/energy ratio in the weaning foods. I am talking about the UK, and possibly Sweden, not about Africa at the moment. The reaction of the subcommittee on weaning foods of the Scientific Committee for Food to the suggestion that the protein/energy ratio in these foods could be reduced is usually a sharp intake of breath. Of course, in these committees there has to be a consensus view to represent many different attitudes, even just as in Europe. So although the Committee has allowed some reduction, the ratios are still pretty high, and perhaps we should pursue that more.

Finally, coming out of Europe, what about Dr. Househam's point? There must certainly be concern about weaning foods that have a very low protein/energy ratio. I see this in the UK, particularly in immigrant groups from Moslem countries, who do not give any of our weaning foods that contain meat because they feel the animals have not been killed in an appropriate fashion. And they often don't like vegetables, so the weaning foods they choose tend to consist of fruits or puddings, and the protein/energy ratio of the solid food in their weaning diets is therefore very low. For this population a milk with a high protein/energy ratio acts as a safety net. I don't think there is any real disagreement among us here. We are saying that we must consider the individual needs of each country before we come up with particular recommendations.

Dr. Lönnerdal: I would like to emphasize what Dr. Wharton has just said, that one should take each country into consideration. There is a tendency among researchers in both developing and industrialized countries to consider the "developing world" as though it were a homogeneous population. In many countries in South America where I have been working, the problem is not the protein level of formulas—they cannot buy them in the first place, so the subject is irrelevant. The problem is much more how to engineer social change. In other parts of the developing world the situation is totally different. I have just met some Chinese pediatricians who told me about the present situation in China. Although there is a low mean income, nevertheless because of the prevailing policy of only one child per family, 25-50% of the family income goes on feeding the infant. This is regarded as a high priority and parents are willing and able to pay for formula feeds and follow the guidelines that we use in Europe and the USA. Therefore, we should never take the view that in the developing world there are so many problems that we don't need to optimize the situation. Far from it—we must adapt to the individual needs of particular communities.

Dr. Marini: Dr. Wharton did not mention the problem of calcium. This is an important point because, with low phosphate, low calcium milk, if you don't add food enriched with calcium after the first 6 months, the calcium intake will be very low indeed. In Italy we use rice powder from Nestlé which is enriched in calcium and is a good supplementary source.

REFERENCES

1. Siimes MA, Salmenperä L, Perheentupa J. Exclusive breast-feeding for 9 months: risk of iron deficiency. *J Pediatr* 1984; 104: 196–8.
2. Aronsson B, *et al.* Is there increased hypertonic dehydration among children with gastroenteritis? *J Swed Med Assoc* 1992; 89: 2268–9.
3. Vis HL. *Assignment children*, Vol 63/64, 1983.
4. Kahn A, Blum D. Hyperkalaemia and UNICEF type rehydration solutions. *Lancet* 1980; i: 1082.

Essential and Non-essential Amino Acids in Neonatal Nutrition

David Keith Rassin

Department of Pediatrics, Children's Hospital C3T16, 301 University Boulevard, University of Texas Medical Branch, Galveston, Texas 77555-0344, USA

There has been an evolution in the appreciation of the functional roles that amino acids play in neonatal nutrition over the last 30 years. Up to the early 1960s primary interest was in the identification of those amino acids that had to be supplied in the diet in order to maintain growth and positive nitrogen balance (Table 1). Most of the experiments that were performed to identify these amino acids were conducted in animals or human adults; thus they often failed to take into account the special needs of the biochemically immature neonate, particularly the preterm neonate.

Appreciation of the special biochemical needs of the developing neonate resulted in the designation of a second group of amino acids as semiessential, conditionally essential, or developmentally essential (Table 2). Studies that have defined the role that the latter compounds play in early development have reflected on properties such as infant growth (histidine, cysteine) (1,2), the developmental pattern of enzymes that catalyze synthesis (cysteine, tyrosine) (3,4), stable isotope metabolism (glycine) (5), presence in human milk coupled with animal and human studies (taurine) (6–8), and prevention of adverse metabolic effects (arginine) (9). The data to support the "essentiality" of these amino acids have often been conflicting, as for cysteine (3,10) and glycine (5,11), making a final determination difficult at best.

The next stage in the understanding of the role that amino acids play in early nutrition evolved from the many functions that they were identified with, other than as precursors for protein synthesis. Some of this impetus came from a growing appreciation of the nutritional importance of the amino acid taurine, despite the fact that it could not be incorporated into protein as it is a sulfonic amino acid. Other functions of amino acids include supporting the synthesis of hormones, neurotransmitters, and bile acids. In addition, many amino acids may themselves function as neurotransmitters (Table 3).

Thus, monitoring the amino acid responses of neonates to different nutritional regimens may be of more functional importance than just determining the sufficiency of the protein synthesis precursor pool. Intakes of individual enteral formulas and total parenteral nutrition solutions are reflected by specific amino acid patterns in

TABLE 1. *Essential amino acids[a]*

Threonine	Phenylalanine
Valine	Lysine
Isoleucine	Methionine
Leucine	Tryptophan

[a] These amino acids have been accepted as the classic indispensable amino acids based on data regarding growth and nitrogen balance.

TABLE 2. *Proposed developmentally essential amino acids[a]*

Cysteine	Histidine
Taurine	Arginine
Tyrosine	Glycine

[a] These amino acids may be conditionally essential during early development due to biochemical immaturity of the infant.

TABLE 3. *Central nervous system functions of amino acids[a]*

Neurotransmitter precursors (neurotransmitters or putative neurotransmitters)
  Phenylalanine (catecholamines)
  Tyrosine (catecholamines)
  Histidine (histamine, carnosine)
  Tryptophan (serotonin)
  Serine (glycine)
  Glutamate (GABA)
  Methionine (cystathionine, cysteinesulfinic acid, taurine)
  Cysteine (cystathionine, cysteinesulfinic acid, taurine)
Neurotransmitter or putative neurotransmitter amino acids
  Glutamate
  GABA
  Aspartate
  Proline
  Taurine
  β-Alanine
  Cysteic acid
  Cysteinesulfinic acid
  Glycine
  Cystathionine

[a] Amino acids that serve as either metabolic precursors of neurotransmitters or which have been suggested to serve as neurotransmitters themselves.

the neonate. These patterns influence protein synthetic rates, but may also regulate future development of various organs, particularly of the central nervous system, via mechanisms other than the synthesis of new proteins.

In this brief review it is intended to present some of the features of plasma amino patterns that reflect specific nutritional regimens in neonates, a discussion of the proposed mechanisms by which these patterns may influence the development of the central nervous system, and some evidence to support the validity of these mechanisms.

PLASMA AMINO ACID RESPONSES TO NUTRITION

The majority of enteral infant formulas fed to neonates are based upon cow's milk proteins, supplied usually as either casein-protein-predominant or whey-protein-predominant preparations. Other available formulas include various hydrolysates (usually of cow's milk casein or whey proteins) and soy preparations. The standard against which these preparations are compared is human milk. This comparison is influenced by the fact that formulas generally contain more protein per volume than does human milk, and that cow's milk proteins contain different amino acids than those of human milk proteins (even though they belong to the same classes of proteins, casein or acid-insoluble and whey or acid-soluble). The result is that formulas generally provide not only more amino acids than human milk but also a different pattern of amino acids, and different formulas provide different ratios or patterns of amino acids, reflecting the primary source of protein (Fig. 1).

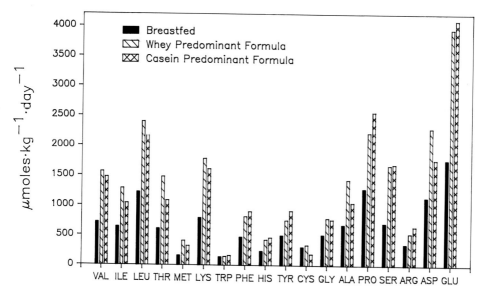

FIG. 1. Intakes of individual amino acids in breast-fed, casein-protein-predominant formula-fed, and whey-protein-predominant formula-fed term infants expressed as μmol intake per kg body weight per day. Developed from data collected in the author's laboratory.

The effect of these varying amino acid intakes is to cause distinct plasma amino acid patterns in the recipient infants, reflecting the source and amount of protein intake. These plasma amino acid patterns have been thoroughly documented in a number of different studies in both term and preterm infants (12–19). Typically, the overall pattern of plasma amino acid concentrations is increased in formula-fed infants compared to breast-fed infants and specific amino acids (such as threonine in infants fed whey-protein-predominant formulas and tyrosine in infants fed casein-protein-predominant formulas) are particularly increased and may actually be utilized as markers in the plasma to determine the protein composition fed to a particular infant.

Total parenteral nutrition (TPN) provides yet another pattern of amino acids to the infant, reflecting a variety of factors, such as chemical stability, solubility, and cost of the individual compounds. First-generation solutions were prepared from casein hydrolysates, but subsequent preparations have been prepared directly from crystalline amino acids. These preparations include second-generation solutions or "adult" preparations and third-generation solutions or "pediatric" preparations. The so-called "pediatric" preparations have been modified to attempt to provide a more balanced solution (the adult solutions contain large percentages of the cheaper amino acids glycine and alanine relative to other amino acids) and to include amino acids such as tyrosine, cysteine, and taurine that are felt to be essential for the neonate.

The plasma amino acid patterns of infants fed total parenteral nutrition also reflect the amounts of the individual compounds included in the solutions. Typically, such infants have low concentrations of tyrosine and cystine, reflecting the low or absent concentrations of these amino acids in the TPN solutions due to the difficulty in solubilizing adequate amounts of tyrosine and in stabilizing cysteine from being oxidized into cystine (which is also insoluble).

The reduced concentrations of tyrosine and cystine observed in the plasma of infants fed TPN occur despite usually very large amounts of their respective precursors, phenylalanine and methionine, in the amino acid solutions. This illustrates the biochemical immaturity of these infants with respect to the aromatic amino acid and sulfur-containing amino acid metabolic pathways.

Route of administration may play an important role in the ability of the infants to synthesize tyrosine. Data on the maturity of the enzymes required to catalyze tyrosine synthesis and catabolism indicate that phenylalanine hydroxylase is far more active than the tyrosine oxidizing system (4,20,21). This finding is compatible with the response of infants to excessive aromatic amino acid enteral intake with large increases in plasma tyrosine (14). However, it does not explain why parenterally fed infants have low plasma tyrosine in the presence of more than adequate phenylalanine intake. Even adult rats, which are reported to have no problems synthesizing tyrosine (22), have low plasma tyrosine concentrations when nourished parenterally (23). Similarly, reduced plasma cystine has been much more evident in parenterally fed (24) than in enterally fed (13) infants.

Taurine, a metabolite of cysteine, is also found in low concentrations in both enterally and parenterally fed infants when it is not included in the feeding regimens. Until about 8 years ago taurine was not added to either enteral formulas (which contained only trace amounts of this compound) or parenteral solutions (which contained none). Human milk contains relatively large amounts of taurine, so breast-fed infants manage to remain taurine-sufficient (6,15). Human infants appear to be dependent on a dietary supply of taurine; in fact, human beings do not ever appear to develop much hepatic activity of the enzyme cysteine sulfinic acid decarboxylase, responsible for catalyzing taurine synthesis, so are probably dependent on a dietary source throughout the life cycle (13).

In summary, plasma amino acid patterns reflect the amount, type, and route of protein intake. The question that has to be asked is how important are these variations in concentrations. As mentioned above (Table 3), the aromatic and sulfur amino acids are precursors for a variety of biologically active compounds that could have consequences for normal development of the central nervous system if they were present in either excess or deficiency.

CONSEQUENCES OF CHANGES IN AMINO ACID PATTERNS

As noted earlier, amino acids subserve a variety of functions that could potentially be affected by changes in concentrations due to diet. Of particular interest is the influence that such changes might have on the developing central nervous system. Such interest is particularly acute in the light of reports that the intellectual outcome of infants, both term and preterm, may reflect their early feeding history (25,26). The nutrients supplied in human milk appear to have a real benefit in respect to such neurologic outcome when breast-fed infants are compared to bottle-fed infants.

One proposed mechanism for such an effect is relevant to the class of large neutral amino acids (which includes valine, isoleucine, leucine, tyrosine, phenylalanine, tryptophan, methionine, threonine, and histidine). This group of amino acids includes several important neurotransmitter precursors (phenylalanine and tyrosine for the catecholamines, tryptophan for serotonin) as well as compounds important to the synthesis and catabolism of amino acids (methionine is a precursor of the methylation donor S-adenosylmethionine utilized in the catabolism of catecholamines, serotonin, and histamine).

The large neutral amino acids are transported across the blood-brain barrier by a common carrier, which has varying affinity constants and velocity maximums for each member of the group. The implication of this system is that any variation in the plasma concentration of one member of the group will affect the precursor pool for all members of the carrier group. The simplest expression of this relationship is as a ratio of the plasma concentration of the amino acid in question (as the numerator) to the sum of all the other members of the transport group (as the denominator) (27,28). The true velocities of each compound may actually be expressed in terms of their influx rates modified for individual affinity and velocity constants (Table 4) (29,30).

The general implication of this system is that if the concentration of one amino

TABLE 4. *Expression of competition of the large neutral amino acids at the blood-brain barrier[a]*

1. Ratio:

$$\frac{[AA]_i}{\sum [AA]}$$

$[AA]_i$ = individual plasma amino acid concentrations
$\sum [AA]$ = sum of large neutral plasma amino acid concentrations

2. Kinetic equation:

$$K^{aa}_{mapp} = K^{aa}_m \left(1 + \sum \frac{[AA]}{K^{aa}_m}\right)$$

$$v = \frac{V^{aa}_{max} [AA]}{K^{aa}_{mapp} + [AA]}$$

aa = individual amino acid
K_m = affinity constant, μM
K_{mapp} = modified affinity constant reflecting amino acid pattern μM
[AA] = plasma amino acid concentration, μM
V_{max} = velocity constant, μmol/min/g
v = brain amino acid influx, μmol/min/g

Adapted from Pardridge WM. *Nutrition and the Brain,* Vol. 1. New York: Raven Press, 1977; 141–204.
[a] The ratio gives all amino acids the same weight, while the kinetic method allows for the difference in affinities of each amino acid for the carrier. For example, tyrosine has a much higher affinity (K_m = 160 μm) than that of valine (K_m = 630 μm).

acid is increased (phenylalanine, for example), transport into the brain of the others in the group will be decreased. The ratio approach gives equal weight to all the amino acids in the group; however, some have lower affinity constants than others (such as tyrosine compared to threonine), so may have a greater actual impact on the transport system.

Such a system may put the synthesis of catecholamines at particular risk in the TPN-fed infant who has a combination of an increased plasma phenylalanine and a decreased plasma tyrosine, particularly limiting tyrosine access to the brain. Such a situation is significant because the brain has little capacity to synthesize tyrosine from phenylalanine and so is dependent on an exogenous source of this amino acid (31). Indeed, tyrosine should be considered an essential amino acid for the brain, which raises the concept of essentiality at the level of individual tissues as well as at that of the whole body.

Limitations on access of tyrosine to the brain (as well as of other neurotransmitter precursors) are important because the enzymes responsible for regulating the catalysis of neurotransmitter synthesis (e.g., tyrosine hydroxylase and tryptophan hydroxylase) have affinity constants that are higher than the endogenous concentrations of their precursors (32,33). The effective result of this situation is that changing brain tyrosine (or tryptophan) concentrations will change the concentrations of the catecholamine (or serotonin) products. Thus a biochemical sequence of events exists to permit modification of central nervous system neuroactive compounds when peripheral concentrations of plasma amino acids are modified.

Animal studies support this general mechanism; the data indicate that varying the protein in the diet causes modifications of plasma amino acid concentrations, which are reflected by changes in brain concentrations (34). These changes appear to reflect particularly the calculated brain influxes of amino acids (29,30) rather than just plasma concentrations (35). It is interesting to note that even peripheral catecholamine concentrations (synthesized in the adrenal gland) may reflect a dietary supply of tyrosine (36).

These animal studies lay a foundation for understanding the results of a number of studies in humans that have associated amino acid modifications due to diet with cognitive outcome. High protein intake in preterm infants [similar to that reported to result in abnormal amino acid patterns (12,14)] has been associated with an increased incidence of low IQ scores (37). Both term and preterm infants exposed to dietary (high protein, casein-predominant formula)-induced tyrosinemia during early infant feeding have been reported to have reduced intellectual outcome (38,39). Preterm infants have also been reported to have reduced behavior indices when exposed to low protein diets (40,41). Such changed behavior appears to be related to the amount of the protein intake as well as to changes in specific plasma amino acid concentrations, particularly of the large neutral amino acids (41).

There is additional experimental evidence that specific short-term infant behaviors may be modified by just such a mechanism as that described above. When healthy infants were fed tryptophan, sleep (modulated by serotonin) was induced more rapidly than when the infants were fed unmodified formula (42). In contrast, when the infants were fed the tryptophan transport competitor valine, sleep was induced much later than when formula was fed alone (42). Secondary dietary effects, such as modification of carbohydrate, which would cause increased branched-chain amino acid uptake into muscle as a result of stimulating insulin secretion, also might affect sleep by reducing competing amino acid concentrations (43). While carbohydrate feeding resulted in changing sleep patterns compared to either water or balanced formula, it did not increase sleep more than formula (43). Thus the situation may be more complex than simple modification of tryptophan concentrations. Human milk and formulas are complex nutrient mixtures and it is probably unreasonable to expect the properties of single nutrients to be unaffected by the presence of all the other nutrients in these preparations.

The sulfur-containing amino acids have not been analyzed in like manner in human infants, but animal data (and some human data) indicate that modifications in concentrations of these compounds also affect brain development. Cysteine has been shown to be neurotoxic in the developing rodent hypothalamus, suggesting that giving excessive amounts to compensate for the lack in situations such as TPN may backfire (44). Taurine, on the other hand, has been shown to be important for retinal and cerebellar development in both cats and monkeys (7,8,45,46), so situations that limit its presence may be of concern. One study of children on long-term total parenteral nutrition lacking taurine found similar retinal changes to those observed in taurine-deficient cats associated with very low plasma taurine concentrations (47). This finding suggests that human infants may be liable to retinal damage when exposed to long-term taurine-free nutrition. Although only minimal experimental evidence support this hypothesis in human infants, the pediatric TPN solutions and all infant formulas are now supplemented with taurine.

CONCLUSION

Appropriate protein nutrition of the infant is important for structural growth and development. However, less appreciated is the importance of such nutrition for the support of other functions, particularly within the central nervous system. The emerging data that demonstrate improved intellectual development in infants fed human milk suggests that characterizing the role of nutrients, such as amino acids, in the central nervous system is of great importance in establishing the most appropriate composition of early nutrition for the support of optimal outcome.

ACKNOWLEDGMENT

The author is extremely grateful for the expert secretarial assistance of Mrs. Deborah LaVictoire.

REFERENCES

1. Snyderman SE, Boyer A, Roitman E, Holt LE. The histidine requirement of the infant. *Pediatrics* 1963; 31: 786–801.
2. Snyderman SE. The protein and amino acid requirements of the premature infant. In: Jonxis JHP, Visser HKA, Troelstra JA, eds. *Metabolic processes in the foetus and newborn infant*. Leiden: Stenfert Kroese, 1971; 128–41.
3. Sturman JA, Gaull GE, Räihä NCR. Absence of cystathionase in human fetal liver. Is cystine essential? *Science* 1970; 169: 74–6.
4. Del Valle JA, Greengard O. Phenylalanine hydroxylase and tyrosine aminotransferase in human fetal and adult liver. *Pediatr Res* 1976; 11: 2–5.
5. Jackson AA, Shaw JCL, Barker A, Golden MHN. Nitrogen metabolism in preterm infants fed donor breast milk: the possible essentiality of glycine. *Pediatr Res* 1981; 15: 1454–61.
6. Rassin DK, Sturman JA, Gaull GE. Taurine and other free amino acids in milk of man and other mammals. *Early Hum Dev* 1978; 2: 1–13.
7. Hayes KC, Carey RE, Schmidt SY. Retinal degeneration associated with taurine deficiency in the cat. *Science* 1975; 188: 949–51.
8. Sturman JA, Wen GY, Wisniewski HM, Neuringer MD. Retinal degeneration in primates raised on a synthetic human infant formula. *Int J Dev Neurosci* 1984; 2: 121–30.
9. Anderson TL, Heird WC, Winters RW. Clinical and physiological consequences of total parenteral nutrition in the pediatric patient. In: Greef JM, Soeterz B, Wesdorp RIC, Phaf CWC, Fischer JE, eds. *Current concepts in parenteral nutrition*. The Hague: Martinus Nijhoff, 1977; 111–27.
10. Zlotkin SH, Anderson GH. The development of cystathionase activity during the first year of life. *Pediatr Res* 1982; 16: 65–8.
11. Pencharz PB, Farri L, Papageorgiou A. The effects of human milk and low-protein formulae on the rates of total body protein turnover and urinary 3-methyl histidine excretion of preterm infants. *Clin Sci* 1983; 64: 611–6.
12. Rassin DK, Gaull GE, Heinonen K, Räihä NCR. Milk protein quantity and quality in low-birth-weight infants. II. Effects on selected essential and non-essential amino acids in plasma and urine. *Pediatrics* 1977; 59: 407–22.

13. Gaull GE, Rassin DK, Räihä NCR, Heinonen K. Milk protein quantity and quality in low-birth-weight infants. III. Effects on sulfur-containing amino acids in plasma and urine. *J Pediatr* 1977; 90: 348–55.
14. Rassin DK, Gaull GE, Räihä NCR, Heinonen K. Milk protein quantity and quality in low-birth-weight infants. IV. Effects on tyrosine and phenylalanine in plasma and urine. *J Pediatr* 1977; 90: 356–60.
15. Järvenpää A-L, Rassin DK, Räihä NCR, Gaull GE. Milk protein quantity and quality in the term infant. II. Effects on acidic and neutral amino acids. *Pediatrics* 1982; 70: 221–30.
16. Rassin DK, Gaull GE, Järvenpää A-L, Räihä NCR. Feeding the low-birth-weight infant. II. Effects of taurine and cholesterol supplementation on amino acids and cholesterol. *Pediatrics* 1983; 71: 179–86.
17. Janas LM, Picciano MF, Hatch TF. Indicies of protein metabolism in term infants fed human milk, whey-predominant formula, or cow's milk formula. *Pediatrics* 1985; 75: 775–84.
18. Kashyap S, Okamoto E, Kanaya S, et al. Protein quality in feeding low-birth-weight infants: a comparison of whey-predominant versus casein-predominant formulas. *Pediatrics* 1987; 79: 748–55.
19. Picone TA, Benson JD, Moro G, et al. Growth and serum biochemistries and amino acids of term infants fed formulas with amino acid and protein concentrations similar to human milk. *J Pediatr Gastroenterol Nutr* 1989; 9: 351–60.
20. Kretchmer N, Levine SZ, McNamara H, Barnett HL. Certain aspects of tyrosine metabolism in the young. I. The development of the tyrosine oxidizing system in human liver. *J Clin Invest* 1956; 35: 236–44.
21. Kretchmer N, Levine SZ, McNamara H. The *in vitro* metabolism of tyrosine and its intermediates in the liver of the premature infant. *Am J Dis Child* 1957; 93: 19–20.
22. Wurtman RJ. Aspartame effects on brain serotonin. *Am J Clin Nutr* 1987; 45: 799–801.
23. Rivera A, Bhatia J, Rassin DK, Gourley WK, Catarau E. *In vivo* biliary function in the adult rat: the effects of parenteral glucose and amino acids. *J Parenter Enteral Nutr* 1989; 13: 240–5.
24. Pohlandt F. Cysteine: a semi-essential amino acid in the newborn infant. *Acta Paediatr Scand* 1974; 63: 801–4.
25. Rodgers B. Feeding in infancy and later ability and attainment: a longitudinal study. *Dev Med Child Neurol* 1978; 20: 421–6.
26. Lucas A, Morley R, Cole TJ, Lister G, Leeson-Payne C. Breast milk and subsequent intelligence quotient in children born preterm. *Lancet* 1992; 339: 261–4.
27. Wurtman RJ, Fernstrom JD. Control of brain neurotransmitter synthesis by precursor availability and nutritional state. *Biochem Pharmacol* 1976; 25: 1691–6.
28. Cohen EL, Wurtman RJ. Nutrition and brain neurotransmitters. In: Winick M, ed. *Human nutrition: a comprehensive treatise*, Vol 1, *Nutrition: pre- and postnatal development*. New York: Plenum Press, 1979; 103–32.
29. Pardridge WM. Regulation of amino acid availability to the brain. In: Wurtman RJ, Wurtman JJ, eds. *Nutrition and the brain*, Vol 1. New York: Raven Press, 1977; 141–204.
30. Pardridge WM. Effects of the dipeptide sweetener aspartame on the brain. In: Wurtman RJ, Wurtman JJ, eds. *Nutrition and the brain*, Vol 7, *Food constituents affecting normal and abnormal behaviors*. New York: Raven Press, 1986; 199–241.
31. Guroff G, Lovenberg W. Metabolism of aromatic amino acids. In: Lajtha A, ed. *Handbook of neurochemistry*, Vol 3. New York: Plenum Press, 1970; 209–23.
32. Gibson CJ, Wurtman RJ. Physiological control of brain norepinephrine synthesis by brain tyrosine concentration. *Life Sci* 1978; 22: 1399–406.
33. Lovenberg W, Jequier E, Sjoerdsma A. A tryptophan hydroxylation in mammalian systems. *Adv Pharmacol* 1968; 6A: 21–36.
34. Voog L, Eriksson T. Relationship between plasma and brain large neutral amino acids in rats fed diets with different compositions at different times of the day. *J Neurochem* 1992; 59: 1868–74.
35. Glanville NT, Anderson GH. The effect of insulin deficiency, dietary protein intake and plasma amino acid concentrations on brain amino acid levels in rats. *Can J Physiol Pharmacol* 1985; 63: 487–94.
36. Morita T, Teraoka K, Oka M, Hamano S. Further studies on the relationship between tyrosine supply and catecholamine production in cultured adrenal chromaffin cells. *Neurochem Int* 1992; 20: 229–35.
37. Goldman HI, Goldman JS, Kaufman I, Liebman OB. Late effects of early dietary protein intake on low-birth-weight infants. *J Pediatr* 1974; 85: 764–9.
38. Mamunes P, Prince PE, Thornton HH, Hunt PA, Hitchcock ES. Intellectual deficits after transient tyrosinemia in the term neonate. *Pediatrics* 1976; 57: 675–80.

39. Menkes JH, Welcher DW, Levi HS, Dallas J, Gretsky NE. Relationship of elevated blood tyrosine to the ultimate intellectual performances of premature infants. *Pediatrics* 1972; 49: 218–24.
40. Tyson JE, Lasky RE, Mize CE, *et al.* Growth, metabolic response, and development in very-low-birth-weight infants fed banked human milk or enriched formula. I. Neonatal findings. *J Pediatr* 1983; 103: 95–104.
41. Bhatia J, Rassin DK, Cerreto MC, Bee DE. Effect of protein/energy ratio on growth and behavior of premature infants: preliminary findings. *J Pediatr* 1991; 119: 103–10.
42. Yogman MW, Zeisel SH. Diet and sleep patterns in newborn infants. *N Engl J Med* 1983; 309: 1147–9.
43. Oberlander TF, Barr RG, Young SN, Brian JA. Short-term effects of feed composition on sleeping and crying in newborns. *Pediatrics* 1992; 90: 733–40.
44. Olney JD, Ho OL, Rhee V. Cytotoxic effects of acidic and sulphur containing amino acids on the infant mouse central nervous system. *Exp Brain Res* 1971; 14: 61–76.
45. Sturman JA, Hayes KC. The biology of taurine in nutrition and development. *Adv Nutr* 1980; 3: 231–99.
46. Sturman JA, Moretz RC, French JH, Wisniewski HM. Postnatal taurine deficiency in the kitten results in a persistence of the cerebellar external granule cell layer: correction by taurine feeding. *J Neurosci Res* 1985; 13: 521–8.
47. Geggel HS, Amant ME, Heckenlively JR, Martin DA, Kopple JD. Nutritional requirement for taurine in patients receiving long-term parenteral nutrition. *N Engl J Med* 1985; 312: 142–6.

DISCUSSION FOLLOWING THE PRESENTATION OF DR. RASSIN

Dr. Räihä: Do you think that the reason why we don't see lesions in humans on a taurine-deficient diet is that if it takes 10 weeks in the cat, it may take much longer in the human? Most human infants are starting to be fed with supplementary food after 3–4 months, so it is possible that the reason that we don't see the lesions is because the infants are getting supplementary foods that contain taurine.

Dr. Rassin: I think that is an extremely important consideration. The only cases in which human infants have been shown to develop abnormal electroretinograms have been in a small cohort of children studied by Ament (1). They were given total parenteral nutrition for more than 6 months because they had no gut, and when they were examined, four out of six had abnormal electroretinograms; they also had low plasma taurine because they were getting no taurine in their TPN solution. When taurine was supplemented I think three of the four abnormal ones responded. It is interesting to note that they found no abnormal electroretinograms among a group of adults on TPN.

Dr. Räihä: The second question relates to the Brazelton score in those infants who were fed various levels of protein intake derived from formula. Do you have any data on the same taurine intakes supplied with human milk? I think this would be a good control.

Dr. Rassin: Unfortunately, unlike in Sweden, we were unable to find human-milk-fed infants in our population. It is a real problem. John Tyson in Dallas has done a study comparing breast-fed and formula-fed preterms, looking at the Brazelton score and relating it to the lower protein in human milk (2). He found lower Brazelton scores in the breast-fed infants, which I think is a matter for concern, but I would remind you that we don't really know what the Brazelton score means in the long term. There are some indications that is related to long-term cognitive outcome, but on the other hand, it may only reflect the current status of the infant.

Dr. Räihä: In the rabbits that received the low tyrosine TPN, was it low plasma tyrosine or brain tyrosine?

Dr. Rassin: Both were low.

Dr. Räihä: Did they have enough phenylalanine hydroxylase to synthesize tyrosine from phenylalanine? The preterm infant has.

Dr. Rassin: Every study I have ever seen that has looked at total parenteral nutrition in infants has shown a low plasma tyrosine and some have shown high phenylalanine. Enterally fed infants on casein formulas have very high tyrosine levels. They seem to be able to convert phenylalanine to tyrosine efficiently. However, when we look at parenterally fed infants, this is not the case.

Dr. Kashyap: Was the amino acid mixture you used one of the older ones without added tyrosine, or was it one of the newer ones where tyrosine is added?

Dr. Rassin: The TPN solution we use is one of the older ones—I have to admit that we use it for selfish reasons, because it has no taurine in it and we were interested in taurine and cysteine metabolism when these compounds were given as supplements. In the babies we have looked at who have been treated with one of the newer formulations containing more tyrosine, plasma tyrosine is still low. I think this is in agreement with other studies.

Dr. Kashyap: Do plasma taurine levels increase with age because the infants are taking additional taurine in their diet or because of maturation of enzymes that convert cysteine to taurine?

Dr. Rassin: I don't think the taurine synthetic enzyme, cysteine sulfinic acid decarboxylase, has any maturational pattern in humans. Everybody in this room has the same activity of this enzyme as preterm babies. We probably have better stores of taurine and so can last longer than babies without becoming deficient, but in fact we cannot synthesize it any better than babies can.

Dr. Kashyap: Does this mean that we should regard taurine as an essential nutrient?

Dr. Rassin: If taurine is indeed important to us, then we have to maintain our supplies. We know that vegetarians become taurine-deficient and the milk of vegetarian mothers is also taurine-deficient because there is very little taurine in non-meat products. Taurine and zinc are similar in that they are both nutrients that we probably need to get from some kind of meat ingestion.

Dr. Fern: By and large, vegetarians live normal lives without complications. We should be careful about what we call "deficient" and what is simply a lowering of the plasma concentrations. Just because taurine levels go down in the plasma does not mean that there is a deficiency state. You said at the beginning of your talk that the cat could make more taurine than humans. We have been doing a lot of work with ^{15}N and we think it is the other way round.

Dr. Rassin: I can only say that our enzymatic data show that this is not true. Perhaps in the adult, vegetarianism does not have any severe consequences, but I would refer you to the literature on babies breast-fed by vegetarian mothers and there seem to have been fairly serious consequences to a cohort of these infants (3,4). I don't think it is entirely benign to be vegetarian; I think it does have consequences and it is consequences for the baby that we are perhaps most interested in.

Dr. Heine: You mentioned that there is no phenylalanine hydroxylase in the brain. Has this any consequences for long-term nutrition?

Dr. Rassin: I don't think that is known. We are very actively looking at catecholamine metabolism. The only studies we have done so far have looked at whole brain catecholamine concentrations, which is probably not a very good way to look at what happens in response to the low brain tyrosine. You really need to look at specific areas, such as the hypothalamus, and that is the direction we are going toward now. If we find low concentrations of those compounds in catecholamine-rich areas, it really raises important concerns about reducing the access of tyrosine into the brain.

Dr. Heine: My other question is related to the essentiality of glycine in premature infants. Alan Jackson claims that it is an essential amino acid and we know it is incorporated in many metabolic processes. What is known about the concentration of glycine in preterm infant formulas? Are these concentrations sufficient to meet the requirements of preterm infants?

Dr. Rassin: We never saw any differences in glycine concentrations among the three feeding groups that I showed you. There is an enormous amount of glycine in parenteral nutrition solutions. We have found that puppies on TPN tend to be somnolent. Glycine is thought to be an inhibitory neurotransmitter and I have sometimes wondered, based on the very large concentrations of glycine that we found in the brains of these puppies, whether we were making them somnolent with millimolar concentrations of cerebral glycine. This illustrates the fact that these compounds do have neurologic activity on their own as well as through some of their metabolic interactions.

Dr. Bremer: I think we are half blind if we measure only amino acids because there are so many steps of degradation. If you analyze urine from premature infants, you find spectra of organic acids in the urine very comparable to inborn errors of metabolism. These are usually combined with coenzyme A, and this certainly has very profound effects on metabolism. We need to extend our techniques into this marginal area.

Dr. Rassin: One of the things we don't talk about much are the implications of vitamin B-6. This is the major cofactor involved in protein and amino acid metabolism and whenever you start loading protein you may be reducing the B-6 pool and changing some of the metabolic pathways and products.

Dr. Cooper: I am interested in whether anyone has looked at taurine in relation to retinopathy of prematurity and whether there is any association of taurine deficiency with that disease.

Dr. Rassin: In the few studies that I have seen there has been no close relationship between taurine and the retinopathy of prematurity, and ROP develops much earlier than the taurine-deficient lesions in cats. The lesions themselves are also different.

Dr. Marini: You have an indirect estimation in the human of amino acids going into the brain, by looking at the levels in venous blood and at the K_m value of the particular amino acid. Was this system validated by measuring arteriovenous extraction and brain blood flow?

Dr. Rassin: No, but we discussed this with Dr. Partridge, who developed this approach. His group has done comparative studies across a variety of species and found that although the individual K_m values may differ somewhat, the relative K_m values don't, and certainly if you go from the rat to the guinea pig to the rabbit you get essentially the same relative influxes by this kind of calculation. I could not tell you for certain whether those K_m values are the true K_ms, but I think that the relative K_m values for each amino acid are probably appropriate, so you get a fairly solid relative competition and influence for each amino acid.

Dr. Marini: Do you think that the amino acid content of red blood cells can be an indicator of what is going on in other tissues?

Dr. Rassin: We have tried to use this because we have been very interested in glutathione metabolism and red blood cells are the primary repository for glutathione in the circulation. We found that the red cell values did not reflect liver or brain glutathione concentration changes that we observed in our animals. I think this is of some concern. Tom Hanson's group has similar findings. They have also been trying to relate glutathione levels in the plasma to what goes on in the tissues, which we haven't tried to do yet. That is difficult because there is very little glutathione in the plasma, but it may be a better reflection of what goes on in the tissues than what is actually in the red cells.

REFERENCES

1. Geggel HS, Ament ME, Heckenlively JR, Martin DA, Kopple JD. Nutritional requirement for taurine in patients receiving long-term parenteral nutrition. *N Engl J Med* 1985; 312: 142–6.
2. Tyson JE, Lasky RE, Mize CE, et al. Growth, metabolic response and development in very-low-birth-weight infants fed banked human milk or enriched formula. I. Neonatal findings. *J Pediatr* 1983; 103: 95–104.
3. Zmora E, Gorodescher R, Bar-Ziv J. Multiple nutritional deficiencies in infants from a strict vegetarian commune. *Am J Dis Child* 1979; 133: 141–4.
4. Higginbotham MC, Sweetman L, Nyhan WL. A syndrome of methylmalonic aciduria, homocystinuria, megaloblastic anemia and neurologic abnormalities in a vitamin B_{12}-deficient breast-fed infant of a strict vegetarian. *N Engl J Med* 1978; 299: 317–23.

Significance of Nucleic Acids, Nucleotides, and Related Compounds in Infant Nutrition

Ricardo Uauy-Dagach and *Richard Quan

*Instituto de Nutrición y Tecnología de los Alimentos (INTA), Universidad de Chile, Casilla 138-11, and Hospital Sótero del Río, Santiago, Chile; and *University of Texas Southwestern Medical Center at Dallas, Dallas, Texas 75235-9063, USA*

Nucleotides are low molecular weight intracellular compounds that play major roles in physiological and biological functions. They act as precursors for nucleic acid synthesis and are also fundamental for intermediary metabolism. Nucleotides and nucleic acids turn over rapidly, especially in growing tissues or those undergoing constant cell renewal. Tissues that grow have a net formation of new DNA and a rapid turnover of RNA (1,2).

Nucleotides are typically conformed by a nitrogenous base (purine or pyrimidine), a pentose (ribose or deoxyribose), and one or more phosphate groups. The nitrogenous bases are derived from two parent heterocyclic molecules. The major purines found in living organisms are adenine and guanine, while cytosine, thymidine, and uracil are the major pyrimidine bases. Nitrogenous bases can be formed *de novo* from amino acid precursors or reutilized after their release from nucleic acid breakdown via the *salvage pathway*. The purine ring carbon atoms formed by *de novo* synthesis are derived from the dispensable amino acids glycine, glutamine, and aspartate. The carbon atoms for the *de novo* synthesis of pyrimidines are derived from carbamoyl phosphate and aspartate. Figure 1 summarizes the structure and precursors needed for the biosynthesis of purine and pyrimidine bases. The energy cost of *de novo* synthesis of nitrogenous bases requires significantly more energy than the salvage process, 5 mol of ATP as opposed to 1 mol of ATP for each nucleotide monophosphate produced; thus from an evolutionary point of view the salvage pathway has a selective advantage for rapidly growing organisms (3,4).

DIETARY NUCLEOTIDES: ABSORPTION AND METABOLISM

Dietary nucleotides are ingested mainly as nucleic acids and nucleoproteins derived from nuclear material. Digestion of nucleoproteins is initiated by proteases. Nucleic acids undergo partial hydrolysis in the stomach and are then subjected to pancreatic nucleases and phosphoesterases to yield nucleotides and nucleosides. Most of the

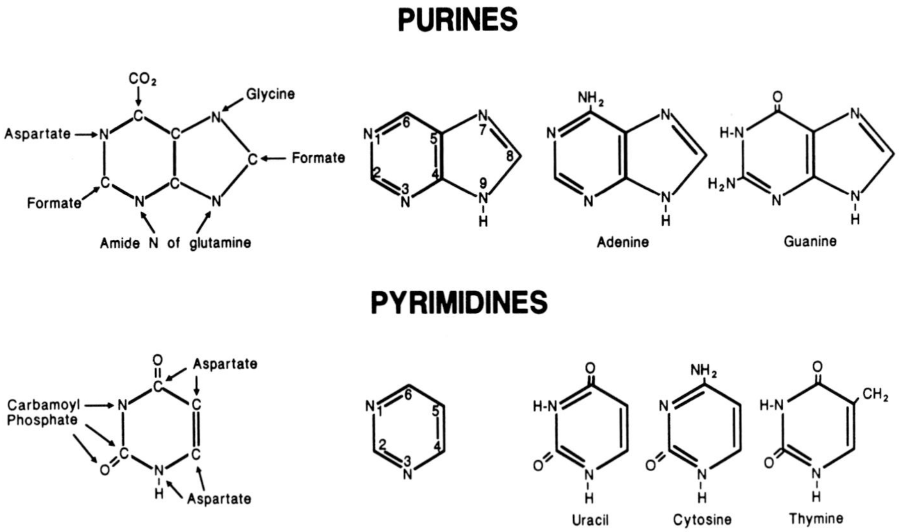

FIG. 1. Major purine and pyrimidine bases and precursors for *de novo* synthesis. Arrows indicate structural carbon and nitrogen atoms provided by precursors. Simplified ring structure for adenine, guanine, uracil, cytosine, and thymine are given on the right side. [Adapted from Uauy R. *Textbook of gastroenterology and nutrition in infancy*, Vol. 2. New York: Raven Press, 1989; 265–280.]

DNA and RNA is fully hydrolyzed to nucleotides in the gut. Alkaline phosphatases in the enterocyte will cleave the phosphate groups in nucleotides to form nucleosides; nucleosidases will release the sugar moiety and hence the free nitrogenous bases (5,6). The evidence suggests that a mixture of nucleosides and nitrogenous bases is offered to the enterocyte for absorption. Transport studies using everted intestinal sacs and isolated perfused loops have demonstrated a highly efficient Na^+-dependent active transport system for nucleosides. Over 90% of the ingested nucleotides are absorbed, yet less than 5% are incorporated into intestinal nucleic acids and a relatively smaller amount appears in hepatic cells (7–9). Most of the absorbed purines are degraded to uric acid in the gut. The studies conducted to date have been performed under relative excess nitrogenous base supply and in mature adult animals. Whether these observations are applicable to earlier developmental stages remains undetermined (5).

The evidence suggests that the intestine not only salvages significant amounts of nitrogenous bases and nucleosides but also oxidizes the excess purines to form uric acid (7–10). A developmental pattern of coordinated expression of key enzymes involved in purine catabolism has been described. A study by Witte *et al.* demonstrated a postnatal rise in the expression and products of the 5′-nucleotidase, alkaline phosphatase, adenosine deaminase, purine nucleoside phosphorylase, xanthine oxidase, and guanase genes in the mouse proximal gastrointestinal tract. It may be speculated that the high purine degradation activity is possibly linked to the need to stabilize the concentrations of adenosine, a potent gut neurohormone that could have potential toxic effects; it has also been suggested that intestinal degradation of nucleotides may be advantageous since it would restrict nucleotides available for microbial or parasite growth [for some organisms, such as protozoa, purine bases are essential nutrients (11)].

The diet-induced increase in the transcription of genes and increased activity of enzymes responsible for the salvage pathway in the upper intestine suggests that evolution has privileged the salvage pathway over *de novo* nucleotide formation in this cell type (12–14). A molecular mechanism that may be responsible for the regulation of the salvage pathway by dietary purines has now been characterized using site-specific mutation of the 5' flanking region of the hypoxanthine guanine phosphoribosyl transferase (HGPRT) gene and transfecting the corresponding DNA constructs into various cell types (15). The regulatory response to dietary purines has been linked to a specific region upstream from the transcriptional start site of the HGPRT gene. Concomitantly, the diet also induces a rise in adenosine deaminase and other key enzymes that regulate the catabolism of purines (11). Figure 2 depicts the regulation of cellular nucleotide pools in the intestine. Synthesis can occur by the two

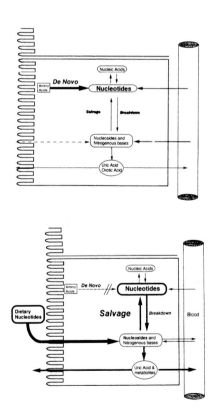

FIG. 2. Regulation of enterocyte nucleotide pools in the absence of nucleotides in the diet. **Top** panel shows enhanced *de novo* synthesis and in the presence of nucleotides; **Bottom** panel shows activation of the salvage pathway and oxidation of purine to form uric acid. [Adapted from Quan R, et al. *J Pediatr Gastroenterol Nutr* 1990;11:429–437.]

alternative paths, depending on what is fed. When purines are fed, a significant proportion of the bases are oxidized rapidly to form uric acid. Thus the intestine plays an important role in defining the metabolic fate of dietary purines; less is known about the regulation of pyrimidine metabolism by the enterocyte, although recent evidence suggests that a greater proportion of these bases are absorbed and incorporated into tissue nucleic acids (16).

Evidence from isotopomer studies in mice and chicken yet unpublished using food sources uniformly labeled with ^{13}C suggests that pyrimidine nucleosides are handled differently from purine nucleosides at the whole body level in terms of their metabolic fate. Dietary adenosine and guanosine are quickly oxidized, and most isotopomers that remain have a mass (m) of m or m + 1, and virtually no intact purine nucleosides, (m + 10) are incorporated into liver nucleic acid pools. In essence, purines are treated like most dispensable amino acids. In contrast, a significant proportion of pyrimidine nucleosides are absorbed intact as m + 9 isotopomers and incorporated into chicken liver nucleic acids. Cytidine and uridine m + 9 isotopomers are conserved in liver nucleoside pools, demonstrating a slow decay in the specific activity of the uniformly labeled isotopomer species with few intermediate mass forms (16). This pattern is similar to what is observed with the disposition of the indispensable amino acids.

A need for nucleotides for the development and maturation of the gastrointestinal tract and the immune system can be suggested on a biochemical basis since they lack or have a limited capacity for *de novo* synthesis of the nitrogenous bases (17,18). The biochemical evidence also indicates that when nucleotides are fed, the gut will activate the transcription and expression of the HGPRT gene, thus enhancing the salvage of preformed nucleosides from the diet (14).

BIOLOGICAL SIGNIFICANCE OF NUCLEOTIDES (ANIMAL AND HUMAN STUDIES)

Effects on the Immune System

The possibility of a role for exogenous nucleotides in the modulation of normal immune response has been suggested by experimental studies. It has been demonstrated that in the Balb C mouse maintained on a nucleotide-free diet, the survival of a cardiac allograft was significantly prolonged as compared to a nucleic acid supplemented group (19). In addition, animals receiving a nucleotide-free diet had significant suppression of lymphoproliferative response to alloantigens. Similarly delayed cutaneous hypersensitivity response upon challenge with purified protein derivative was diminished in the nucleotide-free group (20). These experiments suggest that T lymphocytes are the target of dietary nucleotide deprivation, yet the mechanism responsible for these effects remains undetermined.

The lack of nucleotides in the diet has been shown to decrease resistance to staphylococcal sepsis in the mouse and adversely affects host resistance to *Candida albicans* (21). Macrophages from mice receiving the nucleotide-free diet also showed diminished phagocytic activity (22). The addition of nucleotides to a nucleotide-free diet determined an increase of the phagocytic capacity of mouse macrophages, along with an increase in the natural killer cell activity of spleen cells (23). Carver and Barness also recently showed that 13 infants fed nucleotide-supplemented formula had similar natural killer (NK) activity to that of nine breast-fed infants, and both groups had significantly higher NK activity than 15 infants receiving non-supplemented formula (24). Additional studies on adult hospital inpatients indicate that the enteral nutrition providing nucleic acids as well as arginine and ω-3 fatty acids in the form of a commercial product preserved immune responsivity in surgical patients as compared to a nucleotide-free product given at comparable protein and energy intakes (25,26).

Effects on Gut Microflora

Nucleotides modify the intestinal microflora; *in vitro* experiments have revealed that the addition of nucleotides to bifidobacteria in minimal culture media enhances their growth (27). Young infants fed nucleotide-supplemented formula have higher percentages of fecal bifidobacteria and lower percentages of gram-negative enterobacteria than those of formula-fed infants (28). Thus it is possible that dietary nucleotides may favor the development of fecal flora with a predominance of bifido bacteria, similar to what is observed in the breast-fed infant (2,6). Bifidobacteria have a number of potential benefits for the infant. They lower the pH of the colonic content due to their capacity to hydrolyze sugars to lactic acid, which in turn suppresses the proliferation of pathogenic bacteria that are acid-intolerant. Bifidobacteria also inhibit the growth of some enterobacteria responsible in part for diarrheal disease (2,6,29).

Effects on Intestinal Development and Repair

The evidence to date indicates that nucleotides promote intestinal growth and maturation in the young rat. We studied weanling rats fed either a nucleotide-free semisynthetic diet or the same diet fed isocalorically and supplemented with 0.8% nucleosides over a 2-week period. Villus height, crypt depth, total protein, and DNA content of the proximal gut were higher in nucleotide-supplemented young animals than in those fed the nucleotide-free semisynthetic diet. Maltase-specific activity in the mucosa throughout the gut was also higher in the supplemented animals, while sucrase and lactase activity were less affected (30). The effects were always more pronounced in the duodenum and proximal jejunum. These segments are exposed to higher concentrations of nucleotides derived from the diet and, based on the expression of salvage enzymes, may be more responsive to changes in dietary nucleotide supply. Figure 3 summarizes our observations on changes in disaccharidase activity in nucleotide-supplemented weanling rats. Similar studies conducted by Barness and Barness using scanning electron microscopy demonstrated increased villus density and deeper crypts in the nucleotide-supplemented animals (5).

Recent evidence indicates that recovery from radiation-induced injury is enhanced by the ingestion of nucleotides after exposure to sublethal radiation doses. We have

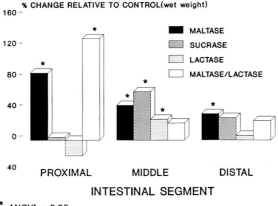

FIG. 3. Change in disaccharidases observed in nucleoside supplemented weanling rats expressed as percent change relative to control animals fed a nucleotide-free chemically defined diet over a 2-week period. Activity expressed as amount of substrate hydrolyzed per minute per gram of wet weight. Maltase/lactase ratio was calculated as an index of intestinal maturation. Two-way ANOVA was used to test for diet and intestinal segment effects. [Adapted from Uauy R, et al. *J Pediatr Gastroenterol* 1990; 10: 497–503.]
* $p < 0.05$ using *post-hoc* Newman-Keuls tests.

reported preliminary results from studies in the rat demonstrating an increase in mortality over the 10 days following radiation in chow-fed animals and a lower mortality in a 0.8% nucleotide-supplemented group relative to controls fed the nucleotide-free semisynthetic diet. Less inflammation was identified in the histology of the intestinal sections in the nucleotide-supplemented group. Maltase and sucrase specific activity were significantly higher in the nucleotide-supplemented groups 5 days after injury (31). The results on differences in mortality were fairly dramatic, but the interpretation of improved survival of the nucleotide-supplemented animals may be confounded by the potential effects of nucleotides in the recovery of the gut-associated lymphoid system and on other immune functions after radiation.

The potential for a beneficial effect of nucleotide supplementation during the recovery from diarrheal illness has also been investigated in the weanling rat using the lactose-induced chronic diarrhea model (32). After 2 weeks of lactose feeding, animals were randomized to receive a control semisynthetic diet or the same diet supplemented with 0.5% nucleotides. After 4 weeks of feeding the study diets, the nucleotide-supplemented animals had higher maltase, lactase, and sucrase activities than those of the controls. Histologic and ultrastructural analysis of the intestine from these animals showed greater improvement in the villous height/crypt depth ratio, fewer intraepithelial lymphocytes, and enhanced recovery of goblet cell population in the nucleotide-supplemented animals relative to those that recovered from the lactose-induced injury on a nucleotide-free diet (32). Mitochondrial matrix density and cristae were also closer to the normal animals in the nucleotide-supplemented group.

A preliminary report from a randomized, blind, controlled study in young infants indicates that feeding a nucleotide-supplemented diet over a 3-month period decreases the incidence of diarrhea while disease severity remains unaltered. After weaning, the group fed a formula supplemented with nucleotides in concentrations similar to human milk had fewer episodes of diarrhea, and the total number of days with the disease was less relative to the control group fed formula without added nucleotides (33).

Effect on Tissue Growth and Repair

In addition to these effects, nucleotides provided intravenously have been shown to ameliorate the injury induced by D-galactosamine, a hepatotoxic agent, as well as to promote the recovery from injury induced by this agent (34). Studies in parenterally fed animals after stress have demonstrated enhanced nitrogen balance with nucleotide supplementation (35). It has been suggested that the use of nucleotides in patients receiving total parenteral nutrition after surgical stress may promote nitrogen retention and thus enhance protein utilization when dietary supply is marginal. The beneficial effects of nucleotides in these circumstances may be explained by improved immunity or alternatively, may be due to the protein sparing effect of nucleotides. A portion of the glycine, aspartate, and glutamine pools would otherwise have been used for *de novo* nitrogenous base synthesis.

Other Effects

Studies in weanling rats and humans have indicated that polyunsaturated fatty acids and lipoprotein metabolism may be modulated by nucleotide supply of the early diet (36–38). Infants fed nucleotide-supplemented formula show, over the initial month of life, higher high-density lipoprotein concentrations and increased concentration of n-3 and n-6 fatty acids of greater than 18 carbon chain length (37,39,40). In conclusion, the results to date suggest that nucleotides may play a potentially significant role in infant nutrition.

IMPLICATIONS FOR INFANT NUTRITION

Increasing attention is being given to the biological effects of nutrients that traditionally are considered non-essential, but which under some conditions may be essential; that is, their endogenous supply may be insufficient for full normal function, but their lack does not lead to a classical nutritional deficiency syndrome (2,5,6,41). The cell or the body may have the biochemical pathways to synthesize these compounds, but regulatory or developmental factors may interfere with the full expression of this capacity. Recent studies suggest that dietary nucleotides (purines and pyrimidine bases) may be conditionally essential for newborn animals. Rapidly growing tissue

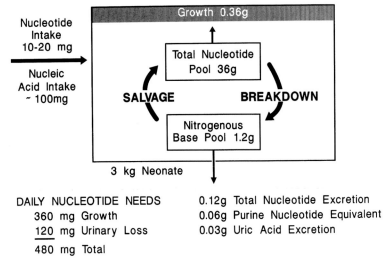

FIG. 4. Whole body nitrogenous base metabolism for a representative 3-kg neonate. Analysis based on assumptions described in detail in Uauy (2). Nucleotide requirements to replace urinary losses are estimated to be 120 mg per day and those for normal growth are estimated to be 360 mg per day. Total nucleotide pool includes soluble cytosolic nucleotides as well as nucleotides forming part of the nucleic acid pool. Nucleotide intake corresponds to free nucleotide content of human milk, depending on stage of lactation. Nucleic acid intake assumes that nucleic acid from the cells in human milk are fully absorbed. [Adapted from Uauy R. *Textbook of gastroenterology and nutrition in infancy, Vol. 2.* New York: Raven Press, 1989; 265–280.]

such as intestinal epithelium and lymphoid cells have an increased demand for purine and pyrimidine bases. This phenomenon may also occur in other tissues during recovery from injury.

Nucleic acids, nucleotides, and their related metabolic products are present in human milk in relatively large amounts: up to 20% of the non-protein nitrogen consists of free nucleotides, while most infant formulas are low in nucleotides and contain nucleotides of different type (2,5,6). Nucleic acid from cells present in human milk may also be considered a further source of available nucleotides for the breast-fed infant. Figure 4 presents a quantitative analysis of whole body nitrogenous base metabolism for a representative neonate. The assumptions for this analysis are described in detail elsewhere (2). We have estimated that a full-term breast-fed infant receives 10–20 mg/d as free nucleotides and 100–150 mg/d as nucleic acids, mainly DNA contained in the cells present in human milk (2,42). The amount of nucleotides required to replace the oxidative losses was estimated to be 120 mg/d, and the amount deposited in new tissues formed, 360 mg/d. Thus a total of 480 mg/d would be needed daily; for a human-milk-fed infant, about a third would come from the salvage of nucleotides and for a formula-fed, over 95% would come from *de novo* synthesis (2).

Is the addition of nucleotides to infant formula within the range of concentrations found in mature human milk safe? The evidence from several controlled clinical trials of nucleotide supplementation indicates that at the present level of addition there are no demonstrable adverse effects. The safety evaluation has included several long-term controlled trials in humans over a 9- to 12-month feeding period, showing similar growth and clinical laboratory findings in the nucleotide-fed group relative to control groups (2,5,6). Comparable studies done in preterm infants showed similar results (2,5,6). These studies demonstrated that nucleotide supplementation is safe. The evidence for efficacy is more tenuous, yet the bulk of the evidence indicates that nucleotide-fed infants are more like human-milk-fed infants: the induced changes in fecal flora enhance the growth of bifidobacteria and inhibit the growth of gram-negative enteric bacteria; in addition, changes in lipid patterns like those in breast-fed infants have been shown during the initial months of life.

It is clear that human milk has a nucleotide profile which is quite different from that of cow's milk formula (43,44). Human milk is particularly rich in cytidine and adenosine monophosphates, while cow's milk formula contains predominantly orotate, a by-product of pyrimidine catabolism. The decision of whether to add nucleotides to infant formula is a matter of choice for infant formula manufacturers and pediatricians. Regulatory agencies in the USA, Europe, Japan, and other countries have accepted the safety of this practice, and since these compounds are present in human milk, there is a high likelihood that they are of benefit to human infants (2,5,6). Several studies have found no adverse effects of feeding nucleotides and we are not aware of any by others in which the feeding of nucleotides within the range found in human milk has caused any problem. We agree with the position that the burden of proof should be on those who propose that artificial formulas that differ from human milk are safe and efficacious for optimal infant growth and development.

The need to demonstrate clinical efficacy is not at present a prerequisite for adding other conditionally essential nutrients. For example, taurine is now added to most formulas destined for full-term infants. No study has ever demonstrated any benefit of feeding taurine to healthy full-term infants. Another example is the addition of carnitine and choline to formulas for healthy infants. This practice is mainly based on human milk content and not on any evidence for their need from clinical trials. The standards that are applied to nucleotides should be no different from those applied to other nutrients. It is noteworthy that several national and international regulatory agencies have considered nucleotides to be appropriate supplements in order to make infant formula more like human milk. Table 1 provides a summary of the potential significance of nucleotides for infant nutrition.

We are of the opinion that human milk is the best source of nucleotides for young infants and that in the absence of breast feeding a nucleotide-containing formula

TABLE 1. *Potential significance of nucleotides to infant nutrition*

Enhance normal development of gastrointestinal tract and immune function
Promote recovery from gut injury (viral, radiation, drug, surgery, vascular, other)
Improve liver function after damage
Higher protein accretion after stress
Improve host resistance in immunocompromised patients

---

patterned after human milk is the best alternative for providing these important compounds artificially. The choice of what to add, how much to add, and for what infant groups is a difficult one to make because the evidence is incomplete, yet animal and human studies suggest that there are definite advantages of including them in the early diet. Presently the best approach is to use the nucleotide profile of human milk. This differs markedly from that of milk formula. We believe that as evidence on total nucleotide content of human milk becomes available, including the substantial amounts of nucleotides in the cells present in human milk, we may in fact suggest that nucleic acids as well as free nucleotides be added to infant formula. As specific results become available we can provide a more definitive answer to whether all infants should be fed nucleotide- and nucleic-acid-containing formula.

REFERENCES

1. Scriver CR, Beaudet AL, Sly WS, Valle D. Purines and pyrimidines. In: Scriver CR, Beaudet AL, Sly WS, Valle D, eds. *The metabolic basis of inherited disease*, Vol 6. New York: McGraw-Hill, 1989; 965–1126.
2. Uauy R. Dietary nucleotides and requirements in early life. In: Lebenthal E, ed. *Textbook of gastroenterology and nutrition in infancy*, Vol 2. New York: Raven Press, 1989; 265–80.
3. Lehninger AL. *Principles of biochemistry*. New York: Worth Publishers, 1982.
4. McGilvery RW. Turnover of nucleotides. In: *Biochemistry: a functional approach*. Philadelphia: WB Saunders, 1983; 674–94.
5. Quan R, Barness LA, Uauy R. Do infants need nucleotide supplemented formula for optimal nutrition? *J Pediatr Gastroenterol Nutr* 1990; 11: 429–37.
6. Gil A, Uauy R. Dietary nucleotides and infant nutrition. *J Clin Nutr Gastroenterol* 1989; 4: 145–53.
7. Savaiano DA, Clifford AJ. Absorption, tissue incorporation and excretion of free-purine bases in rat. *Nutr Rep Int* 1978; 17: 551–6.
8. Sonoda T, Tatibana M. Metabolic fate of pyrimides and purines in dietary nucleic acids ingested by mice. *Biochim Biophys Acta* 1978; 521: 55–66.
9. Jarvis SM. Characterization of sodium-dependent nucleoside transport in rabbit intestinal brush-border membrane vesicles. *Biochim Biophys Acta* 1989; 979: 132–8.
10. Savaiano DA, Ho Cy, Chu V, Clifford AJ. Metabolism of orally and intravenously administered purines in rats. *J Nutr* 1980; 110: 1793–804.
11. Witte DP, Wiginton AD, Hutton JJ, Aronow BJ. Coordinate development regulation of purine catabolic enzyme expression in gastrointestinal and postimplantation reproductive tracts. *J Cell Biol* 1991; 115: 179–90.
12. LeLeiko NS, Bronstein AD, Munro NH. Effect of dietary purines on the novo synthesis of purine nucleotides in the small intestinal mucosa. *Pediatr Res* 1979; 13: 401–3.
13. LeLeiko NA, Bronstein AD, Baliga S, Munro AN. De novo purine nucleotide synthesis in the rat small and large intestine: effect of dietary protein and purines. *J Pediatr Gastroenterol Nutr* 1983; 2: 313–9.
14. LeLeiko NS, Martín BA, Walsh M, Kazlow P, Rabinowitz S, Sterling K. Tissue-specific gene expression results from a purine- and pyrimidine-free diet and 6-mercaptopurine in the rat small intestine and colon. *Gastroenterology* 1987; 93: 1014–20.
15. Walsh MI, Sánchez-Pozo A, Leleiko NS. A regulatory element is characterized by purine-mediated and cell-type-specific gene transcription. *Mol Cell Biol* 1990; 10: 4356–64.
16. Berthold HK, Reeds PJ, Klein PD. Unpublished data.
17. Roux JM. Nucleotide supply of the developing animal: role of the so-called "salvage pathways." *Enzyme* 1973; 15: 361–77.
18. Savaiano DA, Clifford AJ. Adenine, the precursor of nucleic acid in intestinal cells unable to synthesize purines de novo. *J Nutr* 1981; 111: 1816–22.
19. Van Buren CT, Kulkarni AD, Schandle VB, Rudolph FB. The influence of dietary nucleotides on cell-mediated immunity. *Transplantation* 1983; 36: 350–2.

20. Van Buren CT, Kulkarni AD, Rudolph FB. Nucleotide deprivation retards cutaneous hypersensitivity (DCH). *J Parenter Enteral Nutr* 1983; 6: 582.
21. Kulkarni AD, Fanslow WC, Rudolph FB, Van Buren CT. Effect of dietary nucleotides on response to bacterial infections. *J Parenter Enteral Nutr* 1986; 10: 169–71.
22. Kulkarni AD, Fanslow WC, Drath D, Rudolph FB, Van Buren CT. Influence of dietary nucleotide restriction on bacterial sepsis and phagocytic cell function in mice. *Arch Surg* 1986; 121: 169–72.
23. Carver JD, Coc WI, Barness LA. Dietary nucleoside effects upon murine natural killer activity and macrophage activation. *J Parenter Enteral Nutr* 1990; 14: 18–22.
24. Carver JD, Pimentel B, Cox WI, Barness LA. Dietary nucleotide effects upon immune function in infants. *Pediatrics* 1991; 88: 359–63.
25. Pizzini RP, Saroj Kumar BS, Kulkarni AD, Rudolph FB, Van Buren CT. Dietary nucleotides reverse malnutrition and starvation-induced immunosuppression. *Arch Surg* 1990; 125: 86–90.
26. Van Buren CT. Role of dietary nucleotides in non-specific immune function. AIN Symposium on Dietary Nucleotides. *J Nutr* (Suppl) 1994 (in press).
27. Tanaka R, Mutai M. Improved medium for selective isolation and enumeration of bifido bacterium. *Appl Environ Microbiol* 1980; 40: 866–9.
28. Gil A, Coval E, Martínez A, Molina JA. Effects of dietary nucleotides on the microbial pattern of feces of at term newborn infants. *J Clin Nutr Gastroenterol* 1986; 1: 34–8.
29. Quan R. Dietary nucleotides: potential for immune enhancement. In: Paubert-Braquet M, Dupont Ch, Paoletti R, eds. *Foods, nutrition and immunity*, Vol 1. Basel: S Karger, 1992; 13–21.
30. Uauy R, Stringel G, Thomas R, Quan R. Effect of dietary nucleosides on growth and maturation of the developing gut in the rat. *J Pediatr Gastroenterol Nutr* 1990; 10: 497–503.
31. Quan R, Gil A, Uauy R. Effect of dietary nucleosides on intestinal growth and maturation after injury from radiation. (Abstract). *Pediatr Res* 1991; 29: 111A.
32. Núñez MC, Ayudarte MV, Morales D, Súarez MD, Gil A. Effect of dietary nucleotides on intestinal repair in rats with experimental chronic diarrhea. *J Parenter Enteral Nutr* 1990; 14: 598–604.
33. Espinoza J, Araya M, Cruchat S, Pacheco I, Brunser O. Nucleotide-enriched milk and diarrheal disease in infants. (Abstract). *Pediatr Res* 1992; 32: 739A.
34. Ogoshi S, Iwasa M, Kitagawa S, *et al*. Effects of total parenteral nutrition with nucleosides and nucleotide mixture on D-galactosamine-induced live injury in rats. *J Parenter Enteral Nutr* 1988; 12: 53–7.
35. Ogoshi S, Iwasa M, Yonezawa T. Effect of nucleotide and nucleoside mixture on rats given total parenteral nutrition after 70% hepatectomy. *J Parenter Enteral Nutr* 1985; 9: 339–42.
36. Gil A, Pita ML, Martínez A, Molina JA, Sánchez-Medina F. Effects of dietary nucleotides on the plasma fatty acids in at-term neonates. *Hum Nutr Clin Nutr* 1986; 40C: 185–95.
37. Uauy R, Gil A. Fatty acid metabolism in the neonate: effect of age, diet and nucleotide. In: *Proceedings of the III international symposium on infant nutrition and gastrointestinal disease*, Brussels, Belgium, 1985; 65–75.
38. De Lucchi C, Pita ML, Faus MJ, Molina JA, Uauy R, Gil A. Effect of dietary nucleotides on the fatty acid composition of erythrocyte membrane lipids in term infants. *J Pediatr Gastroenterol Nutr* 1987; 6: 568–74.
39. Sánchez-Pozo A, Pita ML, Martínez A, Molina JA, Sánchez-Medina F, Gil A. Effects of dietary nucleotides upon lipoprotein pattern of newborn infants. *Nutr Res* 1986; 6: 763–71.
40. Ramirez M, Hortelano P, Boza J, Jimenez J, Gil A, Pita ML. Effect of dietary nucleotides and orotate on the blood levels of prostacyclin and thromboxane in the weanling rat. *Prostaglandins Leukot Essent Fatty Acids* 1991; 43: 49–54.
41. National Research Council. Other substances in food. In: *Recommended dietary allowances*, 10th ed. Washington, DC: National Academy Press, 1989; 262–70.
42. Sanguansermsri J, Gyorgy P, Zilliken F. Polyamines in human and cow's milk. *Am J Clin Nutr* 1974; 27: 859–65.
43. Gil A, Sánchez-Medina F. Acid-soluble nucleotides of human milk at different stages of lactation. *J Dairy Res* 1982; 49: 301–4.
44. Janas LM, Picciano MF. The nucleotide profile of human milk. *Pediatr Res* 1982; 16: 659–62.

DISCUSSION FOLLOWING THE PRESENTATION OF DR. UAUY

Dr. Tracey: You drew a picture showing the nucleotides coming from the gut lumen, going through the various pathways, and back to the blood, and presented data suggesting that

208   NUCLEIC ACIDS/NUCLEOTIDES/RELATED COMPOUNDS

feeding nucleotides could improve enterocyte histology and DNA content. Have you looked to see if you can confer protection against radiation injury or enhance recovery from radiation injury when you supply nucleotides to the vascular side of your picture, perhaps by including them in TPN?

Dr. Uauy: That has not been tested. We know that if they are supplied to the gut there is active salvage. It would be a very nice experiment to do.

Dr. Schöch: We have done a breakdown balance study in human children and adults and found that on a diet rich in nucleic acids 90–95% of the nucleotides were catabolized and excreted.

Dr. Dhansay: Could you comment on which claims for nucleotides have not been substantiated?

Dr. Uauy: There are many claims that I don't think have been substantiated. For example, I think that it is still quite premature to decide that nucleotides should be give in formula to normal infants. I don't think we can substantiate the suggestion that they potentiate the immune system in the human. I don't think we can say that giving nucleotides reduces the length of hospital stay, as has been claimed. Ninety percent of the data I showed were obtained in animals. This needs to be taken into account when making a decision about giving nucleotides to babies.

Dr. Räihä: In light of the effect on intestinal injury that you showed in the experimental animals, have any studies been done showing that nucleotides could have an effect on necrotizing enterocolitis in very low birthweight infants? We know that this is a problem in formula-fed infants, but we don't see it very often in infants fed on human milk, which contains nucleotides.

Dr. Uauy: There are many other things that I would do to try to prevent NEC before giving nucleotides. However, it might be worth trying nucleotide-enriched products to assist recovery from NEC, since substantial gut damage occurs in this condition.

Dr. Garlick: There is a parallel between the nucleotide-induced improvements in the gut that you showed and what people are now claiming for glutamine. It is said that the enterocyte requires glutamine as an oxidative substrate, but of course glutamine could also be providing part of the nucleic acid. Do you think that may be an explanation for its effect?

Dr. Uauy: That is potentially valid information and obviously labeling studies with labeled glutamine would be useful in determining how much in the label goes into purine.

Dr. Heine: You mentioned the data published by Tanaka from Japan. It is not very astonishing that bifida bacteria grow better when they are provided with nucleotides since this is a very normal pathway. The epithelium in the small bowel serves as a substrate for bacterial growth in the large bowel, and a lot of nucleotide is provided in this way. My question is, is this true only for the bifida bacteria, or do other bacteria, like bacterioides, also grow better when provided with nucleotides?

Dr. Uauy: I am not an expert in microbiology, but probably other bacteria would also do better.

Dr. Rey: I have a query about your very difficult calculations on the requirement for nucleotides. You said that at birth we need 480 mg a day of nucleotides?

Dr. Uauy: For a 3-kg infant, yes.

Dr. Rey: We recommend an intake of approximately 1–2 g of protein per kilogram of body weight daily. Your calculation implies that approximately 5–10% of this protein should be nucleotides, which is 10 times higher than in human milk.

Dr. Uauy: I am saying the 160 mg/kg is what the baby needs to make. This does not mean that it all needs to come from the diet. The liver makes nucleotides *de novo* to make up for the difference.

Dr. Rey: Yes, but you say that the infant formulas you cited from Spain and the USA are supplemented with only very small amounts of nucleotides, but they are just the quantities present in human milk.

Dr. Uauy: The amounts that were added are equivalent to the free nucleotides present in human milk, which are a very small proportion of the total. When the cells in the human milk are digested and absorbed, you come up with a total intake of about 150 mg per day. Nobody has yet produced a formula that has the same nucleotide equivalent as human milk. This would be 10 times higher than presently found in nucleotide supplemented formula.

Dr. Rey: What do you recommend for infant formula?

Dr. Uauy: I provide the data. I am not responsible for decisions about putting nucleotides in formula, but if I was making a formula for healthy full-term infants, I would not include them for now.

Dr. Rey: So you have no idea about what should be recommended for babies?

Dr. Uauy: The situation is the same as for taurine. As David Rassin stated, there are no data proving that taurine should be incorporated in formulas for full-term infants. So there is some other reason for its presence overriding the factual data. My aim is to provide the knowledge base and to try to improve on it where possible. Manufacturers can then use it or not, as they choose, to make their decisions.

Dr. Rey: I do not think you understood my question fully. In the European Community it is forbidden to add any nutrient to infant formula or follow-on milk unless there is a directive that expressly permits it. So for the moment we cannot add nucleotides, although we can add taurine and carnitine. The EC also forbids the export to countries outside the Community of a formula that does not comply with EC regulations. So the infant food industry in the European Community has an obligation to stick very precisely to these rules. The role of the members of the Scientific Committee for Food is to tell the Community authorities whether it is permissible or not to add nucleotides to a formula. We said yes, and we gave as a maximum value the level of free nucleotide in human milk. So my question was, do you think it is reasonable to fix an upper limit to 5–6 mg per 100 kcal, or is this silly? Our problem is not in relation to the food industry; it is to write the rules governing the free circulation of infant formula in the unique market of the European Community. This is a very special problem.

Dr. Uauy: I think your question can be approached in two ways. One approach is to say that anything that is present in human milk potentially has advantages, and in this case it would be reasonable to add free nucleotides up to the values obtained in human milk analyses. The other approach is to demand that it be shown that it makes a clinical or nutritional difference when a substance is added. Which approach is used is up to the Scientific Committee, but as a pediatrician I would want to know whether or not the addition of nucleotides makes a difference. We are still pending an answer on this.

Dr. Beatty: Your model of the radiation-damaged animal is an artificial situation. Is there any comparison to diet-induced damage to the intestinal tract?

Dr. Uauy: The other model that I mentioned is the chronic high lactose ingestion model that induces gut damage in the rat. A paper by Gil's group in the *Journal of Parenteral and Enteral Nutrition* (1) using this method also showed a benefit from providing nucleotides. So those are the two models that I am aware have been tested. I think this is an area that has obvious relevance to the human, but you probably have to use animal models initially and eventually test the findings during recovery from injury in humans.

Dr. Beatty: If you look at the histology of a radiation-damaged gut, is it in any way analogous to nutritional damage?

Dr. Uauy: No. The amount of inflammation that you have with radiation far exceeds anything else that you would find, so it is very different, but it could be a model for what happens in radiation-induced injury in cancer patients. So it is a model that is potentially relevant to human pathology. An important advantage of the model is that you can regulate it: you can give variable amounts of rads to induce specific amounts of damage.

Dr. Beatty: You presented information of the nucleotide content of breast milk and of formula milk. What are the other dietary sources?

Dr. Uauy: Meats and legumes are very high in nucleotides; yoghurt and fermented foods are also high in nucleotides.

Dr. Marini: Is there any difference in nucleotide content between colostrum and mature milk?

Dr. Uauy: The free nucleotides are mostly derived from RNA and DNA originating from milk cells. There is about four times as much nucleotides in colostrum as in mature milk.

Dr. Wharton: You referred very briefly to our work, which is not yet published. I will just summarize by saying that we could not repeat the observation that nucleotides promoted the growth of bifida bacteria. We have been considering various reasons for this. One could be that the formula we originally used had higher concentrations of nucleotides. There is still dissent among workers in this field about the actual content of nucleotides in breast milk. Some of this may be due to degradation during storage. Could you comment?

Dr. Uauy: There is potential for variability in the way you handle the sample (e.g., centrifugation will remove the cells), in the way you store it, and from any sort of thermal treatment the milk receives. If bacterial growth occurs in the sample, nucleotides are an important by-product of microbial fermentation. Gil has also studied various species and has found that there are species differences in nucleotide content of milk. Another point is that the analytic methodology has evolved in the last 15 years, so the more recent data look different from the 1982 studies of Janas & Picciano (2) and of Gil (3).

Dr. Guesry: You spoke about safety. If we would take your calculation of adding 480 mg of nucleic acid nucleotide per day, don't you think that uric acid could increase quite a lot in the baby?

Dr. Uauy: I am not suggesting that we add 480 mg. The model we are following is human milk, so I would propose only adding an amount equivalent to the total nucleotide content (i.e., about one-third of the 480). So far the data adding the free nucleotides content have shown no demonstrable effect on uric acid.

REFERENCES

1. Núñez MC, Ayudarte MV, Morales D, Súarez MD, Gil A. Effect of dietary nucleotides on intestinal repair in rats with experimental chronic diarrhea. *J Parenter Enteral Nutr* 1990; 14: 598–604.
2. Janas LM, Picciano MF. The nucleotide profile of human milk. *Pediatr Res* 1982; 16: 659–62.
3. Gil A, Sánchez-Medina F. Acid-soluble nucleotides of human milk at different stages of lactation. *J Dairy Res* 1982; 49: 301–4.

Inborn Errors of Metabolism: A Model for the Evaluation of Essential Amino Acid Requirements

Jean-Louis Bresson, Françoise Rey, Florence Poggi, Eliane Depondt, Véronique Abadie, Jean-Marie Saudubray, and Jean Rey

Département de Pédiatrie, Hôpital des Enfants Malades, 149 rue de Sèvres, 75743 Paris Cédex 15, France

Inborn errors of metabolism are caused by mutations that alter the functions of physiologically important proteins. These errors can be seen as experiments by nature enabling the clarification of the different steps in a biochemical pathway at which that protein intervenes as an enzyme, transporter, or regulator, as well as the processes mediated by the entire pathway. They thereby serve to shed light on the role the pathway plays in the maintenance of cellular homeostasis (1).

At the turn of the century Archibald Garrod already understood this when he surmised that urinary excretion of large quantities of homogentisic acid, characteristic of alkaptonuria—a disease that had probably been known since the end of the sixteenth century—was the result of partial or complete alteration of an enzyme whose role is to split a benzene ring (2). Garrod considered that this acid, which accumulates at the enzymatic block, could only be an intermediary substance that went unnoticed through the normal degradation pathway. This hypothesis was in fact confirmed 50 years later by demonstrating homogentisic acid oxidase deficiency in the liver of a patient suffering from this disease (1).

Garrod's fundamental idea about the biochemistry of the inborn errors of metabolism was that each condition should be interpreted as a block at some particular point in the normal course of intermediary metabolism due to the congenital deficiency of a specific enzyme (3). *A priori* it seemed to Garrod unlikely that alternative pathways could be used when for any reason the normal one was blocked. To back up this idea he quoted a meaningful sentence from Claude Bernard in his 1880 *Pathologie Expérimentale:* "Et maintenant, oserait-on soutenir qu'il faut distinguer les lois de la vie à l'état pathologique des lois de la vie à l'état normal? Ce serait vouloir distinguer les lois de la mécanique dans une maison qui tombe, des lois de la mécanique dans une maison qui tient debout" (2).

Some inborn errors of amino acid metabolism offer a remarkable model for the study of essential amino acid requirements. There are two opposing schools of thought about this matter: (a) those who believe that essential amino acid needs could be higher in normal infants than in those suffering from a block in the hydroxylation pathway from phenylalanine to tyrosine or a defect in the decarboxylation of branched chain amino acids; (b) those who, along Garrod's line, believe that the calculated needs of amino acids for children with phenylketonuria (PKU) or suffering from leucinosis (maple syrup urine disease; MSUD) constitute the best approximation of the minimum requirement for phenylalanine, leucine, valine, and isoleucine.

METHODS TO COMPUTE THE ESSENTIAL AMINO ACID REQUIREMENTS

The Human Milk Model

The first part of the answer is given by the studies on amino acid composition of human milk. There is general agreement that breast feeding is the most appropriate feeding practice between birth and 4–6 months of age. In fact, it is the reference that is agreed upon by each and every international advisory committee intending to compute nitrogen and amino acid needs during the first months of life (4–6).

Although in industrialized countries growth in height slows slightly after 6 months in exclusively breast-fed babies (7) and breast milk alone is not able to cover the energy needs of many infants after 3 months of age in less developed countries (8), there is nothing to prove that protein needs are not covered during the first months of life in infants breast-fed by well-nourished mothers or that more than 5% of infants receive an inadequate intake in Western countries (9).

Using the average volume of milk absorbed by breast-fed babies and the mean amino acid composition of mature human milk, it is possible to estimate mean daily intake of amino acids during the first months, particularly those relevant to some inborn errors [i.e., phenylalanine and the branched-chain amino acids (leucine, valine, and isoleucine)]. If one accepts an average intake of 800 ml/d (5) and uses the mean amino acid composition of human milk in the UK (10), which is consistent with previous and later reports (Table 1), one can estimate average intakes of phenylalanine, leucine, valine, and isoleucine to be about 285, 960, 700, and 535 mg/d, respectively.

Unfortunately, we have no information on the relation between the volume of milk ingested by infants and the amino acid concentration of human milk. We only know that the variation coefficient between lactating women with regard to the output of human milk is about 15% (14) and that it is about 10% for amino acid composition (12). If we accept that there is no adaptation of the protein composition of human milk in relation to output and if we accept the lower values both for intake and amino acid concentration, the intakes of phenylalanine, leucine, valine, and isoleucine could be as low as 200, 540, 390, and 300 mg/d, respectively. However, as we shall see later,

TABLE 1. *Phenylalanine and branched-chain amino acid concentrations in mature human milk (mg/100 ml)*

Reference	Phe	Leu	Val	Ile
COMA (1977) [10][a]	48	120	87	67
Harzer & Bindels [11][b]	60	134	66	65
Janas & Picciano [12][c]	40 (31–46)	133 (107–162)	54 (42–66)	87 (62–137)
George & DeFrancesca [13][d]	40 (24–58)	97 (65–147)	73 (45–114)	61 (41–92)

[a] Mean of five UK centers (2–9 weeks).
[b] 36 days.
[c] 4 weeks.
[d] Infants 12 weeks to 15 months old.

clinical experience gathered from typical PKUs and MSUDs makes this minimum hypothesis unlikely and suggests that the intake of those amino acids is not lower than 265, 660, 480, and 370 mg/d, respectively (lowest limit of human milk output × mean amino acid concentration).

We can also question the reference proteins whose amino acid content is used to estimate intakes. Human milk proteins do not have the same amino acid composition. Lactoferrin is arginine-rich; α-lactalbumin contains large quantities of isoleucine and lysine; casein has a high glutamic acid and proline content (11). Furthermore, significant quantities of lactoferrin and secretory IgA (IgAS) are excreted in the infant's stools, particularly during the first month of life, and part of these proteins, contributing to the immunological properties of human milk, is recovered intact (15). After 4 weeks of life, native protein excretion still represents 10–35% of ingested IgAS, and 1–2% of the lactoferrin present in the milk (15, 16). The excretion in the stools of significant quantities of these immunological proteins suggests that the amount of nutritionally available proteins in human milk is overestimated (15) even though after 6 weeks of life 95% of human milk protein is digested (16). This factor could therefore be disregarded in the calculations as whichever reference is used (total protein or nutritionally available protein), differences in composition are minimal for phenylalanine and branched-chain amino acids (17).

It is worth stressing that human milk intake accounts for only 10–30% of the interindividual variation in weight gain of infants between birth and 4 months of age (14). Moreover, the important variations in human milk output from well-nourished mothers (550–1100 ml/d with an average of 800 ml/d) (17, 18) could be explained by variations in the baby's requirements rather than by insufficiencies in milk production (19), in keeping with the remarkable ability of breast-fed babies to regulate their intake (14).

Estimation of the Requirements of Artificially Fed Infants

A study of the amino acid (or nitrogen) content of human milk does not really inform us about the physiological requirements of the infant, that is, the amount and the chemical form of a nutrient which is needed systemically to maintain normal health and development without disturbance of the metabolism of any other nutrient and without inducing extreme homeostatic processes or excessive depletion or increase of body stores (20). In any case, the balance between the different nutrients, their chemical form, and the other elements they are eventually associated with is liable to affect their absorption substantially. It would therefore be unsound to deduce the requirements of artificially fed infants from the intake of breast-fed babies.

Two techniques have been employed. The first was to use a synthetic diet in which nitrogen was provided as a mixture of the 18 amino acids in their L form and in the same proportions as in human milk. The intake of one amino acid was varied until the minimum quantity compatible with normal growth and a normal nitrogen retention was reached. The nitrogen intake (non-limiting) was maintained at a constant level by adjustment of a non-essential source of nitrogen (e.g., glycine) (21). Under these conditions, the adequate intake of phenylalanine was estimated at 47–94 mg/kg of body weight per day (22), but four of the six infants studied between 2 and 10 weeks of age had birthweights below 2.5 kg. Moreover, there is no proof that the intakes considered to be inadequate for four of them were so for the other two. The conditions for the study on branched-chain amino acids are less debatable, but the interpretation of the data is still opened to discussion. The estimates for infants and children are essentially selected values based on an inspection of growth curves and nitrogen balances obtained when different levels of the essential amino acids were fed. The results thus obtained depend largely on the judgment of the investigator, and there is no assurance that comparable judgments would be exercised by different investigators (23). This seems particularly true for valine, for which an adequate intake has been estimated at 85–105 mg/kg/d (24), although there was no real difference in growth rate or nitrogen balance when the same infant was fed 85 or 105 mg/d. Similarly, adequate intakes of leucine have been estimated at 76–229 mg/kg/d (25), those for isoleucine between 79 and 126 mg/kg/d (26), values later changed, without any obvious reason, to 102–119 mg/kg/d (21). The ethical issues raised by these experiments led to a 4-year interruption of the study in the state of New York, while it was carried on in Texas (21).

A second set of experiments consisted of varying the quantities of amino acids given to healthy infants by using formulas based on cow's milk or soya. The aim was to determine the lowest quantity of each amino acid able to achieve normal weight and height growth during the first 4 months of life, and maintaining the concentration of plasma albumin within physiological limits at minimal protein intakes (1.5–1.65 g per 100 kcal). The values obtained from this method were 253 ± 22 mg/d, 653 ± 77 mg/d, 389 ± 35 mg/d, and 293 ± 26 mg/d for phenylalanine, leucine, valine, and isoleucine, respectively (27, 28). These values are not very different from our earlier estimation of the breast-fed baby's minimum intakes (Table 2). This result is not surprising since Fomon, for easily understood reasons, did not dare to go far away from the mean composition of human milk. However, there is no proof that the lowest intake achieving "nutritionally adequate performances" is the best estimate of the needs. It makes it unlikely that requirements are higher than these values, but

TABLE 2. *Adequate intakes of phenylalanine and branched-chain amino acids (mg/d)[a]*

	Phe	Leu	Val	Ile
Snyderman et al. [24–26][b]	—	737 ± 371	453 ± 81	305 ± 47
Fomon et al. [27,28]	255 ± 23	670 ± 60	390 ± 35	295 ± 26

[a] m ± SD.
[b] Computed from results in mg/kg/d.

does not preclude the possibility that they might be substantially lower for a certain number of amino acids (28). It is pointless to comment further on the differences between Fomon's results (28) and those of Holt & Snyderman (21), since a valid comparison should be based on accurate measurements performed on infants of similar age and body weight. An estimate of isoleucine requirements at 102 mg/kg/d or more (26) is not really tenable in the face of the much lower intake of Fomon's infant groups (28). It should be stressed that the figures obtained for each amino acid by successive Expert Committee Meetings (FAO, WHO) were derived from the lowest values from one or the other method covering the higher requirement range from birth to 6 months of age (25). It is obvious that this choice can only result in overestimating requirements.

Results by Factorial Methods

The estimation of the protein requirements by the factorial method is still a highly controversial subject. This method implies that maintenance needs are known precisely and that there is an agreement on the way to assess growth needs. Fomon's method for calculating maintenance needs consists of adding urine, stool, and skin nitrogen losses in 4- to 6-month-old infants receiving minimal intakes of protein (i.e., similar to those of breast-fed babies). Starting from this estimation of losses and assuming that they remain constant during the first year of life relative to the protein mass of the child (1.17 g of nitrogen loss per kilogram of protein), Fomon estimates the maintenance requirement at 150 mg N/kg/d or 0.9 g protein/kg/d (29). However, this figure is overestimated since basal losses should be measured on protein-free diets (4).

Another way consists of measuring nitrogen retention with variable protein intakes while energy intake is kept constant. The maintenance requirement is thus calculated by linear fitting of nitrogen retention on intakes (interpolation at zero nitrogen retention) and deducting 10 mg/kg to account for non-measurable losses (30). On this basis the 1985 WHO/FAO Expert Committee arrived at a figure of 120 mg N/kg/d, equivalent to 0.75 g protein/kg/d (5). This same value was also used by the Food and Nutrition Board (FNB) of the American Academy of Sciences in the last issue of its recommendations (31). Beaton and Chery also considered it as a sound estimate, even though they implemented their calculations with two other values: 0.72 and 0.62 g protein/kg/d (9).

There is not uniform agreement on the efficiency of protein utilization for growth. Based on an apparent efficiency of 70% (30), the FAO/OMS Expert Committee (5) and the FNB (31) use a ×1.5 correction factor to estimate growth requirements. Beaton and Chery used the same figure in their calculations (9), whereas Fomon estimates the efficiency of food protein conversion into body proteins at 90% (29). However, the differences in agreement are greatest for the calculation of the increase in body protein mass. Fomon starts from the body composition of "reference children" from birth to 10 years of age (32). Taking into account the rapid decrease in the rate of weight gain during the first year of life, protein accretion slows down from 0.93 g/kg/d during the first month, to 0.5 g/kg/d at 2–3 months, to 0.34 g/kg/d at 3–4 months, to 0.26 g/kg/d at 5–6 months, and to 0.18 g/kg/d at 9–12 months (29). The FAO/WHO Expert Committee and the FNB used the same values but with a security coefficient of 1.5 to cover the variability in weight gain from one day to another (5) and the alleged inability of infants and children to store free amino acids (31). Since they take a protein conversion efficiency of 70%, these two committees arrived at growth requirement estimates more than twice as high as Fomon's calculations (Table 3).

These two differing views on the respective importance of maintenance requirements and growth needs have an influence on more than just the evaluation of total needs. Of course, overestimation or underestimation of one factor has a bearing on the total requirement, but Fomon's divergence goes in opposite directions (overestimation of the maintenance expenses and underestimation of the growth needs) in such a way that his evaluation of total requirements is probably not so far away from reality. On the other hand, the overestimation of growth needs by the FAO/WHO Expert Committee and the FNB leads to values much higher than those computed by Beaton and Chery (9). For 3- to 6-month-old infants the FAO/WHO Expert Committee and the FNB reach a figure of 1.38 g protein/kg/d (5, 31), whereas in 3- to 4-month-old infants, Fomon estimates the average need to be 1.27 g/kg/d (29), and Beaton and Chery to be 1.12 g/kg/d (9). Beaton and Chery also very aptly pointed out that the mean requirement could not be comparable with the mean protein intake

TABLE 3. *Protein, phenylalanine, and leucine requirements estimated by the factorial method*

Age (months)	Δ Weight gain (g/d)	Protein gain[a] (g/d)	Maintenance[b] (g/d)	Σ (g/d)	Phenylalanine (mg/d)	Leucine (mg/d)
0–2	30	4.8	2.9	7.7	355	595
2–6	22	3.5	4.4	7.9	365	610
6–12	12.5	2.0	6.1	8.1	375	625

[a] Assuming that proteins account for 16% of weight gain.
[b] 0.7 g/kg/d.

of breast-fed infants unless an excessive percentage (31–34% according to the different figures used in their calculations) were provided with insufficient protein intakes. This would be in conflict with the hypotheses of the FAO/WHO Expert Committee and the FNB whereby mother's milk should satisfy the needs of all infants up to at least 4 months of age (3).

The respective share of maintenance expenses and growth requirements should also theoretically be taken into account for the computation of amino acid needs. Amino acid requirements for growth are in fact a function of body composition and growth rate, whereas a certain adaptation of the maintenance expenses would be expected, since essential amino acids are likely to be spared if intake is lower than requirements (23). In 1973 the FAO/WHO Expert Committee estimated that essential amino acid needs amount to 37% of infants' protein requirement, a figure much higher than the corresponding 15% in adults (4). However, adult amino acid requirements have been underestimated, even if one accepts that endogenous amino acid recycling (i.e., preferential channeling into body proteins) may reach 90% when protein intake is dramatically reduced or almost nil, with a decrease in obligatory losses through oxidative catabolism to less than 10% of endogenous amino acid appearance rates (33). Therefore, the only way to calculate accretion of phenylalanine and leucine is by using as reference their concentration in muscle proteins (respectively, 46.2 and 73.3 mg/g of muscle protein), as if in this respect there was no difference between maintenance and growth. If we assume maintenance protein expenses of 0.7 g/kg/d and a weight gain protein content of 16%, the figures shown in Table 3 can be derived.

Direct Estimation of the Net *In Vivo* Synthesis

Phenylalanine, leucine, isoleucine, and valine requirements for growth and maintenance can also be estimated by quantitating the metabolism of their analogs (stable isotope labeled) in the organism. Although in use since the 1930s, labeling with ^{15}N is not the best option because an active transamination leads to overestimate the utilization of the perfused tracer (34). Furthermore, calculations based on the so-called "final products" method are made on incorrect assumptions (35). On the other hand, the dilution of the L-$[1-^{13}C]$leucine used as a tracer allows the follow-up of its carbon skeleton and it is further possible to measure the enrichment of intracellular leucine, the true precursor of protein synthesis and of leucine catabolism, from plasma ketoisocaproate (36). Despite this remarkable breakthrough, one should not overlook the fact that these measurements provide information only at the whole body level, disregarding any possible difference between body compartments. However, these results provide an integrated view of protein synthesis, breakdown, and amino acid catabolism and may be compared to the values obtained from energy expenditure measurements which integrate oxygen consumption and carbon dioxide production of individual tissues (37). The net incorporation of leucine could thus be obtained from the difference between protein synthesis and proteolysis rates. However, to make this measurement representative, food intake should be given regularly

TABLE 4. *Phenylalanine and branched-chain amino acid intakes of breast-fed and bottle-fed infants compared to requirements estimated by the factorial method and net protein synythesis measured by tracer dilution*

	Phe	Leu	Val	Ile
Breast-fed infants [10][a]	385	960	700	535
Infant formulas [27,28]	255	670	390	295
Factorial method[b]	365	610		
[1^{13}C]Leucine [38]		490		

[a] Computed by using a mean 800 ml/d milk intake and the mean amino acid concentrations of mature human milk.
[b] From Table 3.

over 24 h. A few studies follow these conditions and make it possible to estimate net incorporation of leucine at 490 mg/d on average for infants from 2 to 8 months of age, steadily growing on total parenteral nutrition (TPN) (38). It is possible to calculate from data obtained in low birthweight infants (2–4 weeks old) on continuous enteral nutrition a net leucine deposition rate ranging from 340 to 560 mg/d (39). Other measurements made on TPN give lower figures, around 350 mg/d (40), but this estimation is based on [^{15}N]glycine dilution without any assay of the isotopic enrichment of plasma leucine. This lower value could thus reflect an underestimation of the rate of protein synthesis by the dilution of labeled glycine (41). Moreover, values varying from 160 to 190 mg/d in low birthweight newborns (42) or 150–160 mg/d in normal healthy newborns (on 24-h feeding) (43) have been calculated by using [^{13}C]leucine from the first hours or days of life. These low values might be accounted for by insufficient intakes (42). In conclusion, net leucine incorporation in newborn infants, under close to physiological conditions, should be about 450–500 mg/d (Table 4).

PHENYLKETONURIA AS A MODEL

Phenylketonuria (PKU) is the result of a defect of the phenylalanine hydroxylase (PAH) gene (44). The consequences of these mutations on the function of the hydroxylase system vary according to their type and the localization of the defect on the polypeptide chain in relation to the catalytic domain.

Classification of the Usual Variants and Correlation Between Genotypes/Phenotypes

More than 70 mutations of the PAH gene had been listed by the end of 1992 (Scriver, the PKU Mutation Analysis Consortium). Some mutations totally abolish the hydroxylase activity. As a consequence, homozygocity or compound heterozygocity are always associated with severe forms of PKU. This holds true for mutations affecting the initiation codon for mRNA translation, "no-sense" mutations resulting in a stop codon leading to incomplete proteins or deletions whether or not they include intron-exon junctions. Many missense mutations turn into severe forms of the disease. Others are associated with more moderate forms of PKU or with moderate hyperphenylalaninemia (45–47).

The distinction between these three phenotypes is, in practical terms, somewhat arbitrary. Actually, the classical indices used to distinguish them—plasma phenylalanine concentration at presentation or during reintroduction of a normal phenylalanine intake, tolerance, *in vivo* kinetics of disappearance, *in vitro* PAH activity—constitute a continuum without any bimodal or trimodal distribution (48). Moreover, the hepatic PAH activity assayed *in vitro* is known for only a few of these mutations, and PAH expression after transfection of eukaryote cells by cDNAs carrying different molecular alterations of the gene has not been performed for all of them. Finally, phenylalanine tolerance (i.e., the maximum quantity of phenylalanine that a child can ingest during a period of normal growth without increase in plasma phenylalanine concentration beyond what is considered as ideal for his treatment) is one of the classical criteria for the classification of "typical PKU," "atypical PKU," and "hyperphenylalaninemias" (48). In practice it is only possible to obtain information on the minimum phenylalanine intake compatible with normal growth and development of phenylketonuric children from studies performed on patients with a complete PAH defect and if the tolerance is not a criteria used to distinguish between "typical" and "atypical" PKU. If this is not the case, the minimum intake would, by definition, be limited to the tolerance that had been fixed *a priori*.

Phenylalanine Intake in Typical PKU

Selma Snyderman appears to have been the first to suggest that the requirement for a single amino acid in the general population could be deduced by estimating intakes of children treated for a metabolic block on this amino acid pathway, as long as their growth and development were normal and that the plasma concentration of the limiting amino acid remained within normal limits. This information was in fact deduced from the efficacy of the treatment. Up to that point it was considered that phenylketonuric children had lower phenylalanine needs than those of normal children (21), but Holt and Snyderman demonstrated that the requirements of phenylketonuric children were of the same order of magnitude (47–90 mg/kg/d) as those of normal infants of comparable age (21). Today we know that Holt and Snyderman overestimated both the amino acid needs of normal infants and the intake of phenylalanine required to maintain plasma phenylalanine concentrations of phenylketonuric children within acceptable limits.

To discuss the matter fairly, two elements should be taken into consideration: whether the enzymatic block is complete or incomplete, and the plasma concentration of the given amino acid at which the children are maintained throughout the study—both can profoundly influence the result. Any residual activity of the enzyme opens the route to amino acid catabolism and as such apparently increases the tolerance beyond the quantity of amino acid necessary to balance obligatory losses and protein deposition. This is demonstrated by comparing the intakes of children suffering from classical PKU to those of children who retain some PAH activity (49). Moreover, the American Collaborative Study (50) clearly showed that the phenylalanine needs were higher during the first year of life in 44 infants for whom the objective of the treatment was to maintain the phenylalaninemia between 5.5 and 9.9 mg per 100 ml (TG2), as compared to 44 others, for whom the objective was to limit the phenylalaninemia to between 1.0 and 5.4 mg per 100 ml (TG1). This study also showed, for the first time, that in the first year of life the total daily phenylalanine intake does not change in relation with age or energy intake. Phenylalanine intake varies from 55 ± 16 mg/kg/d during the first trimester, to 27 ± 8 mg/kg/d during the fourth trimester in the group TG1, and from 62 ± 19 to 34 ± 8 mg/kg/d in the group TG2 (50). The results we obtained in 43 typical phenylketonuric infants—who had a phenylalanine plasma concentration above 25 mg per 100 ml at the time of presentation and above 20 mg per 100 ml at 1 year of age when overloaded with 3 g protein/kg for a few days—fully confirm the results of Acosta et al. (50) and Wendel et al. (51) obtained from 140 infants and children between birth and 6 years of age with a mean phenylalaninemia below 6 mg per 100 ml during the first year of life (Table 5). In our study there is no difference between infants who carry two of the mutations, resulting in complete PAH deficiency, and those who carry only one. Among the patients we studied, the differences between "typical" PKU and the 26 children classified as "atypical" PKU (plasma phenylalanine concentration between 15 and 25 mg per 100 ml at presentation) are not statistically significant during the first year of life but are after 18 months. In this particular case this does not mean that the tolerance for "atypical" PKU is identical to the tolerance of "typical" PKU because no attempt was made to increase the protein intake of the infants above their minimum needs during the first 6–9 months of their treatment (Table 6).

The adequate phenylalanine intake could thus be estimated to be 250 mg/d on average during the first year of life, such values being very close to the lower range of phenylalanine intake of breast-fed babies. This suggests that a sufficient intake is not very different from the requirement (52) and that infant phenylalanine requirement would only represent 60–70% of the values considered up to now to be quite reasonable (49).

TABLE 5. *Phenylalanine intakes of classical PKU infants*

	n	Epoch (months)			
		0–3	3–6	6–9	9–12
Acosta et al. [50][a]	44	253	245	264	254
Wendel et al. [51]	137	251 ± 46		262 ± 52	

[a] Serum phenylalanine levels between 1 and 5.4 mg/dl (TG1).

TABLE 6. *Phenylalanine tolerance of classical and atypical PKU*

	n	3 months	6 months	12 months	18 months
Classical PKU	43	245 ± 50[a]	261 ± 61	283 ± 70	308 ± 77
0/0	6	243 ± 45	248 ± 59	258 ± 57	300 ± 90
0/?	13	248 ± 13	263 ± 47	286 ± 52	305 ± 45
Atypical PKU	26	257 ± 67[b]	289 ± 85	349 ± 151	370 ± 138

[a] Significantly different from 12 and 18 months.
[b] Significantly different from 6, 12, and 18 months.

LEUCINOSIS AS A MODEL

Leucinosis, or maple syrup urine disease (MSUD), is a hereditary disease of branched-chain amino acid metabolism associated with a more or less important reduction in the activity of the enzymatic complex (BCKAD) responsible for the oxidative decarboxylation of their α-ketone acids (53). This activity depends on the expression on the internal mitochondrial membrane of four different proteins. The decarboxylase itself is composed of two subunits (E1 α, which binds the thiamine, and E1 β) associated in an α2, β2 hetero-tetramer. The acyltransferase (E2) constitutes the core protein around which the other components are organized. In addition, there is a lipoamide oxygenase (E3), common to BCKAD and to pyruvate dehydrogenase and α-ketoglutarate dehydrogenase, which is present in the form of a homo-dimer. The expression of these proteins is controlled by nuclear genes distributed over different chromosomes. Moreover, its activity is modulated by a specific kinase and phosphatase. The complexity of the system explains how difficult it is to characterize the unique or multiple defects involved and suggests that specific mutations acting at the level of distinct subunits could lead to phenotypes that are indistinguishable at a biochemical level.

Classification of Usual Variants and Correlations Between Genotypes and Phenotypes

The most common form of the disease is the acute neonatal type, resulting from an almost complete loss of activity (<2% of the normal *in vitro*). The decarboxylation of the L-[1^{13}C]leucine *in vivo* is almost nil in these patients (54). "Intermediary" and "intermittent" forms have also been described, in which the residual activity *in vitro* varies between 5 and 10% of the normal value (55) and could *in vivo* be as high as 30% of the production of $^{13}CO_2$ in normal subjects (56). One of the first correlation trials between genotype and phenotype suggested that the classical and intermediate forms resulted from anomalies in E1 β protein differing in their kinetic properties, whereas the intermittent forms were linked to anomalies in the E2 protein (57). However, more recent studies have shown that the classical forms could indifferently result from anomalies of the E1 α gene (58), E1 β gene (59), or E2 gene (60) and that the defect could either be a deletion of the "targeting leader peptide" of E1 β (probably responsible for its transfer into the mitochondria), a mutation at a splicing site with elimination of an E2 exon, or mutations in E1 α gene impeding the normal assembly of the E1 component of BCKAD. Thiamine-sensitive leucinosis are of more moderate expression than the acute form and may also result from anomalies in the E2 gene (61) or the E1 β gene (62). In fact, the distinction between the classical form and the intermediary or intermittent form appears to be very minor beside the large range of functional consequences of the causative mutations. As in PKU, only the classical type may provide some information about the minimum intake leading to adequate growth.

Leucine Intake in the Classical Leucinose Forms

The leucine intake of one infant suffering from MSUD was estimated at 90 mg/kg/d at 12 months of age and 50 mg/kg/d at 1 year of age. Recalculated over 24 h, intakes are almost constant during the first year of life at 480 mg/d. This patient's isoleucine intake was also stable during the first year, at 280 mg/d, and 325 mg/d for valine (52). Similar results (480 mg/d for leucine at 4 months of age and 500 mg/d at 1 year of age) have also been found in two other patients suffering from classical MSUD (49). Leucine intake estimated from a larger group (19 infants) is about 420 mg/d at 1 year of age (55; personal results), in good agreement with earlier results. However, leucine intake is slightly lower (375 mg/d) in patients treated since the early 1980s because of tighter control of the plasma branched-chain amino acid concentrations. Once again, adequate leucine intake is constant throughout the first year of life, amounting to 350 mg/d at 3 and 6 months of life. As in PKU, the persistence of residual activity gives an apparently higher tolerance (49).

REFERENCES

1. Gargus JJ, Mitas M. Physiological processes revealed through an analysis of inborn errors. *Am J Physiol* 1988; 55: F1047–58.
2. Garrod AE. *Inborn errors of metabolism. The Croonian lectures.* London: Oxford University Press, 1909.
3. Harris H. The "inborn errors" today. In: *Garrod's inborn errors of metabolism.* London: Oxford University Press, 1963; 120–97.
4. Joint FAO/WHO *ad hoc* Expert Committee. *Energy and protein requirements.* FAO Nutrition Meetings Report Series No 52. Rome: FAO/WHO, 1973.
5. Joint FAO/WHO/UNU Expert Consultation. *Energy and protein requirements.* World Health Organization Technical Report Series No 724. Geneva: WHO, 1985.
6. Commission Directive of 14 May 1991 on infant formulae and follow-up formulae (91/321/EEC). Official Journal of the European Communities No L 175/35-49.
7. Salmenperä L, Perheentupa J, Siimes MA. Exclusively breast-fed healthy infants grow slower than reference infants. *Pediatr Res* 1985; 19: 307–12.
8. Waterlow JC, Thomson AM. Observations on the adequacy of breast-feeding. *Lancet* 1979; ii: 238–42.
9. Beaton GH, Chery A. Protein requirements of infants: a reexamination of concepts and approaches. *Am J Clin Nutr* 1988; 48: 1403–12.

10. Committee on Medical Aspects of Food Policy. *The composition of mature human milk*. Report of Health and Social Subjects No 12. London: HMSO, 1977.
11. Harzer G, Bindels JG. Changes in human milk immunoglobulin A and lactoferrin during early lactation. In: Schaub J, ed. *Composition and physiological properties of human milk*. Amsterdam: Elsevier Science Publishers, 1985; 285–93.
12. Janas LM, Picciano MF. Quantities of amino acids ingested by human milk-fed infants. *J Pediatr* 1986; 109: 802–7.
13. George DE, DeFrancesca BA. Human milk in comparison to cow milk. In: Lebenthal E, ed. *Textbook of gastroenterology and nutrition in infancy*, 2nd ed. New York: Raven Press, 1989; 239–61.
14. Butte NF, Garza C, O'Brian Smith E, Nichols BL. Human milk intake and growth in exclusively breast-fed infants. *J Pediatr* 1984; 104: 187–95.
15. Davidson LA, Lönnerdal B. Persistence of human milk proteins in the breast-fed infant. *Acta Paediatr Scand* 1987; 76: 733–40.
16. Prentice A, Ewing G, Roberts SB, *et al*. The nutritional role of breast-milk IgA and lactoferrin. *Acta Paediatr Scand* 1987; 76: 592–8.
17. Georgi G, Sawatzki G. Design of an optimal protein composition for infant nutrition. In: Koletzko B, Okken A, Rey J, *et al*., eds. *Recent advances in infant feeding*. Stuttgart: Georg Thieme Verlag, 1992; 148–55.
18. Dewey KG, Heinig J, Nommsen LA, Lönnerdal B. Maternal versus infant factors related to breast milk intake and residual milk volume: the Darling Study. *Pediatrics* 1991; 87: 829–37.
19. Dewey KG, Lönnerdal B. Infant self-regulation of breast milk intake. *Acta Paediatr Scand* 1986; 75: 893–8.
20. Aggett P. Scientific considerations in the formulation of RDI. *Eur J Clin Nutr* 1990; 44(Suppl 2): 37–43.
21. Holt LE, Snyderman SE. The amino acid requirements of children. In: Nyhan WL, ed. *Amino acid metabolism and genetic variation*. New York: McGraw-Hill, 1967; 381–90.
22. Snyderman SE, Pratt EL, Cheung MW, *et al*. The phenylalanine requirement of the normal infant. *J Nutr* 1955; 56: 253–63.
23. El Lozy M, Hegsted DM. Calculation of the amino acid requirements of children at different ages by the factorial method. *Am J Clin Nutr* 1975; 28: 1052–4.
24. Snyderman SE, Holt LE, Smellie F, *et al*. The essential amino acid requirements of infants: valine. *Am J Dis Child* 1959; 97: 186–91.
25. Snyderman SE, Roitman EL, Boyer A, Holt LE. Essential amino acid requirements of infants: leucine. *Am J Dis Child* 1961; 102: 157–62.
26. Snyderman SE, Boyer A, Norton PM, *et al*. The essential amino acid requirements of infants. IX. Isoleucine. *Am J Clin Nutr* 1964; 15: 313–21.
27. Fomon SJ, Filer LJ. Amino acid requirements for normal growth. In: Nyhan WL, ed. *Amino acid metabolism and genetic variation*. New York: McGraw-Hill, 1967; 381–90.
28. Fomon SJ, Thomas LN, Filer LJ, *et al*. Requirements for protein and essential amino acids in early infancy. *Acta Paediatr Scand* 1973; 62: 33–45.
29. Fomon SJ. Requirements and recommended dietary intakes of protein during infancy. *Pediatr Res* 1991; 30: 391–5.
30. Huang PC, Lin CP, Hsu JY. Protein requirements of normal infants at the age of about 1 year: maintenance nitrogen requirements and obligatory nitrogen losses. *J Nutr* 1980; 110: 1727–35.
31. Food and Nutrition Board–National Research Council. *Recommended dietary allowances*, 10th ed. Washington, DC: National Academy Press, 1989.
32. Fomon SJ, Haschke F, Ziegler EE, Nelson SE. Body composition of reference children from birth to age 10 years. *Am J Clin Nutr* 1982; 35: 1169–75.
33. Young VR, Bier DM, Pellett PL. A theoretical basis for increasing current estimates of the amino acid requirements in adult man, with experimental support. *Am J Clin Nutr* 1989; 50: 80–92.
34. Matthews DE, Bier DM, Rennie MJ, *et al*. Regulation of leucine metabolism in man: a stable isotope study. *Science* 1981; 214: 1129–31.
35. Bier DM. Intrinsically difficult problems: the kinetics of body proteins and amino acids in man. *Diabetes Metab Rev* 1989; 5: 111–32.
36. Horber FF, Horber-Feyder CM, Krayer S, *et al*. Plasma reciprocal pool specific activity predicts that of intracellular free leucine for protein synthesis. *Am J Physiol* 1989; 257: E385–99.
37. Waterlow JC. Protein turnover in the whole animal. *Invest Cell Pathol* 1980; 3: 107–19.
38. Bresson JL, Bader B, Rocchiccioli F, *et al*. Protein-metabolism kinetics and energy-substrate utilization in infants fed parenteral solutions with different glucose-fat ratios. *Am J Clin Nutr* 1991; 54: 370–6.

39. Beaufrère B, Fournier V, Salle B, Putet G. Leucine kinetics in fed low-birthweight infants: importance of splanchnic tissues. *Am J Physiol* 1992; 263: E214–20.
40. Pencharz P, Beesley J, Sauer P, et al. Total-body protein turnover in parenterally fed neonates: effects of energy source studies by using [^{15}N]glycine and [1-^{13}C]leucine. *Am J Clin Nutr* 1989; 50: 1395–400.
41. Golden MHN, Waterlow JC. Total protein synthesis in elderly people: a comparison of results with ^{15}N glycine and ^{14}C leucine. *Clin Sci Mol Med* 1977; 53: 277–88.
42. Mitton SG, Garlick PJ. Changes in protein turnover after the introduction of parenteral nutrition in premature infants: comparison of breast milk and egg protein-based amino acid solutions. *Pediatr Res* 1992; 32: 447–54.
43. Denne SC, Rossi EM, Kalhan SC. Leucine kinetics during feeding in normal newborns. *Pediatr Res* 1991; 30: 23–7.
44. Scriver CR, Kaufman S, Woo SLC. The hyperphenylalaninemias. In: Scriver CR, Beaudet AL, Sly WS, Valle D, eds. *The metabolic basis of inherited disease*, 6th ed. New York: McGraw-Hill, 1989; 495–546.
45. Konecki DS, Lichter-Konecki U. The phenylketonuria locus: current knowledge about alleles and mutations of the phenylalanine hydroxylase gene in various populations. *Hum Genet* 1991; 87: 377–88.
46. Rey F, Abadie V, Lyonnet S, et al. Expression phénotypique de 12 mutations du gène de la phénylalanine hydroxylase. *Arch Fr Pediatr* 1992; 49: 705–10.
47. Scriver CR, John SMW, Rozen R, et al. Associations between populations, phenylketonuria mutations and RFLP haplotypes at the phenylalanine hydroxylase locus: an overview. *Dev Brain Dysfunct* 1993; 6: 11–25.
48. Rey F, Munnich A, Lyonnet S, Rey J. Classification et hétérogénéité des hyperphénylalaninémies liées à un déficit en phénylalanine hydroxylase. *Arch Fr Pediatr* 1987; 44: 639–42.
49. Ruch T, Kerr D. Decreased essential amino acid requirements without catabolism in phenylketonuria and maple syrup urine disease. *Am J Clin Nutr* 1982; 35: 217–28.
50. Acosta PB, Wenz E, Williamson M. Nutrient intake of treated infants with phenylketonuria. *Am J Clin Nutr* 1977; 30: 198–208.
51. Wendel U, Ullrich K, Schmidt H, Batzler U. Six-year follow up of phenylalanine intakes and plasma phenylalanine concentrations. *Eur J Pediatr* 1990; 149: S13–6.
52. Kindt E, Halvorsen S. The need of essential amino acids in children. An evaluation based on the intake of phenylalanine, tyrosine, leucine, isoleucine, and valine in children with phenylketonuria, tyrosine amino transferase defect, and maple syrup urine disease. *Am J Clin Nutr* 1980; 33: 279–96.
53. Danner DJ, Elsas LJ. Disorders of branched chain amino acid and keto acid metabolism. In: Scriver CR, Beaudet AL, Sly WS, Valle D, eds. *The metabolic basis of inherited disease*, 6th ed. New York: McGraw-Hill, 1989; 671–92.
54. Thompson GN, Bresson JL, Pacy PJ, et al. Protein and leucine metabolism in maple syrup urine disease. *Am J Physiol* 1990; 258: E654–60.
55. Saudubray JM, Amédée-Manesme O, Munnich A, et al. Hétérogénéité de la leucinose. Corrélations entre l'aspect clinique, la tolérance protéique et le déficit enzymatique. *Arch Fr Pediatr* 1982; 39: 735–40.
56. Elsas LJ, Ellerine NP, Klein PD. Practical methods to estimate whole body leucine oxydation in maple syrup urine disease. *Pediatr Res* 1993; 33: 445–51.
57. Indo Y, Akaboshi I, Nobukuni Y, et al. Maple syrup urine disease: a possible biochemical basis for the clinical heterogeneity. *Hum Genet* 1988; 80: 6–10.
58. Fisher CR, Chuang JL, Cox RP, et al. Maple syrup urine disease in Mennonites. Evidence that the Y393N mutation in E1α impedes assembly of the E1 component of branched-chain α-keto acid dehydrogenase complex. *J Clin Invest* 1991; 88: 1034–7.
59. Nobukuni Y, Mitsubuchi H, Akaboshi I, et al. Maple syrup urine disease. Complete defect of the E$_1$β subunit of the branched chain α-ketoacid dehydrogenase complex due to a deletion of an 11-bp repeat sequence which encodes a mitochondrial targeting leader peptide in a family with the disease. *J Clin Invest* 1991; 87: 1862–6.
60. Mitsubuchi H, Nobukuni Y, Akaboshi I, et al. Maple syrup urine disease caused by a partial deletion in the inner E$_2$ core domain of the branched chain α-keto acid dehydrogenase complex due to aberrant splicing. A single base deletion at a 5'-splice donor site of an intron of the E$_2$ gene disrupts the consensus sequence in this region. *J Clin Invest* 1991; 87: 1207–11.
61. Fisher CN, Lau KS, Fisher CR, et al. A 17-bp insertion and a Phe215 → Cys missense mutation in the dihydrolipoyl transacylase (E$_2$) mRNA from a thiamine-responsive maple syrup urine disease patient WG-34. *Biochem Biophys Res Commun* 1991; 174: 804–9.

62. Zhang B, Wappner RS, Brandt IK, et al. Sequence of the E1 α subunit of branched-chain α-ketoacid dehydrogenase in two patients with thiamine-responsive maple syrup urine disease. *Am J Hum Genet* 1990; 46: 843–6.

DISCUSSION FOLLOWING THE PRESENTATION OF DR. REY

Dr. Bremer: If one looks at a large number of phenylketonuric patients, most of them certainly have a requirement of between 300 and 350 mg/d, but there are some with a much lower requirement and others with a much higher one. This means that on the basis of the average minimum requirement, there are certainly some individuals who need more phenylalanine. This is probably also true for normal children. So one should try to give a range with respect to the amino acid and probably also with respect to total protein. Do you agree?

Dr. Rey: Yes, and I think the range is between 200 and 400 mg/d.

Dr. Guesry: I am puzzled because the requirement for phenylalanine and leucine, based on the study of inborn errors of metabolism, seems to be approximately half of the amount calculated from the intake of breast-fed babies. You also say that the needs expressed in milligrams per day do not vary with age, which means that we have been overestimating the requirement for protein in our previous discussions. You say that your assumption is based on the fact that Garrow does not believe that babies with phenylketonuria or maple syrup urine disease have higher than normal requirements. But how strong is the evidence for this? You seem to be giving us results that contradict what we have been discussing on the previous days of this workshop.

Dr. Rey: In breast milk we have nitrogen and amino acids, or protein and non-protein nitrogen, which is the same. It seems to me that it is an advantage for the species to have enough nitrogen in the milk and a margin of safety for essential amino acids. You can convert all these essential amino acids to non-essential amino acids if you need to, if all the enzymatic steps are mature. So it seems to me that we should have an excess of phenylalanine and leucine in human milk because there is an advantage to this, but this is not a reason to believe that we *need* such a quantity of leucine or phenylalanine.

To return to Garrow, I did not say that the requirement for phenylalanine is higher in phenylketonuria; I said that there is no reason to believe that the requirement should be different depending on whether or not there is a block somewhere. When Archibald Garrow published his Croonian lecture in 1905, he quoted Claude Bernard, a very well known French physiologist. He wrote: "Il n'y a pas de raison de penser que les lois de la physique soient différentes dans une maison qui s'écroule et dans une maison qui tient debout." In other words, there is no reason to believe that the laws of physics are different in a house that is in good order and in a house that is destroyed. It is probable that the requirements are exactly the same in phenylketonuric and normal infants.

Dr. Rassin: I am a little bothered by this concept because some phenylalanine has to go to tyrosine to provide a supply of tyrosine for protein metabolism; that portion of the requirement is completely undefined in phenylketonuric persons. In the phenylketonuric patients the usual metabolic pathway has been removed. I think that without being able to describe what goes on in a normal individual in terms of the flux of phenylalanine → tyrosine → protein as well as of phenylalanine → protein, it is hard to say what goes on in the phenylketonuric individual in comparison with the normal individual.

Dr. Rey: You are right if we consider a situation where phenylalanine is in excess, because hydroxylation is the main pathway and transamination is very weak below the concentration of phenylalanine in the blood of around 10–15 mg/dl. But I am not sure that a lot of phenylalanine is hydroxylated to tyrosine if you give the minimum requirement of phenylalanine and tyrosine. If we return to the model of human milk with 3.3% of protein as phenylalanine and approximately the same amount of tyrosine, you could say that part of the phenylalanine is transferred to tyrosine and the requirement is therefore less because of this hydroxylation, but you have no proof of this *in vivo*. Have you any data?

Dr. Rassin: No, I don't think there are any data but I am not sure that you can evaluate phenylalanine in complete isolation from what goes on with the tyrosine. I would like to see some data on tyrosine requirements for protein synthesis to see how important that is and to compare the incorporation into protein and the metabolism and excretion of phenylalanine with those of tyrosine. How can you divorce tyrosine from phenylalanine when considering requirements because the two are so closely related to one another metabolically and both of them are certainly required for protein synthesis, even if we forget everything else.

Dr. Rey: Perhaps you are right. This is a good question.

Dr. Uauy: Do you have follow-up data on infants being fed this low intake of phenylalanine showing that their development is normal?

Dr. Rey: Yes. We have been treating phenylketonuria for 25 years now in our center and we have managed 100 patients. We have followed growth curves and IQ very carefully. Physical growth in these children is absolutely normal. At 5 years of age the IQ is around 100, but this is 8 points less than in the normal siblings. The data are exactly the same as in the US PKU collaborative study. Some of these children are now adults and most attended normal schools. Before 10 years of age about 50% have a 1-year delay in their studies, but some have more. But if they are very well treated with strictly limited phenylalanine and protein during the first year of age and up until 5 years, the results are really good. So we have ample proof that treatment with this low quantity of phenylalanine is successful, and many centers around the world have the same experience.

Dr. Wharton: I presume that in recent years your therapeutic diets have been supplemented with tyrosine, in rather greater amounts than you would get in breast milk. This has certainly been a trend elsewhere and relates to Dr. Rassin's comments. My second question is in relation to brain development. Isn't there some evidence now of abnormal NMR patterns in the brain in apparently normal people who have been treated for phenylketonuria for many years? I wonder if we should be a little bit cautious before adopting this figure for phenylalanine intake as being the correct one. I know there may be many possible reasons for abnormal NMR scans, but it is at least possible that the phenylalanine level has been set too low.

Dr. Rey: In relation to your first point, I would confirm that dietetic products intended for use by phenylketonuric patients are supplemented with tyrosine for many years. We have no experience of magnetic resonance imaging in our patients, but we have a program to look at this. I only know the work published recently by British investigators and summarized by Isabelle Smith in a recent paper in the *British Medical Journal* (1). It is clear from these data that myelin is not normal in phenylketonuric patients even if they are treated very early and very well. The lesions seem nevertheless much more significant if the children are treated late. The difficulty with PKU, or with leucinosis, is that if we don't give enough essential amino acids, growth is disturbed, and this may include deficient myelin formation; on the other hand, if we give too much, the phenylalanine levels increase in the blood and compete with other amino acids for transfer into the brain, resulting in decreased IQ, motor disability, and the many other consequences of excessive phenylalanine. So there has to be a compromise between not enough and too much.

Dr. Wharton: There is another possible explanation for the abnormal NMR patterns. The formulas used for phenylketonuria 20 years ago would have had very primitive fat blends by today's standards. The NMR lesions, assuming that they are significant, could well be related to deficiencies of essential lipids.

Dr. Rey: Phenylketonuric patients receive practically no protein, and the oil used in their formulas has for many years been corn oil. The plasma cholesterol in phenylketonuric children is lower than in normal infants. We have studied the medium-term consequences of this very low cholesterol intake on the regulation of blood cholesterol at 10 years of age but we found no adverse effect. However, it is true that the long-chain polyunsaturated fatty acids are missing in this mixture.

Dr. Bremer: We have done NMR scans on 35 patients of different ages and we have only found changes in older children and adults with phenylalanine levels above 8 mg/dl, not below.

Dr. Heine: These patients have to eat additional hydrolysates every day which are free of phenylalanine. How good is compliance? Hydrolysates have a bad taste. How do you control the intake? Is this a problem? Additional disturbances might be due to a reduced intake of these hydrolysates.

Dr. Rey: During the first 6 months of life we use Lophenalac, which contains a small quantity of phenylalanine. This is not a problem in the first 6 months. Lophenalac is a casein hydrolysate mixture with the aromatic amino acids removed by charcoal, and the taste is not too bad. After this we have to move to mixtures of amino acids and it is very difficult for the children to eat these mixtures, so we have now decided to stop the diet at 5 years of age. This is very unusual in the treatment of these patients. For the moment we don't know the consequences of the increase in phenylalanine level after 5 years of age. However, it is increasingly difficult for the children to continue to eat these mixtures, and if you look at the mean phenylalanine level in young patients you can see that it increases steadily after infancy. Children do their best to avoid these terrible mixtures.

Dr. Bremer: We started a collaborative study of nearly 160 patients in Germany in the 1950s. The patients have been followed since that time. Our results differ from the American collaborative study in that the average phenylalanine values were much lower. We had different classes of phenylalanine level during treatment, and 78 patients from the total of 160 fell into the class with values below 5 mg/dl during treatment. Children with phenylalanine values between 6 and 10 mg/dl did not reach their expected IQ and performance, whereas the children with values below 5 mg/dl did achieve them. It is almost certain that the impaired performance was not a problem of reduced energy intake. I think the results from your study are certainly valid.

Dr. Rey: Phenylketonuric women should be treated during pregnancy if we wish to avoid microcephaly and other fetal malformations. It is recommended that these women should resume a very low phenylalanine intake before conception. We shall probably obtain interesting data on the minimum phenylalanine intake required by these adults to maintain their phenylalanine level below 6 mg/dl, and I think we shall find that the value is not very different from 300–500 mg/d.

REFERENCE

1. Smith I, Cook B, Beasley M. Review of neonatal screening programme for phenylketonuria. *Br Med J* 1991; 303: 33–5.

Protein Metabolism During Infancy, edited by
Niels C. R. Räihä. Nestlé Nutrition Workshop
Series, Vol. 33. Nestec Ltd., Vevey/
Raven Press, Ltd., New York © 1994.

Role of Tumor Necrosis Factor in Protein Metabolism

Kevin J. Tracey

Department of Surgery, Division of Neurosurgery, North Shore University Hospital–Cornell University Medical College, Manhasset, New York 11030, USA

The relentless catabolism of body protein during chronic disease may kill. Early investigators presumed that the origins of cachexia lay in the underlying neoplasia or organism usurping the host's energy stores. More recent evidence indicates that immunological and neuroendocrinological mediators produced in response to invasion amplify the catabolism of lean body tissue, diverting substrate from peripheral tissues to liver as part of an acute-phase response. Anorexia and decreased food intake invariably accompany catabolic illnesses, but unlike the metabolic adaptation that occurs during unstressed fasting, the cachectic host fails to downregulate muscle protein catabolism and urinary nitrogen losses, and persistently excretes up to 15 g of nitrogen per day. Since mammals lack a protein storage depot, loss of body protein is associated with loss of function, resulting in immunosuppression, weakness, delayed wound healing, decreased tolerance for surgical and chemotherapeutic regimens, and death.

Cytokines, an expansive family of immunological mediators, have been implicated in mediation of the catabolic response to illness. The number of cytokines that have been identified and isolated continues to increase; at least 30 have been purified, sequenced, and cloned. These protein or glycoprotein mediators, produced in response to invasive stimulae, possess potent activities capable of modulating biochemical and molecular changes in nearly every known cell type. Cytokine-induced responses augment host defense, immune responsiveness, and the mobilization of energy stores during invasion. High doses of some cytokines are also toxic, however, so that the biological effects of an individual cytokine may be either beneficial or injurious to the host, depending on the amount produced. For instance, an acute overproduction of one cytokine, tumor necrosis factor-α (TNF), triggers lethal shock and tissue injury during septicemia (reviewed in 1). Antibodies that prevent TNF toxicity in sepsis are currently being evaluated in clinical trials. Moreover, early toxicity studies revealed that TNF induces the release of catabolic hormones (2) and mediates the development of a protein catabolic state with anorexia and anemia (3).

Subsequently, a great deal of research has been devoted to identifying a role for TNF and other cytokines in the biochemical basis of cachexia.

This chapter is a discussion of the role of TNF in the mediation of whole body protein catabolism. Since TNF mediates catabolic responses through a cascade of secondary cytokines, hormones, and metabolic responses, this brief review presents one model for potential interaction between cytokines and the neuroendocrinological system in mediating the systemic manifestations to injury, infection, or invasion.

TNF PRODUCTION AND BIOCHEMISTRY

Human TNF is synthesized by a variety of cells from hematopoietic and non-hematopoietic lineage, including macrophages, T cells, B cells, natural killer (NK) cells, mast cells, eosinophils, astrocytes, and Kupffer cells. The TNF gene, which lies on chromosome 6 in humans, encodes a cell-membrane-associated prohormone which is subsequently cleaved to yield the mature 17-kDa protein. The regulation of TNF expression has been most extensively studied in macrophage systems after stimulation with endotoxin (LPS), enterotoxin, or a variety of antigens derived from parasites, viruses, or tumors. Within minutes after exposure to LPS, macrophages upregulate the transcription and translation of TNF, leading to a 10,000-fold increase in the release of mature protein. Glucocorticoids inhibit and interferon-γ increases LPS-induced TNF expression (4, 5). Serum TNF levels peak within 90 min after experimental endotoxemia in both human volunteers and laboratory animals, then rapidly return to baseline (undetectable) levels (6). This fall of serum TNF levels occurs because cellular biosynthesis ceases shortly after induction (5), and the serum half-life is brief (6–15 min). Mature h-TNF circulates as a non-covalently bound trimer that interacts with two types of TNF receptor present on most cells. These receptors are cleaved *in vivo* by an unidentified process that yields TNF-binding peptides (7–9). These TNF-binding fragments are present in serum, where they function at low doses to enhance TNF effects (10), but at high doses to inhibit TNF effects (11, 12).

METABOLIC EFFECTS OF CHRONIC TNF EXPOSURE

Animals chronically exposed to TNF develop cachexia characterized by anorexia, weight loss, protein and lipid catabolism, insulin resistance, and anemia (Table 1). Some of these metabolic effects are directly attributable to the interaction of TNF with its cellular receptor, but others are attributable to secondary mediators the appearance of which is upregulated by TNF.

Protein Loss

Body composition analysis of laboratory rats subjected to twice daily injection of rh-TNF for 7–10 days shows that whole body protein is depleted (3). This net protein

TABLE 1. *Biological responses to TNF that have been implicated in the development of catabolic illness*

Tissue or cell type	Biological response
Brain	Anorexia
	Increase adrenocorticotropin, growth hormone, prolactin
	Decrease of thyroid stimulating hormone
Muscle	Net whole body protein catabolism
	Increased efflux of amino acids from isolated extremity
	Reduction of resting transmembrane potential
	Suppression of GLUT-4
	Glycogenolysis
	Lactate efflux
Adipose	Suppression of lipoprotein lipase
	Increased free fatty acid efflux
	Increased lipolysis
	Decreased lipogenesis
Liver	Acute-phase protein biosynthesis
	Decreased albumin biosynthesis
	Increased lipogenesis
	Enhanced glucagon-mediated amino acid transport
	Increased gluconeogenesis

catabolism was not attributable to decreases in food intake in the TNF-treated animals, since pair-fed controls retained protein despite similar body weight losses (3). Although skeletal muscle mass and protein content were depleted after TNF, there was a simultaneous increase in hepatic mass and protein content (3, 13). Skeletal muscle protein synthesis is suppressed, as indicated by quantitative analysis of expressed mRNA for myofibrillar proteins (13). Net losses of skeletal muscle protein were also observed when TNF was infused continuously rather than intermittently in freely feeding rats (14). Rats injected with TNF and analyzed by [^{14}C]leucine tracer methodology had increased dilution of labeled leucine, suggesting that systemic exposure to TNF promoted skeletal muscle proteolysis (15). Provision of parenteral nutrition during continuous TNF infusion failed to prevent urinary nitrogen loss and depletion of whole body protein (15, 16). By contrast, concurrent provision of insulin during 5 days of twice daily TNF administration prevented the net losses of whole body nitrogen (17). TNF-mediated protein loss is enhanced by glucocorticoids, but steroids alone do not account for all of the catabolized protein (18, 19).

Anorexia

Animals exposed to TNF develop anorexia dependent on the dose given (3, 20, 21). This observation was first made in studies of systemically administered TNF; a direct role for the central nervous system was suggested by studies employing trace quantities of TNF injected intracerebrally (22). The onset of anorexia induced by intracerebral TNF was associated with suppressed activity of glucose-sensitive neurons in the lateral hypothalamus, a region that contributes to the regulation of feeding (22). Exposure of the brain to even higher TNF levels by implantation of a genetically engineered TNF-secreting tumor in the forebrain caused lethal anorexia (23). The development of anorexia in this study was mildly attenuated by insulin, but not by depletion of brain serotonin (23). These studies suggest that TNF is capable of mediating anorexia by a direct effect on the hypothalamus, but leave open the possibility that the mechanism may involve the induction of secondary anorexigenic mediators (e.g., interleukin-1) in the hypothalamus.

Other Cytokines

The recognition that a catabolic state develops after exposure to TNF was followed by the realization that the mechanisms underlying net protein catabolism involved other catabolic cytokines as well. An abbreviated list of other factors that are produced in response to TNF is given in Table 2. TNF occupies a pivotal position in triggering the appearance of other cytokines during inflammation, injury, and invasion. IL-1, IL-2, IL-3, IL-6, and interferon-γ all appear in the circulation after the administration of TNF, and these cytokines have in turn been implicated in the development of catabolic illness. The pivotal role of TNF in mediating this cascade is suggested by the observation that inhibiting TNF during overwhelming septicemia prevents the appearance of the other cytokines (24). Once the humoral cytokine cascade is induced, it is self-propagating, so that the subsequent catabolic responses may be dissociated in time from the initial triggering event (e.g., TNF or LPS).

Adding to this already complex picture, cytokines produced during this cascade are capable of increasing or decreasing the biological activities of other cytokines,

TABLE 2. *Abridged list of factors implicated in catabolic illness that are produced in response to tumor necrosis factor (TNF)*

Cytokines	TNF
	IL-1, -2, -3, -6
	Interferon-γ
	Transforming growth factor-β
Hormones	Cortisol
	Glucagon
	Epinephrine
	Norepinephrine
	Growth hormone
	Insulin-like growth factor-1
Eicosanoids	Prostaglandins
	Leukotrienes
	Platelet activating factor
Miscellaneous	Nitric oxide
	Selectins
	Endothelial procoagulant activity

including TNF. For instance, IL-1, interferon-γ, and LPS each have protein catabolic activities when administered chronically to animals. Moreover, each of these factors synergistically enhances the toxicity and catabolic activities of TNF itself, such that sublethal doses of these factors lower the LD50 of TNF (25–27). Thus, when assessing the catabolic mechanisms underlying the biological response to TNF, consideration must be given to the activities of other cytokines in the cellular milieu.

Catabolic Hormones

TNF administration is followed by the release of glucose counterregulatory hormones that may participate in the mediation of net protein loss. Early studies of dogs infused with increasing doses of h-TNF showed rising serum levels of cortisol, epinephrine, norepinephrine, and glucagon (2); more recently, these observations have been extended to humans (28, 29). TNF interacts with the hypothalamic-pituitary axis to alter hormone production with resultant increases of ACTH, and decreases in growth hormone, prolactin, and TSH (reviewed in 30). A direct glucocorticoid stimulating effect of TNF on the adrenal cortex has been demonstrated (31). In addition to altering the hormonal profile, TNF is also capable of altering the magnitude of cellular responsiveness to these hormones. For example, TNF enhances glucagon-mediated amino acid transport in hepatocytes, alters muscle cell responsiveness to insulin by downregulating glucose transporters, and is synergistic with glucocorticoids in enhancing muscle protein breakdown (32–34). Since glucocorticoids are capable of suppressing TNF biosynthesis, the emerging picture is that protein metabolic interactions between TNF and the neuroendocrinological system occur at the level of cytokine production, hormone release, and cellular responsiveness.

Insulin Resistance

TNF has been implicated in the development of insulin resistance in laboratory animals and human subjects. Serum glucose increases after infusion of increasing doses of TNF, with a TNF dose-dependent influence on rising glucose levels (2, 35). A similar observation has been made in animals receiving total parenteral nutrition during a continuous infusion of TNF (16). The physiological basis for this hyperglycemia lies in both reduced peripheral clearance and reduced suppression of hepatic glucose output (36). The cellular basis for reduced glucose uptake has been studied in adipocytes and myocytes co-cultured with increasing concentrations of TNF (33). These results indicate that the expression of the glucose transporter GLUT-4 is reduced in TNF-treated cells, suggesting that decreased transport underlies the reductions of insulin-mediated glucose uptake (33). Relative insulin resistance in the adipocyte also occurs because TNF suppresses the expression of lipogenic enzymes, effectively preventing the incorporation of glucose into lipid even in the presence of high insulin levels (37). Insulin resistance is known to develop in both cachexia and obesity, and a recent study implicated TNF in the mediation of insulin resistance in genetically obese animals (38). The mechanisms by which insulin resistance influences body weight are unknown, but it is plausible that TNF-mediated insulin resistance may participate in both cachexia and obesity, depending on the concurrent food intake (decreased or increased).

Tolerance

Chronic exposure to TNF increases the tolerance of the animal to subsequent TNF administration, so that progressively higher TNF doses must be administered to maintain the same level of biological response (3). The molecular basis of this tolerance is unknown, but antibody production against exogenous TNF has been excluded as the explanation (39). Other cytokines, including IL-1, IL-6, interferon-γ, and D-factor, induce tolerance that protects experimental animals against the lethal effects of endotoxemia. Less is known, however, about catabolic responses in tolerant animals. The results presently available suggest that several factors may be involved, including (a) suppressed production of the cytokine cascade, (b) suppressed cellular responsiveness to cytokine effects, (c) altered activity or clearance of released cytokines, (d) altered expression of cytokine receptors, or (e) altered expression of cytokine inhibitors.

Tumors

Tumorigenic cell lines have been genetically engineered to constitutively secrete TNF, IL-6, or interferon-γ. When implanted into nude mice, these cells form tumors that continuously produce a single cytokine, providing models for the effects of chronic cytokine overexposure. These studies provide conclusive evidence that chronic overproduction of a single gene product (either TNF, IL-6, or interferon-γ) is capable of inducing net protein catabolism, as determined by analysis of body composition (23, 40–43). In each case animals developed increased serum cytokine levels relative to the transfected cytokine gene, developed some degree of anorexia, and had carcass composition analyses that verified net protein catabolism. A similar observation has been made in rats bearing a sarcoma that induced local macrophage activation and TNF production; the severity of cachexia correlated with the quantity of TNF produced (44). Administration of anti-TNF antibodies to animals bearing either this sarcoma (44), another transplantable tumor (45), or a genetically engineered TNF-secreting tumor (46) partially ameliorated the development of anorexia, weight loss, and lipid and protein depletion, suggesting that TNF was involved in triggering the catabolic state, but that it acts via indirect mechanisms.

CELL BIOLOGY OF CYTOKINE-INDUCED PROTEIN CATABOLISM

Despite the weighty evidence from *in vivo* studies implicating TNF in the mediation of net protein catabolism, the molecular basis of TNF-induced protein catabolism is enigmatic. TNF, IL-1, interferon-γ, and LPS fail to induce protein breakdown when incubated with skeletal muscle *in vitro*, suggesting that *in vivo* protein loss occurs via some unknown indirect mechanism (47). Myocytes *in vitro* respond directly to TNF with a variety of other biological effects, including reduction of resting transmembrane potential (48, 49), enhanced glycogenolysis (50), increased lactate release (50), and decreased expression of glucose transporters (33). Hepatocytes co-cultured with TNF are triggered to upregulate acute-phase protein biosynthesis (51), downregulate albumin production (41), increase lipogenesis (52), and increase glucagon-mediated amino acid uptake (32). Extrapolating from these *in vitro* data, it appears that the net result of these TNF-mediated cellular responses *in vivo* would contribute to a redistribution of amino acids from peripheral tissue to the hepatic compartment for incorporation into acute phase proteins, carbohydrates, and urea.

METABOLIC EFFECTS OF TNF IN HUMANS

TNF Administration

Recombinant human TNF has been administered to cancer patients as an experimental antineoplastic drug. Although these studies have not yet identified clinical efficacy for TNF in the treatment of cancer, they have provided a great deal of information about the biological effects of this cytokine in humans.

TNF infusion is followed by the development of fever, myalgia, tachycardia, headache, hypotension, and a capillary leakage syndrome that is dose-dependent (53–55). Forearm efflux of amino acids is increased, primarily because of increased rates of alanine and glutamine efflux (28). Muscle amino acid efflux occurred in association with increases in circulating levels of cortisol and glucagon, and a fall in serum insulin (28). Resting energy expenditure increased because of increases in oxygen consumption and carbon dioxide production, and lactate levels rose by 50% (29). Whole body protein turnover, assessed by the [^{15}N]glycine method of Picou and Taylor-Roberts, was increased, and synthesis was decreased. Whole body protein catabolism was increased, but the magnitude of the increase did not reach statistical significance (29). A similar trend toward negative nitrogen balance was observed with the higher doses of TNF used in this study (29). Taken together, these data indicate that exposure to a single dose of TNF is capable of triggering metabolic responses that are typical of the catabolic response to invasive illness. What remains unclear is whether these metabolic effects persist during prolonged TNF administration, or whether tolerance develops.

TNF Production in Disease

Increased TNF levels have been detected in body fluids obtained from patients with a variety of catabolic illnesses, including parasite infection, sepsis, malnutrition, cancer, and AIDS (reviewed in 56). Attempts to correlate circulating TNF levels with the magnitude of nutritional depletion have been thwarted, however, by a number of factors that confound the analysis. These factors include (a) trace amounts of other cytokines synergistically increasing or decreasing the effects of TNF; (b) the fact that fragments of the TNF receptor may increase or decrease the biological response to TNF; (c) TNF acting locally in vital organs (e.g., brain or liver) to exert metabolic responses to the whole body without attaining detectable serum levels; and (d) the fact that TNF is produced sporadically and appears only transiently in the serum during chronic catabolic illness [e.g., burn injury (57)], so that periodic serum sampling may miss the TNF peak. These difficulties of interpreting serum levels do not exclude a role for TNF in protein catabolism. Rather, this approach points out the complexity of the catabolic response induced by TNF, and propagated by other humoral factors, that ultimately leads to protein loss. Perhaps assigning a role for TNF in cachexia by measuring serum TNF is analogous to assigning a role for insulin in diabetes by assaying serum insulin levels (which may be increased, decreased, or normal). Unlike hemoglobin-A1c, which is useful for monitoring the chronic effect of insufficient insulin action in diabetics, there are no available tests to assess the impact of chronic TNF excess in cachexia.

SUMMARY AND FUTURE

The search for a humoral factor that mediates protein catabolism has been under way since the demonstration that normal animals develop cachexia when parabiotically joined to a tumor-bearing animal (58). TNF has been implicated as a mediator that is capable of participating in this response because (a) it is produced during protein catabolic illness in humans; (b) administration of highly purified recombinant human TNF induces anorexia and a protein catabolic state in humans and animals; (c) the cellular biology of TNF suggests a role in promoting a transfer of amino acids from peripheral tissue to liver for use in acute-phase protein synthesis and gluconeogenesis; and (d) blocking TNF with antibodies attenuates the development of protein catabolism and insulin resistance. Much less is known about the precise molecular mechanisms underlying TNF-induced protein catabolism. It is hoped that the identification of the mechanisms by which TNF triggers protein losses in cachexia will foster the development of novel therapies to prevent protein catabolism and its complications.

REFERENCES

1. Tracey KJ, Lowry SF. The role of cytokine mediators in septic shock. *Adv Surg* 1990; 23: 21–56.
2. Tracey KJ, Lowry SF, Fahey TJ, et al. Cachectin/tumor necrosis factor induces lethal shock and stress hormone responses in the dog. *Surg Gynecol Obstet* 1987; 164: 415–22.
3. Tracey KJ, Wei H, Manogue KR, et al. Cachectin/tumor necrosis factor induces cachexia, anemia, and inflammation. *J Exp Med* 1988; 167: 1211–27.
4. Beutler B, Tkacenko V, Milsark I, Krochin N, Cerami A. Effect of gamma interferon on cachectin expression by mononuclear phagocytes. Reversal of the lpsd (endotoxin resistance) phenotype. *J Exp Med* 1986; 164: 1791–6.

5. Beutler B, Krochin N, Milsark IW, Luedke C, Cerami A. Control of cachectin (tumor necrosis factor) synthesis: mechanisms of endotoxin resistance. *Science* 1986; 232: 977–80.
6. Hesse DG, Tracey KJ, Fong Y, et al. Cytokine appearance in human endotoxemia and primate bacteremia. *Surg Gynecol Obstet* 1988; 166: 147–53.
7. Brockhaus M, Schoenfeld H-J, Schlaeger E-J, Hunziker W, Lesslauer W, Loetscher H. Identification of two types of tumor necrosis factor receptors on human cell lines by monoclonal antibodies. *Proc Natl Acad Sci USA* 1990; 87: 3127–31.
8. Engelmann H, Novick D, Wallach D. Two TNF-binding proteins purified from human urine. Evidence for immunological cross-reactivity with cell surface tumor necrosis factor receptors. *J Biol Chem* 1990; 265: 1531–6.
9. Seckinger P, Vey E, Turcatti G, Wingfield P, Dayer J-M. Tumor necrosis factor inhibitor: purification, NH_2-terminal amino acid sequence and evidence for anti-inflammatory and immunomodulatory activities. *Eur J Immunol* 1990; 20: 1167–74.
10. Aderka D, Engelmann H, Maor Y, Brakebusch C, Wallach D. Stabilization of the bioactivity of tumor necrosis factor by its soluble receptors. *J Exp Med* 1992; 175: 323–9.
11. van Zee KJ, Kohno T, Fischer E, Rock CS, Moldawer LL, Lowry SF. Tumor necrosis factor soluble receptors circulate during experimental and clinical inflammation and can protect against excessive tumor necrosis factor-a *in vitro* and *in vivo*. *Proc Natl Acad Sci USA* 1992; 89: 4845–9.
12. Ashkenazi A, Marsters SA, Capon DJ. Protection against endotoxic shock by a tumor necrosis factor receptor immunoadhesin. *Proc Natl Acad Sci USA* 1991; 88: 10535–9.
13. Fong Y, Moldawer LL, Marano MA, et al. Cachectin/TNF or IL-1 alpha induces cachexia with redistribution of body proteins. *Am J Physiol* 1989; 256: R659–65.
14. Hoshino E, Pichard C, Greenwood CE, et al. Body composition and metabolic rate in rat during a continuous infusion of cachectin. *Am J Physiol Endocrinol Metab* 1991; 260: E27–36.
15. Flores EA, Bistrian BR, Pomposelli JJ, Dinarello CA, Blackburn GL, Istfan NW. Infusion of tumor necrosis factor/cachectin promotes muscle catabolism in the rat: a synergistic effect with interleukin-1. *J Clin Invest* 1989; 83: 1614–22.
16. Matsui J, Cameron RG, Kurian R, Kuo GC, Jeejeebhoy KN. Nutritional, hepatic, and metabolic effects of cachectin/TNF in rats receiving total parenteral nutrition. *Gastroenterology* 1993; 104: 235–43.
17. Fraker DL, Merino MJ, Norton JA. Reversal of the toxic effects of cachectin by concurrent insulin administration. *Am J Physiol* 1989; 256: E725–31.
18. Angeras MH, Angeras U, Zamir O, Hasselgren PO, Fischer JE. Effect of the glucocorticoid receptor antagonist RU 38486 on muscle protein breakdown in sepsis. *Surgery* 1991; 109: 468–73.
19. Angeras MH, Angeras U, Zamir O, Hasselgren PO, Fischer JE. Interaction between corticosterone and TNF stimulated protein breakdown in rat skeletal muscle, similar to sepsis. *Surgery* 1990; 108: 460–6.
20. Moldawer LL, Andersson C, Gelin J, Lonnroth C, Lundholm K. Regulation of food intake and hepatic protein metabolism by recombinant-derived monokines. *Am J Physiol* 1988; 254: G450–6.
21. Socher SH, Friedman A, Martinez D. Recombinant human tumor necrosis factor induces acute reductions in food intake and body weight in mice. *J Exp Med* 1988; 167: 1957–62.
22. Plata-Salaman CR, Oomura Y, Kai Y. Tumor necrosis factor and interleukin 1-beta: suppression of food intake by direct action in the central nervous system. *Brain Res* 1988; 448: 106–14.
23. Tracey KJ, Morgello S, Koplin B, et al. Metabolic effects of cachectin/tumor necrosis factor are modified by site of production: cachectin/tumor necrosis factor-secreting tumor in skeletal muscle induces chronic cachexia, while implantation in brain induces predominately acute anorexia. *J Clin Invest* 1990; 86: 2014–24.
24. Fong Y, Tracey KJ, Moldawer LL, et al. Antibodies to cachectin/TNF reduce interleukin-1-β and interleukin-6 appearance during lethal bacteremia. *J Exp Med* 1989; 170: 1627–33.
25. Rothstein JL, Schreiber H. Synergy between tumor necrosis factor and bacterial products causes hemorrhagic necrosis and lethal shock in normal mice. *Proc Natl Acad Sci USA* 1988; 85: 607–11.
26. Doherty GM, Lange JR, Langstein HN, Alexander HR, Buresh CM. Evidence for IFN-gamma as a mediator of the lethality of endotoxin and tumor necrosis factor-alpha. *J Immunol* 1992; 149: 1666–670.
27. Waage A, Espevik T. Interleukin 1 potentiates the lethal effect of tumor necrosis factor-α/cachectin in mice. *J Exp Med* 1988; 167: 1987–92.
28. Warren RS, Starnes HF, Gabrilove JL, Oettgen HF, Brennan MF. The acute metabolic effects of tumor necrosis factor administration in humans. *Arch Surg* 1987; 122: 1396–400.
29. Starnes HF, Warren RS, Jeevanandam M, et al. Tumor necrosis factor and the acute metabolic response to tissue injury in man. *J Clin Invest* 1988; 82: 1321–5.

30. Martin SB, Tracey KJ. Tumor necrosis factor-α (TNF) in neuroimmunology. *Adv Neuroimmunol* 1992; 2: 125–38.
31. Darling G, Goldstein DS, Stull R, Gorschboth CM, Norton JA. Tumor necrosis factor: immune endocrine interaction. *Surgery* 1989; 106: 1155–60.
32. Warren RS, Donner DB, Starnes HF, Brennan MF. Modulation of endogenous hormone action by recombinant human tumor necrosis factor. *Proc Natl Acad Sci USA* 1987; 84: 8619–22.
33. Stephens JM, Pekala PH. Transcriptional repression of the GLUT4 and C/EBP genes in 3T3-L1 adipocytes by TNF-a. *J Biol Chem* 1991; 266: 21839–45.
34. Hall-Angeras M, Hasselgren P-O, Dimlich RVW, Fischer JE. Myofibrillar proteinase, cathepsin B, and protein breakdown rates in skeletal muscle from septic rats. *Metabolism* 1991; 40: 302–6.
35. Ciancio MJ, Hunt J, Jones SB, Filkins JP. Comparative and interactive in vivo effects of tumor necrosis factor α and endotoxin. *Circ Shock* 1991; 33: 108–20.
36. Fukushima R, Saito H, Taniwaka K, et al. Different roles of IL-1 and TNF on hemodynamics and interorgan amino acid metabolism in awake dogs. *Am J Physiol* 1992; 2625: E275–81.
37. Torti FM, Dieckmann B, Beutler B, Cerami A, Ringold GM. A macrophage factor inhibits adipocyte gene expression: an *in vitro* model of cachexia. *Science* 1985; 229: 867–9.
38. Hotamisligil GS, Shargill NS, Spiegelman BM. Adipose expression of tumor necrosis factor—a direct role in obesity-linked insulin resistance. *Science* 1993; 259: 87–91.
39. Fraker DL, Stovroff MC, Merino MJ, Norton JA. Tolerance to tumor necrosis factor in rats and the relationship to endotoxin tolerance and toxicity. *J Exp Med* 1988; 168: 95–105.
40. Oliff A, Defeo-Jones D, Boyer M, et al. Tumors secreting human TNF/cachectin induce cachexia in mice. *Cell* 1987; 50: 555–63.
41. Brenner DA, Buck M, Feitelberg SP, Chojkier M. Tumor necrosis factor-α inhibits albumin gene expression in a murine model of cachexia. *J Clin Invest* 1990; 85: 248–55.
42. Matthys P, Dukmans R, Proost P, et al. Severe cachexia in mice inoculated with interferon-γ producing tumor cells. *Int J Cancer* 1991; 49: 77–82.
43. Strassmann G, Fong M, Kenney JS, Jacob CO. Evidence for the involvement of interleukin 6 in experimental cancer cachexia. *J Clin Invest* 1992; 89: 1681–4.
44. Stovroff MC, Fraker DL, Travis WD, Norton JA. Altered macrophage activity and tumor necrosis factor: tumor necrosis and host cachexia. *J Surg Res* 1989; 46: 462–9.
45. Sherry BA, Gellin J, Fong Y, et al. Anticachectin/tumor necrosis factor-α antibodies attenuate development of cachexia in tumor models. *FASEB J* 1989; 3: 1956–62.
46. Teng MN, Park BH, Koeppen HKW, Tracey KJ, Fendly BM, Schreiber H. Long-term inhibition of tumor growth by tumor necrosis factor in the absence of cachexia or T-cell immunity. *Proc Natl Acad Sci USA* 1991; 88: 3535–9.
47. Moldawer LL, Svaninger G, Gelin J, Lundholm KG. Interleukin 1 and tumor necrosis factor do not regulate protein balance in skeletal muscle. *Am J Physiol* 1987; 253: C766–73.
48. Tracey KJ, Lowry SF, Beutler B, Cerami A, Albert JD, Shires GT. Cachectin/tumor necrosis factor mediates changes of skeletal muscle plasma membrane potential. *J Exp Med* 1986; 164: 1368–73.
49. Kagan BL, Baldwin RL, Munoz D, Wisnieski BJ. Formation of ion-permeable channels by tumor necrosis factor-α. *Science* 1992; 255: 1427–30.
50. Lee MD, Zentella A, Pekala PH, Cerami A. Effect of endotoxin-induced monokines on glucose metabolism in the muscle cell line L6. *Proc Natl Acad Sci USA* 1987; 84: 2590–4.
51. Perlmutter DH, Dinarello CA, Punsal Pl, Colten HR. Cachectin/tumor necrosis factor regulates hepatic acute-phase gene expression. *J Clin Invest* 1986; 78: 1349–54.
52. Feingold KR, Soued M, Staprans I, et al. Effect of tumor necrosis factor (TNF) on lipid metabolism in the diabetic rat: evidence that inhibition of adipose tissue lipoprotein lipase activity is not required for TNF-induced hyperlipidemia. *J Clin Invest* 1989; 83: 1116–21.
53. Abbruzzese JL, Levin B, Ajani JA, et al. A phase II trial of recombinant human interferon-gamma and recombinant tumor necrosis factor in patients with advanced gastrointestinal malignancies: results of a trial terminated by excessive toxicity. *J Biol Response Mod* 1990; 9: 522–7.
54. Schiller JH, Storer BE, Witt PL, et al. Biological and clinical effects of intravenous tumor necrosis factor-α administered three times weekly. *Cancer Res* 1991; 51: 1651–8.
55. Spriggs DR, Sherman ML, Frei E, Kufe DW. Clinical studies with tumour necrosis factor. In: Bock G, Marsh J, eds. *Tumour necrosis factor and related cytotoxins*. CIBA Foundation Symposium No 131. Chichester, West Sussex, England: Wiley, 1987; 206–27.
56. Tracey KJ. The acute and chronic pathophysiological effects of TNF: mediation of septic shock and wasting (cachexia). In: Beutler B, ed. *Tumor necrosis factors: the molecules and their emerging role in medicine*. New York: Raven Press, 1992; 255–73.

57. Marano MA, Fong Y, Moldawer LL, et al. Serum cachectin/TNF in critically ill patients with burns correlates with infection and mortality. *Surg Gynecol Obstet* 1990; 170: 32–8.
58. Norton JA, Moley JF, Green MV, Carson RE, Morrison SD. Parabiotic transfer of cancer anorexia/cachexia in male rats. *Cancer Res* 1985; 45: 5547–52.

DISCUSSION FOLLOWING THE PRESENTATION OF DR. TRACEY

Dr. Uauy: Were your experiments conducted in growing animals; in other words, were you demonstrating stunting, or were you demonstrating cachexia in fully grown animals?

Dr. Tracey: My experiments were done in animals with a starting body weight of about 200–220 g, so they were still growing but they were adolescent. Other studies have looked at older animals and the results appear to be fairly reproducible. In a more general sense your question is asking whether cytokines participate in stunting. I don't think anybody knows the answer to this. It is an important area that has to be investigated. A few studies have looked at the effects of toxic amounts of cytokines on embryos and neonates, but I don't think the specific question of stunting has been addressed. There is a transgenic mouse that overproduces tumor necrosis factor, and those animals become stunted or runted and seem to have abnormal thymus function as well.

Dr. Guesry: You show that the end result of these TNF and interleukin studies is cachexia, and we know that the origin of this is partly anorexia. But is the effect on muscle mass due primarily to excessive muscle breakdown or is it due to impaired protein synthesis?

Dr. Tracey: The problem with answering your question is that in the animal model the data suggest that there is an effect on both synthesis and breakdown. This has not been confirmed in tissue culture, so it is fair to say that since we don't understand the molecular basis of the net loss, we don't know the answer. But we do know that there seems to be both decreased synthesis and increased breakdown *in vivo*.

Dr. Garlick: It is a complex issue as to whether or not protein synthesis is altered by these agents *in vivo*. Kevins has showed that when either TNF or IL-1 was injected over periods of days there was no reduction in muscle protein synthesis, although I think there was an increase in liver protein synthesis (1). Grimble, in Southampton, gave injections of TNF but measured the effects at much shorter time intervals, within hours after the injection, when the animals showed fever. At that stage muscle protein synthesis did fall quite substantially after 9 h and liver protein synthesis increased (2). This is very similar to the effect of any acute-phase response.

In our laboratory, Peter Ballmer injected rats with IL-1β and got exactly the same effect as you get with TNF. In other words, at about 9 h there was a big fall in muscle protein synthesis and a large rise in liver protein synthesis. In acute experiments we were unable to measure protein breakdown.

The question I would like to ask is this. If you inject these cytokines in large amounts, there appears to be an acute-phase response. Can the final response be mimicked by other agents that induce acute-phase response: for example, stress hormones, trauma, and so on?

Dr. Tracey: You can reproduce some of the aspects of the acute-phase response, as you suggest, by giving infusions or injections of classical hormones, but these hormones don't induce the entire spectrum of cachexia. They don't induce the anemia and they don't induce the severity of the muscle protein loss that are found after administration of cytokines. Cytokines by themselves seem able to induce the whole spectrum of cachexia, particularly TNF and IL-1.

It is hard to study cachexia during sepsis, and hard to treat it because it goes on for so long. The situation is confused by antigenic responses and it is difficult to do studies over a sufficiently long period of time. People therefore try to extrapolate from the acute sepsis syndrome to the chronic situation. During the acute sepsis syndrome, many cytokines can give you changes in blood pressure or catabolic rate, but so far TNF is the only one that can of its own induce the entire spectrum of septic shock syndrome. If you look at the effects of each cytokine, they suggest that these agents play a role in the host responses that may perhaps give rise to cachexia. So it is difficult to assign priorities to these factors. However, by understanding that there is a cascade effect which, once started, becomes self-perpetuating, we shall hopefully be better able to target therapies.

Dr. Räihä: Do you have any information on cytokines during the neonatal period? I ask this because we know that normal breast-fed newborn infants show very little decrease in weight after birth, whereas formula-fed infants usually lose quite a lot and take longer before they regain their birthweight. Do you think that cytokines could have anything to do with this? We are inducing proteins into these infants that are not species specific.

Dr. Tracey: This is a reasonable hypothesis. Malnourished adults have raised levels of TNF in the blood. This has been reported by two different groups. This is suggestive that malnutrition or aberrant nutrition may somehow trigger a cytokine response. Maybe the cytokines are involved in mobilizing energy as a beneficial response. There has been a great deal of interest in cytokine production in the neonatal period in relation to necrotizing enterocolitis. Babies with this condition seem to have a reduced cytokine response, at least initially, but they also develop a different syndrome of septic shock from the adult, so there is much speculation in this area.

Dr. Yamashiro: Do you think that the stunting that occurs with chronic steroid therapy has something to do with cytokines?

Dr. Tracey: That is a very important point. The mechanism by which glucocorticoids produce their anti-inflammatory action has been an enigma since they were first developed. A great deal of work has now been done on the fact that glucocorticoids seem, as a group, specifically to turn off the production of many of these pro-inflammatory cytokines. So you could be describing a sort of rebound effect. This is an important area for further investigation.

Dr. Beatty: In the process of development of cachexia, do you see a rise in immunoglobulins and acute-phase proteins in your mouse model?

Dr. Tracey: Yes, in all studies on cachexia where indices of acute-phase proteins have been sought they have been found to be increased.

Dr. Rassin: TNF and some of the IL cytokines are present in human milk. What function do you think they might have when they enter the baby by the enteral route?

Dr. Tracey: I don't know. But I was struck by the data showing lactoferrin-specific receptors and I wonder what the cytokine receptor profile is. Many of the cytokines are fairly tough biochemically and would probably survive gastric acid. Some would probably also survive proteolysis in the upper gastrointestinal tract and get absorbed.

Dr. Rassin: We have been working very hard on the mechanism of TPN-associated cholestasis and one of the mechanisms we have been looking at is the fact that light-exposed parenteral nutrition mixtures produce toxic products capable of stimulating TNF release. We have not yet been able to pin down exactly what it is that stimulates TNF release. When this substance gets in the liver it appears to affect amino acid transport at the bile formation site. However, it is hard to understand the sequence of events.

Dr. Tracey: There are now at least two books on TNF both of which have excellent chapters on the regulation of TNF synthesis, each with 100–200 references. This is a very large field because of the obvious medical relevance. The list of things that turn on TNF synthesis keeps growing. I don't know the answer to your question but any kind of pro-inflammatory stimulator may turn on TNF.

Dr. Pettifor: Do all peripheral tissues have receptors for TNF? And how is it catabolized?

Dr. Tracey: The red cell is the only somatic cell type that does not have TNF receptors. All other cells in the body have two TNF receptors, type 1 (55 kDa molecular weight) and type 2 (75 kDa molecular weight). It appears that the type 1 isoform may be the important one for mediating shock and death in the septic shock syndrome. The catabolism of TNF is another area that is not well understood. The half-life of TNF in humans varies between 6 and 20 minutes and a lot of the TNF is secreted through the kidney. Some of it is not accounted for and it is not clear whether it is bound to cell receptors or bound by circulating fragments.

Dr. Guesry: Did you say that the cancer cells keep the TNF receptor?

Dr. Tracey: Many cancer cells have a TNF receptor, and many cancer cells make TNF. The problem with understanding the biology of TNF in cancer cells is that, depending on type, some tumor cells respond to TNF by dying, while others respond by growing. So there does not appear to be 1:1 correlation between biological effect and tumor or tumor receptor.

Dr. Guesry: At one time people were thinking that they could use TNF as an anti-cancer drug. Of course, if the cancer cells are using the receptor, there is no reason to continue this type of investigation because we know that TNF may have deleterious side effects.

Dr. Tracey: Loss of the receptor does not explain cancer resistance or lack of efficacy of TNF. The lack of efficacy in cancer appears to be closely correlated to the development of toxicity, particularly toward endothelial cells. The problem is that the endothelial cells around the tumor are not more sensitive to TNF than endothelial cells in the kidney, or the heart, or the liver, so systemic toxicity is the limiting problem.

Dr. Uauy: Considering that this is a nutrition meeting, do you know of any specific nutritional manipulations that modulate TNF responsiveness—zinc, vitamin A, other specific nutrients?

Dr. Tracey: There is evidence that the intake of ω-3 fatty acids may alter cytokine production. I believe that the largest effect was on IL-1, while the effect on TNF was to attenuate it without reducing it to zero. It would be reasonable to consider these approaches for nutritional modulation of cytokines.

Dr. Wharton: All the conditions we have discussed where there is high TNF in association with wasting are also associated with quite severe curtailment of intake. I have been trying to think of conditions where you get wasting combined with a high intake and wonder what happens to TNF in this situation. Has thyrotoxicosis been studied, or Cushing's syndrome, where you have a high intake and deposition of fat but there is wasting of muscles?

Dr. Tracey: I have not seen any data on TNF in thyrotoxicosis or Cushing's syndrome. I think you are raising an important point. In the animal model—the mouse with prolonged weight loss—there is not a profound degree of anorexia. It would be within the bounds of the typical patients with cancer or AIDS, where food intake is reduced. If you look at Jeejeeboy's data, the animals he studied were getting well in excess of their requirements, but they continued to have negative muscle protein balance. We have worked with other tumor models of cachexia characterized by protein and lipid loss but with normal or increased food intake and we have not found TNF to be important in those models.

Dr. Wharton: Take another clinical example, such as Crohn's disease. We know that some children with Crohn's disease will start to grow if they are fed adequately, whether or not hydrolyzed protein is given. There are also some patients in whom this does not happen, despite what seems to be a very adequate intake. Can you relate this kind of clinical course to TNF levels?

Dr. Tracey: Several companies and several research centers have spent a great deal of time and money investigating inhibitors of TNF and IL-1, particularly the IL-1 receptor antagonist, in Crohn's disease and ulcerative colitis. There is marked upregulation of pro-inflammatory cytokine production in these conditions, but the effects of this over a long period of time in a developing child are largely unknown. It is certainly possible that an overproduction of these pro-inflammatory factors could have deleterious effects on nutritional development.

Dr. Beatty: To follow on what Dr. Wharton was saying, the diencephalic syndrome is characterized by excessive intake and poor weight gain. Is anything known about cytokines in this condition?

Dr. Tracey: In babies with diencephalic syndrome no TNF was found in the blood or in the spinal fluid. Energy expenditure was measured and was found to be in excess of predicted for both age and weight.

Dr. Yamashiro: One established example of cytotoxicity of TNFα in the pediatric field is Kawasaki disease. In patients with Kawasaki disease, the serum level of TNFα is significantly higher than in controls in the initial acute phase of the illness, and those with high levels of TNFα are more likely to have cardiac involvement. TNFα attacks the endothelium of vessels and damages the coronary vasculature.

Dr. Kirsten: Does TNF cross the placenta, and do you know any disease in the mother where TNF is secreted and could have an influence on the fetus?

Dr. Tracey: TNF levels are very high in the amniotic fluid of mothers delivering at the time of chorioamnionitis. It is difficult to diagnose chorioamnionitis and attempts are being made to establish whether there is a possible diagnostic correlation with cytokines locally. I don't know if they cross the placenta. They don't cross the blood-brain barrier very well. There has been renewed interest in the last couple of years on the role of local cytokine production in the pathogenesis of infertility.

REFERENCES

1. Flores EA, Bistrian BR, Pomposelli JJ, Dinarello CA, Blackburn GL, Istfan NW. Infusion of tumor necrosis factor/cachectin promotes muscle catabolism in the rat—a synergistic effect with IL-1. *J Clin Invest* 1989; 83: 1614–1622.
2. Charters Y, Grimble RF. Effect of recombinant human tumour necrosis factor-α on protein synthesis in liver, skeletal muscle and skin of rats. *Biochem J* 1989; 258: 493–497.

Subject Index

A

N-Acetylneuraminic acid, 110–111
  human milk, lactation changes, 111
Adapted whey formula, postprandial amino acid curve, 62
Alkaptonuria, 211
Allergy, 178
Alpha-lactalbumin, 128–129
  cystine, 125
  infant formula, 125–126
  isoleucine, 125
  leucine, 125
  lysine, 125
  methionine, 125
  phenylalanine, 125
  proteolysis, 58
  threonine, 125
  tryptophan, 125
  valine, 125
American Academy of Pediatrics, protein requirement, recommendations, 177–178
Amino acid. *See also* Specific type
  central nervous system functions, 184
  changing view of role, 1
  cognitive outcome, 189
  essentiality, 27
  fetus, 100
  functions, 3–4
  large neutral, 187, 188
  muscle, 130
  nutritionally relevant, 1–4
  oral administration, 46
  pathways, 14
  pattern changes, 187–190
  protein requirement, 75–77
  recommendations, 77
  red blood cell, 194
  sulfur-containing, 189–190
  total parenteral nutrition, 100, 186
    cystine, 186–187
    tyrosine, 186–187
  transport rate, 13
  whole body protein turnover, 44
Amino acid conservation, 8
Amino acid homeostasis, principal metabolic systems, 4–13
  integration, 13–23
Amino acid metabolism
  biochemistry, cellular processes, 5
  energy dependency, 20–23
    qualitative aspects, 20, 21
    quantitative aspects, 20–23
  Michaelis constant, 9
  physiology, cellular processes, 5
  protein nutrition
    biochemistry, 1–23
    physiology, 1–23
Amino acid oxidation, 8–10
  amino acid intake, 8–10
  rate, 8
  substrate availability, 8–10
  tissue enzymes, 9–10
Amino acid requirement, 77
  kinetic data, 83–84
Amino acid transport, 12–13
Amino alcohol, non-protein nitrogen, 113–114
Amino nitrogen, 11
Ammonia
  human milk, 114
  non-protein nitrogen, 114
  urea cycle enzyme, 14
Anabolic drive, 6–7
Anorexia, tumor necrosis factor, 231–232
Anoxia, protein degradation, 47
alpha-Antitrypsin, human milk protein, 54
Arginine, 27
  citrulline, 16–19
  urea cycle enzyme, 14
Arginine kinetics, metabolic cooperativity, 13–19
Arginine metabolism
  functional specialization, 17
  intracellular compartmentation, 17
  in vivo, 15

B

Baby food, protein content, 171
Bacterial flora, urea, 64
Balance study, 149
Bile salt binding, casein, 78
Binding protein, non-protein nitrogen, 117
Biologically active peptide
  human milk, 112–113
  non-protein nitrogen, 112–113
Body composition, 147
Bovine milk protein, proteolysis, 58
Bovine serum albumin, 129

SUBJECT INDEX

Brain, breast feeding, 131
Branched-chain amino acid
 adequate intake, 215
 breast feeding, 218
 infant formula, 218
 mature milk, 213
Branched-chain amino acid catabolism, 9–10
Branched-chain keto acid dehydrogenase, 10
Brazelton score
 human milk, 192
 infant formula, 192
Breast feeding. *See also* Breast milk
 brain, 131
 branched-chain amino acid, 218
 consequences, 173–175
 diabetes, 131
 fecal flora, 78–79
 insulin, 131
 phenylalanine, 218
 protein intake effect, 172–173
 renal function, 131
Breast milk
 N-acetylneuraminic acid, lactation changes, 111
 ammonia, 114
 biologically active peptide, 112–113
 Brazelton score, 192
 carnitine, 114
 choline, 113–114
 consequences, 173–175
 creatine, 114
 creatinine, 114
 essential amino acid, 225
 fetal growth retardation, 102–103
 fortifier, 64
 IgA, 103
 lactoferrin, 103
 lysosyme, 103
 maternal diet, 102
 nitrogen, 225
 non-protein nitrogen
  characterization, 105–107
  total nitrogen, 105
 nucleic acid, 112
 nucleotide, 112
 parity, 130
 polyamine, 112
 preterm infant, 147–148
  growth rates, 147–148
 protein, 53–55, 87–103
  amino acid content, 94–95
  amino acid pattern, 121–127, 128–132
  alpha-antitrypsin, 54
  content variability, 163
  cystine, 125
  digestibility, 53–55
  intake effect, 172–173
  isoleucine, 125

 lactoferrin, 54
 leucine, 125
 lysine, 125
 methionine, 125
 nutritional value, 91–94
 by phases of lactation, 89–90
 phenylalanine, 125
 preterm infant, digestibility, 55
 proteolysis, 58
 qualitative aspects, 121–127, 128–132
 secretory IgA, 53
 threonine, 125
 tryptophan, 125
 valine, 125
 urea, 101–102
 uric acid, 114
Breast milk protein, secretory IgA, 54

C

Calcium, 182
 casein, 78
Cancer patient, protein degradation, 46
Carbohydrate
 large bowel, 64
 nitrogen-containing, 110–112
Carbon-13 method, whole body protein turnover, 33–35
Carbon dioxide, urea cycle enzyme, 14
Carboxypeptidase, casein, 78
Carnitine, 118, 119
 human milk, 114
 non-protein nitrogen, 114
Casein
 bile salt binding, 78
 calcium, 78
 carboxypeptidase, 78
 enzyme, 63
 gastric emptying, 78
 infant formula, 116
 iron, 78
 lactobezoar formation, 78
 milk, 91, 92
 postprandial amino acid curve, 62
 proteolysis, 58
Catabolic hormone, 232, 233
Catch-up growth
 fat, 78
 nitrogen, 78
 protein requirement, 74–75, 141–143
Catecholamine metabolism, 193–194
Catecholamine synthesis, 188
Chicken-based formula, 119–120
Chlorine, 86
Cholesterol, phenylketonuria, 227
Choline
 human milk, 113–114
 non-protein nitrogen, 113–114

SUBJECT INDEX

Chronic high lactose ingestion model, 209
Citrulline, 18
  arginine, 16–19
Cognitive outcome, amino acid, 189
Colostrum, 89–90
  IgA, 103
  lactoferrin, 103
  lysosyme, 103
  nucleotide, 210
Compartmentation, 27
Congenital heart disease, nucleotide, 47
Cow's milk formula, 118
Cow's milk protein
  adaptation to human milk, 121–122
  modifying, 129
Creatine
  human milk, 114
  non-protein nitrogen, 114
Creatinine, 162, 181
  human milk, 114
  non-protein nitrogen, 114
Cystein sulfinic acid decarboxylase, 193
Cysteine, 118, 189, 193
  protein requirement, 75
Cystine, 186–187
  alpha-lactalbumin, 125
  human milk protein, 125
  whey protein, 125
  whole bovine milk protein, 125
Cytokine, 229

D

Developing country, protein needs, 179–180, 182
Diabetes
  breast feeding, 131
  small for gestational age infant, 150
Diet protein, protein turnover, 43
Dietary nucleotide
  absorption, 197–200
  metabolism, 197–200
Dietary protein allowance, whole body protein synthesis, by age, 19
Digestibility
  gastric emptying, 64
  human milk protein, 53–55
Disaccharidase, nucleoside, 201, 202
Downregulation, 28

E

End product method, whole body protein turnover, 30–33
Energy turnover rate
  protein degradation rate, 49–52
    background, 49
    methodological aspects, 49
  RNA degradation rate, 49–52
    background, 49
    methodological aspects, 49
Enzyme, casein, 63
Epidermal growth factor
  gut function, 117
  non-protein nitrogen, 117
ESPGAN, protein requirement, recommendations, 177–178
Essential amino acid, 183–190, 192–194
  breast milk, 225
  proposed developmentally, 184
Essential amino acid requirement
  computation methods, 212–218
  direct estimation of net in vivo synthesis, 217–218
  estimation for artificially fed infants, 213–215
  factorial methods, 215–217
  human milk model, 212–213
  inborn errors of metabolism, 211–222, 225–227

F

Fat, 148–149
  catch-up growth, 78
  milk, 88
Fat globule protein, 118
Fat store, 146–147
Fecal excretion
  lactoferrin, 92, 93
  secretory IgA, 92, 93
Fecal flora
  breast-fed baby, 78–79
  infant formula, 78–79
Fetal growth retardation
  breast milk, 102–103
  insulin-like growth factor, 118
Fetus
  amino acid, 100
  serine, 101
Flooding method, protein turnover, 38–39
Follow-on formula, 83, 178
  weaning, recommendations, 168–170

G

Gastric acid, output by age, 57
Gastric emptying
  casein, 78
  digestibility, 64
Gene transcription, 5
Genotype, phenylketonuria, 218–219
Glutamine, 208
  parenteral nutrition, 46–47
  septic shock, 46–47

SUBJECT INDEX

Glutathione, 118
  red blood cell, 194
Glycine, premature infant, 194
Glycine turnover, neonate, 44
Glycosylation precursor, non-protein
    nitrogen, 117
Growth, variability, 161, 162
Growth data, protein requirement, 159–160
  factorial method, 160
  growth curves, 160
  WHO/FAO, 160–161
Gut closure, 118
Gut function
  epidermal growth factor, 117
  metabolic factors, 79–80
Gut microflora, 85
  nucleotide, 201

H
Heart rate, 162
Heat treatment
  lysine, 63
  serine, 63
  threonine, 63
Histidine, 1, 130
Human whey protein, 53–55
Hydrolysate formula, urea, 116
Hydrolyzed formula, 63–64
Hypercatabolic state, protein requirement, 74
Hypernatremic dehydration, 179

I
IgA
  breast milk, 103
  colostrum, 103
  mature milk, 103
Immune system, nucleotide, 200–201
Inborn errors of metabolism, essential
    amino acid requirement, 211–222, 225–227
Infant. *See also* Sick infant
  protein
    absorption, 53–65
    digestibility, 53–65
    infant nutrition, 59–60
    international recommendations, 67–86
Infant formula, 62. *See also* Specific type
  alpha-lactalbumin, 125–126
  branched-chain amino acid, 218
  Brazelton score, 192
  casein, 116
  consequences, 173–175
  cost, 131
  fecal flora, 78–79
  improved biological value, 125–126

label claim, 83
lower current level, 164
non-protein nitrogen, 107–109, 116
nucleotide, 209
protein, 100
  amino acid pattern, 121–127, 128–132
  digestibility, 56–59
  heat treatment, 57
  intake effect, 172–173
  level, 56
  qualitative aspects, 121–127, 128–132
protein requirement
  clinical studies, 156–158
  recommendations, 156
  reduced protein formulas, 156–158
  supplementation, 118
  Swedish experience, 158
  taurine, 209
  true protein level, 128
  urea, 116
  weaning, recommendations, 168–170
Infection
  neonate, 46
  nitrogen, 83
  protein requirement, 84
Insulin, breast feeding, 131
Insulin-like growth factor, fetal growth
    retardation, 118
Insulin resistance, tumor necrosis factor, 233–234
Interorgan metabolic flow, membrane
    transport, 12
Intestinal development, nucleotide, 201–203
Intestinal injury, nucleotide, 209
Intestinal repair, nucleotide, 201–203
Intravenous feeding, whole body protein
    turnover, 44
Iron, casein, 78
Iron deficiency, 178
Isoleucine
  alpha-lactalbumin, 125
  human milk protein, 125
  whey protein, 125
  whole bovine milk protein, 125

K
Kidney, 181

L
Lactobacillus bifidus, 110–112
Lactobezoar formation, casein, 78
Lactoferrin
  breast milk, 103
  colostrum, 103
  fecal excretion, 92, 93
  human milk protein, 54

mature milk, 103
milk, 92, 93
proteolysis, 58
beta-Lactoglobulin, 129
Lactose, milk, 88
Lamb-based formula, 119
Large bowel, carbohydrate, 64
Leucine
  alpha-lactalbumin, 125
  genotype-phenotype correlation, 221–222
  human milk protein, 125
  kinetic method, 83–84
  polysome profile, 43
  variant classification, 221–222
  whey protein, 125
  whole bovine milk protein, 125
Leucine oxidation, 9, 10
Leucine requirement, 83–84, 225
Leucine turnover, neonate, 44
Leucinosis, 221–222
  genotype-phenotype correlation, 221–222
  leucine, 222
  variant classification, 221–222
Linear growth, 161
Lipid protein interaction, 65
Liver, protein synthesis, 35, 36
Long-chain fatty acid, 65
Lophinalac, 227
Low birthweight infant, protein requirement, 133–141
Low protein diet, 162–163
Low protein intake, adaptation, 85–86
Lysine, 1
  alpha-lactalbumin, 125
  heat treatment, 63
  human milk protein, 125
  whey protein, 125
  whole bovine milk protein, 125
Lysosyme
  breast milk, 103
  colostrum, 103
  mature milk, 103

M

Malnutrition, protein turnover, 45
Mammalian milk, protein, comparative aspects, 87–89
Maternal diet
  breast milk, 102
  human milk, 102
  milk volume, 102
  taurine, 102
Mature milk, 89–90
  branched-chain amino acid, 213
  IgA, 103
  lactoferrin, 103
  lysosyme, 103

nucleotide, 210
phenylalanine, 213
Medium-chain triglyceride, 65
Membrane transport
  interorgan metabolic flow, 12
  substrate utilization, 12
Metabolic cooperativity, arginine kinetics, 13–19
Metalloprotein, 131–132
Methionine
  alpha-lactalbumin, 125
  human milk protein, 125
  whey protein, 125
  whole bovine milk protein, 125
Michaelis constant, amino acid metabolism, 9
Milk. See also Specific type
  casein, 91, 92
  fat, 88
  lactoferrin, 92, 93
  lactose, 88
  protein, 88
  secretory IgA, 92, 93
  whey protein, 91, 92
Milk protein fraction, changes during lactation, 90–91
Mineral absorption, 131–132
Muscle
  amino acid, 130
  protein synthesis, 35, 36, 44

N

Necrotizing enterocolitis, nucleotide, 208
Neonate
  glycine turnover, 44
  infection, 46
  leucine turnover, 44
  phenylalanine turnover, 44
Neuropsychological development, preterm infant, 130
Neurotransmitter precursor, 187
Nitrogen. See also Non-protein nitrogen
  breast milk, 225
  catch-up growth, 78
  infection, 83
  protein intake, 12
  protein requirement, 77–78
  net nutrient absorption, 77–78
Nitrogen-15 method, 116
  whole body protein turnover, 30–33
    ammonia, 32
    end product average, 32
    glycine, 31–32
    in infants, 31
    single dose of isotope, 31
    urea, 32
Nitrogen balance study, 162

Nitrogen-containing carbohydrate, 110–112
Non-essential amino acid, 183–190, 192–194
Non-protein nitrogen, 163
  amino alcohol, 113–114
  ammonia, 114
  binding protein, 117
  biological activity, 116–117
  biologically active peptide, 112–113
  breast milk
    characterization, 105–107
    total nitrogen, 105
  carnitine, 114
  choline, 113–114
  creatine, 114
  creatinine, 114
  epidermal growth factor, 117
  glycosylation precursor, 117
  infant formula, 107–109, 116
  nucleic acid, 112
  nucleotide, 112
  nutritional importance, 105–114, 116–120
  polyamine, 112
  urea, 109–110
  uric acid, 114
  whey formula, 116
Nucleic acid, 203–206
  human milk, 112
  non-protein nitrogen, 112
Nucleoside, disaccharidase, 201, 202
Nucleotide, 203–206
  absorption, 197–200
  bacteria, 208
  biological significance, 200–203
  colostrum, 210
  congenital heart disease, 47
  dietary sources, 210
  gut microflora, 201
  human milk, 112
  immune system, 200–201
  infant formula, 209
  intestinal development, 201–203
  intestinal injury, 209
  intestinal repair, 201–203
  mature milk, 210
  metabolism, 197–200
  necrotizing enterocolitis, 208
  non-protein nitrogen, 112
  radiation injury, 208, 210
  tissue growth, 203
  tissue repair, 203
  unsubstantiated claims, 208
  uric acid, 210
Nucleotide requirement, 208–209

O

Oral rehydration fluid, 179
Ornithine, 18

urea cycle enzyme, 14
Ornithine-urea cycle, 11
Oxygen consumption
  protein degradation rate, 49–52
  RNA degradation rate, 49–52

P

Parenteral nutrition, glutamine, 46–47
Parity, breast milk, 130
Partially hydrolyzed formula, 63–64
Pepsin, output by age, 57
Phenylalanine, 130 131, 193
  adequate intake, 215
  alpha-lactalbumin, 125
  breast feeding, 218
  enrichments, 38
  human milk protein, 125
  mature milk, 213
  phenylketonuria, 218, 219–220, 221, 225
  tyrosine, 225–226
  whey protein, 125
  whole bovine milk protein, 125
Phenylalanine requirement, 225
Phenylalanine turnover, neonate, 44
Phenylketonuria, 218–220, 221
  cholesterol, 227
  genotypes/phenotypes, 218–219
  long-term treatment, 226
  phenylalanine, 218, 219–220, 221, 226
  pregnancy, 227
  variant classification, 218–219
Plasma amino acid, 185–187
  concentration, 8–9
Plasma arginine flux, 16–19
Plasma arginine kinetics, 13–19
Plasma citrulline flux, 16–19
Polyamine, 118
  human milk, 112
  non-protein nitrogen, 112
Polysome profile, leucine, 43
Postprandial amino acid curve
  adapted whey formula, 62
  casein, 62
Potassium, 86
Precursor method
  protein synthesis, 44
  whole body protein turnover, 33–35
Pregnancy, phenylketonuria, 227
Preterm breast milk, protein, 95–98
Preterm infant, 100
  glycine, 194
  human milk, 147–148
    growth rates, 147–148
  human milk protein, digestibility, 55
  neuropsychological development, 130
  protein degradation, 46

SUBJECT INDEX

tyrosine, 130
whole body protein turnover, 44
Protein
　functional, 62–63
　high intakes, 163
　　in early life, 122–124
　human milk, 53–55, 87–103
　　amino acid content, 94–95
　　amino acid pattern, 121–127, 128–132
　　nutritional value, 91–94
　　by phases of lactation, 89–90
　　qualitative aspects, 121–127, 128–132
　infant
　　absorption, 53–65
　　digestibility, 53–65
　　infant nutrition, 59–60
　　international recommendations, 67–86
　infant formula, 100
　　amino acid pattern, 121–127, 128–132
　　digestibility, 56–59
　　heat treatment, 57
　　level, 56
　　qualitative aspects, 121–127, 128–132
　intake
　　nitrogen, 12
　　type of milk feeding effects, 172–173
　low intake in early life, 122–124
　mammalian milk, comparative aspects, 87–89
　milk, 88
　preterm breast milk, 95–98
　protective, 63
　quality, 85
　quantitative assessment, 62–63
　true nutritional, 63
　urea cycle enzyme, differing consumption levels, 11
Protein accretion, 146–147
Protein degradation, 7–8, 45
　anoxia, 47
　cancer patient, 46
　major systems, 7
　metabolic energy, 7–8
　nutritional aspects, 7
　premature infant, 46
　refeeding after protein-energy malnutrition, 45
　regulation, 7
　sick infant, 46
　site, 45
　turnover, 7
Protein degradation rate
　energy turnover rate, 49–52
　　background, 49
　　methodological aspects, 49
　oxygen consumption, 49–52
Protein/energy ratio, weaning, 182
Protein/fat ratio, 148–149

Protein gap, 180
Protein homeostasis, principal metabolic systems, 4–13
　integration, 13–23
Protein hydrolysate, 85
Protein loss, tumor necrosis factor, 230–231
Protein metabolism
　biochemistry, cellular processes, 5
　energy dependency, 20–23
　　qualitative aspects, 20, 21
　　quantitative aspects, 20–23
　physiology, cellular processes, 5
　protein nutrition
　　biochemistry, 1–23
　　physiology, 1–23
　tumor necrosis factor, 229–236, 239–242
Protein nutrition
　amino acid metabolism
　　biochemistry, 1–23
　　physiology, 1–23
　protein metabolism
　　biochemistry, 1–23
　　physiology, 1–23
Protein requirement, 67–80
　American Academy of Pediatrics, recommendations, 177–178
　amino acid, 75–77
　background, 67
　catch-up growth, 74–75, 141–143
　cysteine, 75
　denominator, 69–75
　energy of food, 69–70
　environmental stress factors, 85
　ESPGAN, recommendations, 177–178
　foods, 67
　growth data, 159–160
　　factorial method, 160
　　growth curves, 160
　　WHO/FAO, 160–161
　hypercatabolic state, 74
　individuals, 69
　infant formula
　　clinical studies, 156–158
　　recommendations, 156
　　reduced protein formulas, 156–158
　infection, 84
　intestinal events, 77–80
　label claim, 80
　low birthweight infant, 133–141
　minimum, 149
　multiples of nitrogen, 68
　nitrogen, 77–78
　　net nutrient absorption, 77–78
　non-protein nitrogen, 68–69
　numerator, 67–69
　populations, 67
　protein adequacy/excess indices, 137–141

Protein requirement (contd.)
  protein/energy ratio
    in pathological states, 74–75
    in physiological states, 70, 71, 72
    in weaning diets, 70–74
  Scientific Committee for Food,
      recommendations, 177–178
  small for gestational age infant, 144–145,
      150
  term infant
    factorial method for estimation,
        153–155
    first four months, 153–158, 159–164
    protein intake in growing breast-fed
        infants, 153
    safe intake, 155
    study methods, 153–155
  tryptophan, 76–77
  very low birthweight infant, 135, 143–144
    blood urea nitrogen, 135
    parenteral nutrition, 135
    plasma amino acid, 136
    urinary nitrogen excretion, 135
  volume of food, 69–70
  weaning, 165–175, 177–182
    recommendations, 165–168
  weight, 69
  weight of food, 69–70
Protein synthesis
  cellular origin of amino acids, 37
  downregulating transport precursors, 43
  enzyme, 28
  flooding, 43
  initiation complex, 5, 6
  liver, 35, 36
  modifications in phosphorylation state,
      5–6
  muscle, 35, 36, 44
  precursor method, 44
  regulation site, 5–6
  site, 44
  transcriptional phase, amino acid defect, 6
  translational aspects, 5–7
Protein turnover
  constant infusion of labeled amino acids,
      36–38
  diet protein, 43
  flooding method, 38–39
  individual tissues, 35–39
  isotopic study methods, 29–40
  levels, 42–43
  malnutrition, 45
  nutritional implications, 19–20
  quantitative aspects, 19–20
  rate, 28
  sick infant, 46
  urea labeling study, 46

Proteolysis
  bovine milk protein, 58
  casein, 58
  human milk protein, 58
  alpha-lactalbumin, 58
  lactoferrin, 58
Purine, 208
Purine base precursor, 198
Pyrimidine base precursor, 198

R
Radiation injury, nucleotide, 208, 210
Recombinant glycosetylated protein, 117
Red blood cell
  amino acid, 194
  glutathione, 194
Refeeding after protein-energy malnutrition,
    protein degradation, 45
Renal acid excretion, 147
Renal disease, 181
Renal function, breast feeding, 131
Resting metabolic rate, whole body protein
    turnover, 22, 23
Retinopathy of prematurity, taurine, 194
RNA degradation rate
  energy turnover rate, 49–52
    background, 49
    methodological aspects, 49
  oxygen consumption, 49–52

S
Scientific Committee for Food, protein
    requirement, recommendations,
    177–178
Secretory IgA
  fecal excretion, 92, 93
  human milk protein, 53, 54
  milk, 92, 93
Selenium, 118
Selenocysteine, 1–2
Septic shock, glutamine, 46–47
Serine
  fetus, 101
  heat treatment, 63
Serine synthesis, 27–28
Sialic acid, 119
Sick infant
  protein degradation, 46
  protein turnover, 46
Skeletal muscle, enzyme, 28
Small for gestational age infant
  diabetes, 150
  protein requirement, 144–145, 150
Soy formula, 119

SUBJECT INDEX

Substrate utilization, membrane transport, 12
Supplementary food, weaning, 170–172

T
Taurine, 118, 129, 189, 192–193
  infant formula, 209
  maternal diet, 102
  retinopathy of prematurity, 194
  vegetarian diet, 193
Term infant, protein requirement
  factorial method for estimation, 153–155
  first four months, 153–158, 159–164
  protein intake in growing breast-fed infants, 153
  safe intake, 155
  study methods, 153–155
Threonine, 44
  alpha-lactalbumin, 125
  heat treatment, 63
  human milk protein, 125
  whey protein, 125
  whole bovine milk protein, 125
Tissue growth, nucleotide, 203
Tissue repair, nucleotide, 203
Total parenteral nutrition, amino acid, 100, 186
  cystine, 186–187
  tyrosine, 186–187
Tracer substance, 45–46
Transitional milk, 89–90
Tryptophan, 1, 118, 189
  alpha-lactalbumin, 125
  human milk protein, 125
  measurements, 85
  protein requirement, 76–77
  supplementation, 128
  whey protein, 125
  whole bovine milk protein, 125
Tumor necrosis factor
  anorexia, 231–232
  biochemistry, 230
  insulin resistance, 233–234
  metabolic effects of chronic exposure, 230–234
  production, 230
  protein loss, 230–231
  protein metabolism, 229–236, 239–242
  tolerance, 234
  tumors, 234
Tyrosine, 186–187, 188
  phenylalanine, 225–226
  preterm infant, 130
  route of administration, 186
Tyrosinosis, 130

U
UGA codon, 1–2
Upregulation, 28
Urea
  bacterial flora, 64
  human milk, 101–102
  hydrolysate formula, 116
  infant formula, 116
  non-protein nitrogen, 109–110
  urea cycle enzyme, 14
Urea cycle arginine, 18
Urea cycle enzyme, 11–12
  ammonia, 14
  arginine, 14
  carbon dioxide, 14
  ornithine, 14
  protein, differing consumption levels, 11
  urea, 14
Urea labeling study, protein turnover, 46
Urea production, 11–12
Uric acid
  human milk, 114
  non-protein nitrogen, 114
  nucleotide, 210
Urinary potassium excretion, 147

V
Valine, 189
  alpha-lactalbumin, 125
  human milk protein, 125
  whey protein, 125
  whole bovine milk protein, 125
Vegetarian diet
  taurine, 193
  zinc, 193
Very low birthweight infant, protein requirement, 135, 143–144
  blood urea nitrogen, 135
  parenteral nutrition, 135
  plasma amino acid, 136
  urinary nitrogen excretion, 135
Vitamin B-6, 194
Vitamin D, 86

W
Weaning
  follow-on formula, recommendations, 168–170
  food, 84
  infant formula, recommendations, 168–170
  protein/energy ratio, 182
  protein requirement, 165–175, 177–182
  recommendations, 165–168
  solid foods, 178
  supplementary food, 170–172

Weight gain, 150
Whey formula, 62
  non-protein nitrogen, 116
Whey protein
  cystine, 125
  isoleucine, 125
  leucine, 125
  lysine, 125
  methionine, 125
  milk, 91, 92
  phenylalanine, 125
  threonine, 125
  tryptophan, 125
  valine, 125
Whole body protein metabolism, 30
Whole body protein synthesis, dietary protein allowance, by age, 19
Whole body protein turnover
  amino acid, 44
  carbon-13 method, 33–35
  end product method, 30–33
  intravenous feeding, 44
  nitrogen-15 method, 30–33
    ammonia, 32
    end product average, 32
    glycine, 31–32
    in infants, 31
    single dose of isotope, 31
    urea, 32
  precursor method, 33–35
  preterm infant, 44
  resting metabolic rate, 22, 23
  studies, 29–35
Whole bovine milk protein
  cystine, 125
  isoleucine, 125
  leucine, 125
  lysine, 125
  methionine, 125
  phenylalanine, 125
  threonine, 125
  tryptophan, 125
  valine, 125

Z

Zinc, 86
  vegetarian diet, 193